Universitext

Universitext

Universitext is a series of textbooks that presents material from a wide variety of mathematical disciplines at master's level and beyond. The books, often well class-tested by their author, may have an informal, personal, even experimental approach to their subject matter. Some of the most successful and established books in the series have evolved through several editions, always following the evolution of teaching curricula, into very polished texts.

Thus as research topics trickle down into graduate-level teaching, first textbooks written for new, cutting-edge courses may make their way into Universitext.

For further volumes:
www.springer.com/series/223

Adam Bowers • Nigel J. Kalton

An Introductory Course in
Functional Analysis

 Springer

Adam Bowers
Department of Mathematics
University of California, San Diego
La Jolla, CA
USA

Nigel J. Kalton (deceased)
Department of Mathematics
University of Missouri, Columbia
Columbia, MO
USA

ISSN 0172-5939   ISSN 2191-6675(electronic)
Universitext
ISBN 978-1-4939-1944-4   ISBN 978-1-4939-1945-1 (eBook)
DOI 10.1007/978-1-4939-1945-1

Library of Congress Control Number: 2014955345

Springer New York Heidelberg Dordrecht London
© Springer Science+Business Media, LLC 2014

Printed on acid-free paper

Springer is part of Springer Science+Business Media (www.springer.com)

To the memory of Nigel J. Kalton (1946–2010)

Foreword

Mathematicians are peculiar people who spend their life struggling to understand the great book of Mathematics, and find it rewarding to master a few pages of its daunting and ciphered chapters. But this book was wide open in front of Nigel Kalton, who could browse through it with no apparent effort, and share with his colleagues and students his enlightening vision.

That book is now closed and we are left with the grief, and the duty to follow Nigel's example and to keep working, no matter what. Fortunately, his mathematical legacy is accessible, which includes the last graduate course he taught in Columbia during the Academic Year 2009–2010. Adam Bowers, a post-doctoral student during that year, gathered very careful notes of all classes and agreed with Nigel that these notes would eventually result in a textbook. Fate had in store that he completed this work alone.

Adam Bowers decided that this tribute to Nigel should be co-authored by the master himself. All functional analysts should be grateful to Adam for his kind endeavour, and for the splendid textbook he provides. Indeed this book is a smooth and well-balanced introduction to functional analysis, constantly motivated by applications which make clear not only how but why the field developed. It will therefore be a perfect base for teaching a one-semester (or two) graduate course in functional analysis. A cascade falling from so high is a powerful force, and a beautiful sight. Please open this book, and enjoy.

Paris, France   Gilles Godefroy
January 2014

Preface

During the Spring Semester of 2010, Nigel Kalton taught what would be his final course. At the time, I was a postdoctoral fellow at the University of Missouri-Columbia and Nigel was my mentor. I sat in on the course, which was an introduction to functional analysis, because I simply enjoyed watching him lecture. No matter how well one knew a subject, Nigel Kalton could always show something new, and watching him present a subject he loved was a joy in itself.

Over the course of the semester, I took notes diligently. It had occurred to me that someday, if I had the good fortune to teach a functional analysis course of my own, Nigel Kalton's notes would make the best foundation. When I happened to mention to Nigel what I was doing, he suggested we turn the notes into a textbook. Sadly, Nigel was unexpectedly taken from us before the text was complete. Without him, this work—and indeed mathematics itself—suffers from a terrible loss.

About this book

This book, as the title suggests, is meant as an introduction to the topic of functional analysis. It is not meant to function as a reference book, but rather as a first glimpse at a vast and ever-deepening subject. The material is meant to be covered from beginning to end, and should fit comfortably into a one-semester course. The text is essentially self-contained, and all of the relevant theory is provided, usually as needed. In the cases when a complete treatment would be more of a distraction than a help, the necessary information has been moved to an appendix.

The book is designed so that a graduate student with a minimal amount of advanced mathematics can follow the course. While some experience with measure theory and complex analysis is expected, one need not be an expert, and all of the advanced theory used throughout the text can be found in an appendix.

The current text seeks to give an introduction to functional analysis that will not overwhelm the beginner. As such, we begin with a discussion of normed spaces and define a Banach space. The additional structure in a Banach space simplifies many proofs, and allows us to work in a setting which is more intuitive than is necessary for the development of the theory. Consequently, we have sacrificed some generality for the sake of the reader's comfort, and (hopefully) understanding. In Chap. 2, we meet the key examples of Banach spaces—examples which will appear again and again throughout the text. In Chap. 3, we introduce the celebrated Hahn–Banach Theorem and explore its many consequences.

Banach spaces enjoy many interesting properties as a result of having a complete norm. In Chap. 4, we investigate some of the consequences of completeness, including the Baire Category Theorem, the Open Mapping Theorem, and the Closed Graph Theorem. In Chap. 4, we relax our requirements and consider a broader class of objects known as locally convex spaces. While these spaces will lack some of the advantages of Banach spaces, considerable and interesting things can and will be said about them. After a general discussion of topological preliminaries, we consider topics such as Haar measure, extreme points, and see how the Hahn–Banach Theorem appears in this context.

The origins of functional analysis lie in attempts to solve differential equations using the ideas of linear algebra. We will glimpse these ideas in Chap. 6, where we first meet compact operators. We will continue our discussion of compact operators in Chap. 7, where we see an example of how techniques from functional analysis can be used to solve a system of differential equations, and we will encounter results which allow us to do unexpected things, such as sum the series $\sum_{n=1}^{\infty} \frac{1}{n^4}$.

We conclude the course in Chap. 8, with a discussion of Banach algebras. We will meet the spectrum of an operator and see how it relates to the seemingly unrelated concept of maximal ideals of an algebra. As a final flourish, we will prove the Wiener Inversion Theorem, which provides a nontrivial result about Fourier series.

At the end of each chapter, the reader will find a collection of exercises. Many of the exercises are directly related to topics in the chapter and are meant to complement the discussion in the textbook, but some introduce new concepts and ideas and are meant to expose the reader to a broader selection of topics. The exercises come in varying degrees of diffculty. Some are very straightforward, but some are quite challenging.

It is hoped that the reader will find the material intriguing and seek to learn more. The inquisitive mind would do well with the classic text Functional Analysis by Walter Rudin [34], which covers the material of this text, and more. For further study, the reader might wish to peruse A Course in Operator Theory by John B. Conway [8] or (moving in another direction) Topics in Banach Space Theory by Albiac and Kalton [2].

About Nigel Kalton

Nigel Kalton was born on 20 June 1946 in Bromley, England. He studied mathematics at Trinity College Cambridge, where he took his Ph.D. in 1970. His thesis was awarded the Rayleigh Prize for research excellence. He held positions at Lehigh University, Warwick University, University College of Swansea, the University of

Illinois, and Michigan State University before taking a permanent position in the mathematics department at the University of Missouri in 1979.

In 1984, Kalton was appointed the Luther Marion Defoe Distinguished Professor of Mathematics and then he was appointed to the Houchins Chair of Mathematics in 1985. In 1995, he was appointed a Curators' Professor, the highest recognition bestowed by the University of Missouri.

Among the many honors Nigel Kalton received, he was awarded the Chancellor's Award for outstanding research (at the University of Missouri) in 1984, the Weldon Springs Presidential Award for outstanding research (at the University of Missouri) in 1987, and the Banach Medal from the Polish Academy of Sciences in 2005 (the highest honor in his field).

During his career, he wrote over 270 articles and books, mentored 14 Ph.D. students, served on many editorial boards, and inspired countless mathematicians. For more information about Nigel Kalton, please visit the the Nigel Kalton Memorial Website developed by Fritz Gesztesy and hosted by the University of Missouri:

  http://kaltonmemorial.missouri.edu/

A very nice tribute to Nigel Kalton appeared in the Notices of the American Mathematical Society with contributions from Peter Casazza, Joe Diestel, Gilles Godefroy, Aleksander Pełczyński, and Roman Vershynin:

A Tribute to Nigel J. Kalton (1946–2010), Peter G. Casazza, Coordinating Editor, Notices of the AMS, Vol. 59, No. 7 (2012), pp. 942–951.

The article (which is [6] in the references) is a good starting point to learn about the life and work of Nigel Kalton.

Acknowledgements

I owe much to those who have helped me during the creation of this text, including friends, family, the staff at Springer, and the anonymous reviewers. In particular, I offer my sincere thanks to Greg Piepmeyer, who thoroughly read the first draft and provided invaluable suggestions and corrections. I also wish to thank Minerva Catral, Nadia Gal, Simon Cowell, Brian Tuomanen, and Daniel Fresen for many helpful comments. I am very grateful to Professor Gilles Pisier for sharing his knowledge of the Approximation Problem and providing insight into a source of great confusion. I am indebted to Jennifer Kalton and Gilles Godefroy for their kind support, without which this book would not exist, and I am deeply grateful to Professor Godefroy for all of his advice and generosity.

Of course, I owe my deepest gratitude to Professor Nigel Kalton, who was both an inspiration and mentor to me. Since I cannot show him the gratitude and admiration I feel, I offer this text as my humble tribute to his memory.

I assure the reader that if there are any errors in this book, they belong to me.

La Jolla, California   Adam Bowers
January 2014

Contents

Chapter 1
Introduction

Functional analysis is at its foundation the study of infinite-dimensional vector spaces. The goal of functional analysis is to generalize the well-known and very successful results of linear algebra on finite-dimensional vector spaces to the more complicated and subtle infinite-dimensional spaces. Of course, in infinite dimensions, certain issues (such as summability) become much more delicate, and accordingly additional structure is imposed. Perhaps the most natural structure imposed upon a vector space is that of a *norm*.

Definition 1.1   A *normed space* is a real or complex vector space X together with a real-valued function $x \mapsto \|x\|$ defined for all $x \in X$, called a *norm*, such that

(N1) $\|x\| \geq 0$ for all $x \in X$, and $\|x\| = 0$ if and only if $x = 0$,
(N2) $\|x + y\| \leq \|x\| + \|y\|$ for all x and y in X, and
(N3) $\|\lambda x\| = |\lambda|\,\|x\|$ for all $x \in X$ and scalars λ.

Property (N1) is known as *non-negativity* or *positive-definiteness*, Property (N2) is called *subadditivity* or the *triangle inequality*, and Property (N3) is called *homogeneity*.

A norm on a vector space X is essentially a way of measuring the size of an element in X. If x is in $\mathbb{R}$ (or $\mathbb{C}$), the norm of x is given by the absolute value (or modulus) of x, which is denoted (in either case) by $|x|$.

Definition 1.2   A *metric space* is a set X together with a map $d : X \times X \to \mathbb{R}$ that satisfies the following properties:

(M1) $d(x, y) \geq 0$ for all x and y in X, and $d(x, y) = 0$ if and only if $x = y$,
(M2) $d(x, y) \leq d(x, z) + d(z, y)$ for all x, y, and z in X, and
(M3) $d(x, y) = d(y, x)$ for all x and y in X.

The function d is said to be a *metric* on X. As was the case with a norm, (M1) is known as *non-negativity* and (M2) is called the *triangle inequality*. The final property, (M3), is called *symmetry*.

A metric is a measure of distance on the set X. Any normed space is *metrizable*; that is, we can always introduce a metric on a normed space by

$$d(x, y) = \|x - y\|, \quad (x, y) \in X \times X.$$

© Springer Science+Business Media, LLC 2014
A. Bowers, N. J. Kalton, *An Introductory Course in Functional Analysis*,
Universitext, DOI 10.1007/978-1-4939-1945-1_1

This metric measures the size of the displacement between x and y. While every norm determines a metric, not every metric space can have a norm. (We will see some examples of this in Sect. 5.3.)

Definition 1.3 Let X be a metric space with metric d. A sequence $(x_n)_{n=1}^{\infty}$ in X is said to *converge* to a point x in X if for any $\epsilon > 0$ there exists an $N \in \mathbb{N}$ such that $d(x, x_n) < \epsilon$ whenever $n \geq N$. In such a case, we say the sequence is *convergent* and write $\lim_{n \to \infty} x_n = x$. A sequence $(x_n)_{n=1}^{\infty}$ in X is called a *Cauchy sequence* if for any $\epsilon > 0$ there exists an $N \in \mathbb{N}$ such that $d(x_n, x_m) < \epsilon$ whenever $n \geq N$ and $m \geq N$.

It is well-known that a scalar-valued sequence converges if and only if it is a Cauchy sequence. This is not always true in infinite-dimensional normed spaces. A convergent sequence will always be a Cauchy sequence, but there may be Cauchy sequences that do not converge to an element of the normed space. Such a space is called *incomplete*, because we imagine it lacks certain desirable points. For this reason, we are generally interested in normed spaces that are *complete*; that is, spaces in which every Cauchy sequence does converge. For these spaces we have a special name.

Definition 1.4 A normed space X is called a *Banach space* if it is a complete metric space in the metric given by $d(x, y) = \|x - y\|$ for all $(x, y) \in X \times X$.

Suppose X and Y are Banach spaces (or simply normed spaces) over a scalar field $\mathbb{K}$ (which is $\mathbb{R}$ or $\mathbb{C}$). A basic goal of functional analysis is to solve equations of the form $Tx = y$, where $x \in X$, $y \in Y$, and $T : X \to Y$ is a given linear map (e.g., a differential or integral operator). Consequently, much study is made of linear maps on Banach spaces. Of particular interest are bounded linear maps.

Definition 1.5 Let X and Y be normed spaces. A linear map $T : X \to Y$ is called *bounded* if there exists a constant $M \geq 0$ such that $\|Tx\| \leq M$ for all $\|x\| \leq 1$. We denote the smallest such M by $\|T\|$; that is, if T is a bounded linear map, then $\|T\| = \sup\{\|Tx\| : \|x\| \leq 1\}$. A linear map that is not bounded is called *unbounded*.

An early and important result in functional analysis is the following proposition.

Proposition 1.6 *Let X and Y be normed spaces and suppose $T : X \to Y$ is a linear map. The following are equivalent:*

(i) *T is continuous,*
(ii) *T is continuous at zero, and*
(iii) *T is bounded.*

Proof The implication *(i)* $\Rightarrow$ *(ii)* is clear. To show *(ii)* $\Rightarrow$ *(iii)*, assume T is continuous at zero, but is unbounded. If T is unbounded, then for every $n \in \mathbb{N}$, there exists an element $x_n \in X$ such that $\|x_n\| \leq 1$ and $\|Tx_n\| \geq n$. By the choice of x_n, we have that

$$\left\| \frac{x_n}{n} \right\| \leq \frac{1}{n}, \quad n \in \mathbb{N}.$$

It follows that $\left\|\dfrac{x_n}{n}\right\| \to 0$ as $n \to \infty$, and consequently $\dfrac{x_n}{n} \to 0$ in X as $n \to \infty$. By continuity at zero, it must be that

$$\lim_{n \to \infty} \left\| T\left(\frac{x_n}{n}\right) \right\| = \|T(0)\| = 0.$$

However,

$$\left\| T\left(\frac{x_n}{n}\right) \right\| = \left\| \frac{T x_n}{n} \right\| \geq 1, \quad n \in \mathbb{N}.$$

This is a contradiction. Therefore, T must be bounded.

Now we show *(iii)* $\Rightarrow$ *(i)*. Assume that T is bounded and let $x \in X$. Then $\|T(\frac{x}{\|x\|})\| \leq \|T\|$, because $\frac{x}{\|x\|}$ has norm 1. By the linearity of T,

$$\|T(x)\| = \left\| T\left(\frac{x}{\|x\|}\right) \cdot \|x\| \right\| \leq \|T\| \|x\|, \quad x \in X.$$

To prove the continuity of T, we use the above inequality together with linearity:

$$\|Tx - Ty\| = \|T(x - y)\| \leq \|T\| \|x - y\|, \quad (x, y) \in X \times X.$$

This completes the proof. $\qquad\qquad\qquad\qquad\qquad\qquad\qquad\qquad\qquad\quad\square$

Remark 1.7. In the proof of Proposition 1.6, when showing *(iii)* $\Rightarrow$ *(i)*, we showed that $\|Tx - Ty\| \leq \|T\| \|x - y\|$ for all x and y in X. This allowed us to conclude that T was continuous. In fact, this property is stronger than continuity. A function $f : X \to Y$ that satisfies the condition $\|f(x) - f(y)\| \leq K\|x - y\|$ for a constant $K > 0$ is called *Lipschitz continuous* with *Lipschitz constant* K. Any bounded linear map T between Banach spaces is Lipschitz continuous (with Lipschitz constant $\|T\|$), but a Lipschitz continuous map between Banach spaces need not be linear.

The proof of Proposition 1.6 provides an alternative method for computing $\|T\|$.

Corollary 1.8 *Let X and Y be normed spaces. If $T : X \to Y$ is a linear map, then*

$$\|T\| = \inf\{K : \|Tx\| \leq K\|x\|, \ x \in X\}.$$

Furthermore, $\|Tx\| \leq \|T\| \|x\|$ for all $x \in X$.

Definition 1.9 If X and Y are normed spaces, then $\mathcal{L}(X, Y)$ denotes the set of all bounded linear maps (or *operators*) from X to Y. The set $\mathcal{L}(X, X)$ of bounded linear maps from X to itself is often denoted by $\mathcal{L}(X)$.

The word *operator* implies both boundedness and linearity. Frequently, however, we say $T \in \mathcal{L}(X, Y)$ is a *bounded linear operator*, even though this is redundant.

The set $\mathcal{L}(X, Y)$ is a vector space over the scalar field $\mathbb{K}$ with vector space operations given by:

$$(\alpha S + \beta T)(x) = \alpha S(x) + \beta T(x),$$

where α and β are in $\mathbb{K}$, S and T are in $\mathcal{L}(X, Y)$, and x is in X.

Proposition 1.10 *If X and Y are normed spaces, then*

$$\|T\| = \sup\{ \|Tx\| : \|x\| \leq 1\}, \quad T \in \mathcal{L}(X,Y),$$

defines a norm on $\mathcal{L}(X,Y)$.

Proof We verify the triangle inequality. Let $x \in X$ be such that $\|x\| \leq 1$. Then

$$\|(S+T)(x)\| = \|Sx + Tx\| \leq \|Sx\| + \|Tx\| \leq \|S\| + \|T\|.$$

Therefore,

$$\|S+T\| = \sup\{\|(S+T)(x)\| : \|x\| \leq 1\} \leq \|S\| + \|T\|.$$

The other conditions are also straightforward and are left to the reader. (See Exercise 1.5.)   □

Proposition 1.11 *If X is a normed space and Y is a Banach space, then $\mathcal{L}(X,Y)$ is a Banach space.*

Proof Suppose $(T_n)_{n=1}^{\infty}$ is a Cauchy sequence in $\mathcal{L}(X,Y)$. For each $x \in X$, the sequence $(T_n x)_{n=1}^{\infty}$ is a Cauchy sequence in Y because

$$\|T_n x - T_m x\| \leq \|T_n - T_m\| \, \|x\|,$$

for all m and n in $\mathbb{N}$. It follows that $\lim_{n \to \infty} T_n x$ exists for each $x \in X$. Denote this limit by Tx, so that

$$Tx = \lim_{n \to \infty} T_n x, \quad x \in X.$$

We will show that T is linear and bounded, and so $T \in \mathcal{L}(X,Y)$.

The linearity of T follows from the continuity of the vector space operations. To see this, observe that for all x and y in X,

$$T(x+y) = \lim_{n \to \infty} T_n(x+y) = \lim_{n \to \infty} T_n(x) + \lim_{n \to \infty} T_n(y) = Tx + Ty.$$

We now show that T is bounded. For each m and n in $\mathbb{N}$, we have (from Exercise 1.3)

$$\big| \|T_m\| - \|T_n\| \big| \leq \|T_m - T_n\|.$$

Thus, since $(T_n)_{n=1}^{\infty}$ is a Cauchy sequence in $\mathcal{L}(X,Y)$, it follows that $(\|T_n\|)_{n=1}^{\infty}$ is a Cauchy sequence of scalars. Consequently, $\sup_{n \in \mathbb{N}} \|T_n\| < \infty$, and so, for each $x \in X$,

$$\|Tx\| \leq \sup_{n \in \mathbb{N}} \|T_n x\| \leq \Big(\sup_{n \in \mathbb{N}} \|T_n\| \Big) \|x\|.$$

Therefore, T is bounded and $\|T\| \leq \sup_{n \in \mathbb{N}} \|T_n\|$.

It remains to show that $T = \lim_{n \to \infty} T_n$ in the norm on $\mathcal{L}(X,Y)$. Let $x \in X$ be such that $\|x\| \leq 1$. For $n \in \mathbb{N}$,

$$\|(T - T_n)(x)\| \leq \sup_{m \geq n} \|(T_m - T_n)(x)\| \leq \sup_{m \geq n} \|T_m - T_n\|,$$

by Corollary 1.8. Taking the supremum over all $x \in X$ with $\|x\| \leq 1$, we have

$$\|T - T_n\| \leq \sup_{m \geq n} \|T_m - T_n\|.$$

The sequence $(T_n)_{n=1}^{\infty}$ is a Cauchy sequence, and hence the result. $\square$

When X and Y are normed spaces, a (not necessarily linear) function $f : X \to Y$ is called an *isometry* if $\|f(x) - f(y)\| = \|x - y\|$ for all x and y in X. (In this case, we say f *preserves distances*.) When f is linear, this condition is equivalent to stating that $\|f(x)\| = \|x\|$ for all $x \in X$. (In which case, we say f *preserves the norm*.) If there exists an isometry between the spaces X and Y, they are said to be *isometric*.

We remind the reader that a one-to-one map is called an *injection* and an onto map is called a *surjection*. An injective surjection is also called a *bijection*.

Definition 1.12 Let X and Y be normed spaces. A bijective linear map $T : X \to Y$ is called an *isomorphism* if both T and T^{-1} are continuous. In such a case, we say X and Y are *isomorphic*. If in addition $\|Tx\| = \|x\|$ for all $x \in X$, then T is called an *isometric isomorphism*.

Let X be a vector space. If $\| \cdot \|_\alpha$ and $\| \cdot \|_\beta$ are two norms on X, then they are called *equivalent norms* provided there are constants $c_1 > 0$ and $c_2 > 0$ such that

$$c_1 \|x\|_\alpha \leq \|x\|_\beta \leq c_2 \|x\|_\alpha, \quad x \in X.$$

In such an event, there is a linear isomorphism between the spaces $(X, \| \cdot \|_\alpha)$ and $(X, \| \cdot \|_\beta)$. (See Exercise 1.13.)

The next definitions are central to all that follows, and even indicate the origins of the subject we are studying.

Definition 1.13 Let X be a normed space over the scalar field $\mathbb{K}$ (which is either $\mathbb{R}$ or $\mathbb{C}$). The *dual space* of X is the space $\mathcal{L}(X, \mathbb{K})$, which is denoted X^*. Elements of X^* are called *(bounded) linear functionals*.

We remark that for any normed space X, the dual space X^* is a Banach space with the norm given by $\|x^*\| = \sup\{|x^*(x)| : \|x\| \leq 1\}$ for $x^* \in X^*$. (See Exercise 1.6.) We will see that much can be learned about a Banach space by studying the properties of its dual space.

We close this section by recalling some topological notions that will appear frequently throughout the remainder of this text. We will encounter these concepts in greater generality in Chap. 5, but for now we will restrict our attention to metric spaces.

Definition 1.14 Let M be a metric space with metric d. For $x \in M$, the *open ball of radius δ about x* is the set $B(x, \delta) = \{y \in M : d(x, y) < \delta\}$. The *closed ball of radius δ about x* is the set $\overline{B}(x, \delta) = \{y \in M : d(x, y) \leq \delta\}$.

A subset U of M is called *open* if for every $x \in U$ there exists a $\delta > 0$ such that $B(x, \delta) \subseteq U$. A set is called *closed* if its complement is open. For any subset E of M, the *interior* of E, denoted int(E), is the union of all open sets that are subsets of E. The *closure* of E is denoted $\overline{E}$ and is the intersection of all closed sets which contain E as a subset. If $\overline{E} = M$, then E is called *dense* in M. If there exists a countable dense subset of M, then M is called *separable*.

We remark that a set E in a metric space is closed if and only if any convergent sequence $(x_n)_{n=1}^{\infty}$, where $x_n \in E$ for all $n \in \mathbb{N}$, converges to some $x \in E$. (See Exercise 1.10.)

For a normed space, we have special terminology and notation for the open ball $B(0, 1)$ and the closed ball $\overline{B}(0, 1)$.

Definition 1.15 If X is a normed space, then the *closed unit ball* of X is the set $B_X = \{x \in X : \|x\| \leq 1\}$ and the *open unit ball* of X is the set $U_X = \{x \in X : \|x\| < 1\}$.

Observe that $U_X = \text{int}(B_X)$. We will use both notations for the open unit ball in X. Also, if X is a Banach space (or simply a normed vector space), and consequently has addition and scalar multiplication, we can write the closed and open balls in X (respectively) as:

$$\overline{B}(x, \delta) = x + \delta B_X \quad \text{and} \quad B(x, \delta) = x + \text{int}(\delta B_X).$$

This is the notation we will generally use when working in a Banach space (or normed vector space), in order to emphasize the underlying linear structure.

A set K in a metric space M is called *compact* if any collection $\{U_\alpha\}_{\alpha \in A}$ of open sets for which $K \subseteq \bigcup_{\alpha \in A} U_\alpha$ contains a finite subcollection $\{U_{\alpha_1}, \dots, U_{\alpha_n}\}$ such that $K \subseteq U_{\alpha_1} \cup \cdots \cup U_{\alpha_n}$. In particular, any compact set in a metric space can be covered by finitely many open balls of radius δ for any $\delta > 0$.

Compact sets have many desirable properties. For example, if K is a compact set and $f : K \to \mathbb{R}$ is a continuous function, then f is bounded and attains its maximum and minimum values. (This is the content of the Extreme Value Theorem.)

If M_1 and M_2 are metric spaces, then a map $f : M_1 \to M_2$ is called a *homeomorphism* if it is a continuous bijection with continuous inverse. If such a homeomorphism exists, we say that M_1 and M_2 are *homeomorphic*. Homeomorphic spaces are considered identical from a topological point of view.

Notice that an isomorphism is a linear homeomorphism. Later, we will show that any bounded linear bijection between Banach spaces is necessarily an isomorphism. (See Corollary 4.30.)

Digression: Historical Comments

Functional analysis has its beginnings in the work of Fourier (1768–1830) and the study of differential equations. Fourier's goal was to find solutions to differential equations such as

$$y + \frac{d^2 y}{dx^2} = g(x),$$

where g is some prescribed function. To this end, he applied to this setting the techniques of the already well-developed theory of linear algebra. In 1902, Fredholm applied similar techniques to solve integral equations. Fredholm was trying to find continuous functions f on the unit interval $[0, 1]$ that satisfied equations such as

$$f(x) + \int_0^1 K(x, y) f(y) \, dy = g(x), \quad x \in [0, 1],$$

where K and g were given.

The space of continuous functions on $[0, 1]$ is denoted $C[0, 1]$. There is a natural norm on this space, called the *supremum norm*:

$$\|f\| = \max_{t \in [0,1]} |f(t)|, \quad f \in C[0, 1].$$

Note that this maximum is attained, because f is continuous and $[0, 1]$ is compact. (It is for this reason we compute the *maximum* over t in $[0, 1]$ instead of the *supremum*.) The space $(C[0, 1], \|\cdot\|)$ is a Banach space, and was the first Banach space considered (although the term was not applied until much later).

The space $C[0, 1]$ can be generalized. If K is any compact Hausdorff space, the space $C(K)$ denotes the collection of all scalar-valued continuous functions on K. When equipped with the norm

$$\|f\| = \max_{x \in K} |f(x)|, \quad f \in C(K),$$

this space is a Banach space.

With the advent of Lebesgue's thesis in 1903, and the subsequent development of measure theory, more examples of Banach spaces were discovered. In 1908, Hilbert and Schmidt studied $L_2(0, 1)$, and F. Riesz studied the more general spaces $L_p(0, 1)$ for $p \in [1, \infty)$. (See Appendix A for the relevant definitions.) The notion of a measurable function was critical in the understanding of these spaces—when restricted to continuous functions they are not complete. (However, the continuous functions are dense in $L_p(0, 1)$ for all $p \in [1, \infty)$, by Lusin's Theorem (Theorem A.36).)

Norbert Wiener (in 1919) and Stefan Banach (in his thesis in 1920) independently introduced the axioms of what are now called Banach spaces. Banach and his school at Lwów made many advances in the area, but their work was cut short by the Second World War. Banach himself died shortly after the war in 1945. Much of the work of the Lwów school survived in what is now called the *Scottish Book* [24].

Exercises

Exercise 1.1 Show that in a metric space a convergent sequence is a Cauchy sequence.

Exercise 1.2 Let X be a normed space. A sequence $(x_n)_{n=1}^\infty$ in X is said to be *bounded* if there exists a $M > 0$ such that $\|x_n\| \leq M$ for all $n \in \mathbb{N}$. Show that in a normed space a convergent sequence is bounded. Is the converse true? That is, must a bounded sequence necessarily converge?

Exercise 1.3 Let X be a normed space. Show that $|\,\|x\| - \|y\|\,| \leq \|x - y\|$ for all x and y in X. (This is sometimes called the *reverse triangle inequality*.)

Exercise 1.4 Let X and Y be normed vector spaces. A function $f : X \to Y$ is said to be *bounded on X* if there exists a $M > 0$ such that $\|f(x)\| \leq M$ for all $x \in X$. Show that a nonzero bounded linear operator $T : X \to Y$ is not bounded on X.

Exercise 1.5 Finish the proof of Proposition 1.10.

Exercise 1.6 Let X be a normed space. Prove that X^* is a Banach space.

Exercise 1.7 Let $C[0, 1]$ be the space of continuous functions on $[0, 1]$ and define a norm $\| \cdot \|$ on $C[0, 1]$ by $\|f\| = \max_{t \in [0,1]} |f(t)|$. Show that the map

$$A(f) = \int_0^1 f(x)\, dx$$

defines a bounded linear functional on the normed vector space $(C[0, 1], \| \cdot \|)$.

Exercise 1.8 Let $C^{(1)}[0, 1]$ be the space of all continuous functions on $[0, 1]$ that have continuous derivative. (This space is called the space of *continuously differentiable* functions.) Define a norm $\| \cdot \|$ on $C^{(1)}[0, 1]$ by $\|f\| = \max_{t \in [0,1]} |f(t)|$. Show that the map

$$B(f) = f'(0)$$

defines a linear functional on the normed vector space $(C^{(1)}[0, 1], \| \cdot \|)$ that is not bounded.

Exercise 1.9 Let X be a normed vector space and suppose T and S are in X^*. If $T(x) = 0$ implies that $S(x) = 0$, show that there is a constant c such that $S(x) = c\, T(x)$ for all $x \in X$.

Exercise 1.10 Show that a subset of a metric space M is closed if and only if it is *sequentially closed*. That is, show that E is a closed subset of M if and only if every convergent sequence in E has its limit in E.

Exercise 1.11 Let X and Y be normed vector spaces and suppose $T : X \to Y$ is a linear surjection. Show that T is an isomorphism if and only if there are constants $c_1 > 0$ and $c_2 > 0$ such that

$$c_1 \|x\| \leq \|Tx\| \leq c_2 \|x\|, \quad x \in X.$$

Exercise 1.12 Let X and Y be Banach spaces and let $T \in \mathcal{L}(X, Y)$. Suppose T is not bounded below; that is, there does not exist a $c > 0$ such that $\|T(x)\| \geq c\|x\|$ for all $x \in X$. Show there exists a sequence $(x_n)_{n=1}^{\infty}$ of norm one elements in X such that $\lim_{n \to \infty} \|T(x_n)\| = 0$.

Exercise 1.13 Suppose $\| \cdot \|_\alpha$ and $\| \cdot \|_\beta$ are equivalent norms on a vector space X. Show that there is an isomorphism from $(X, \| \cdot \|_\alpha)$ onto $(X, \| \cdot \|_\beta)$.

Exercise 1.14 Let $(X, \| \cdot \|_X)$ and $(Y, \| \cdot \|_Y)$ be normed vector spaces and assume the map $T \in \mathcal{L}(X, Y)$ is an isomorphism. Define a scalar-valued function $\| \cdot \|_T$ on X by

$$\|x\|_T = \|T(x)\|_Y, \quad x \in X.$$

Prove that $\| \cdot \|_T$ is a norm on X and show that it is equivalent to the original norm $\| \cdot \|_X$.

Exercise 1.15 Suppose $\| \cdot \|_\alpha$ and $\| \cdot \|_\beta$ are two norms on a vector space X. Show that the two norms are equivalent if and only if the normed spaces $(X, \| \cdot \|_\alpha)$ and $(X, \| \cdot \|_\beta)$ have the same open sets.

Exercise 1.16 Suppose $\| \cdot \|_\alpha$ and $\| \cdot \|_\beta$ are two equivalent norms on a vector space X. Show that $(X, \| \cdot \|_\alpha)$ is a complete normed space if $(X, \| \cdot \|_\beta)$ is a complete normed space.

Exercise 1.17 Let X and Y be isomorphic normed vector spaces. Use the preceding exercises to conclude that X is a Banach space if and only if Y is a Banach space.

Chapter 2
Classical Banach Spaces and Their Duals

In the next two sections, we will consider the classical sequence and function spaces. The main purpose of these sections is to make the necessary definitions and to identify the dual spaces for these classical spaces. We will therefore take for granted that the various Banach spaces are indeed Banach spaces—putting off until Sect. 2.3 the proofs that they are complete in the given norms.

2.1 Sequence Spaces

In the context of sequence spaces, we denote by e_n the sequence with 1 in the nth coordinate, and 0 elsewhere, so that $e_n = (0, \dots, 0, 1, 0, \dots)$ for all $n \in \mathbb{N}$. Also, we let $e = (1, 1, 1, \dots)$ be the constant sequence with 1 in every coordinate (not to be confused with the base of the natural logarithm $e \approx 2.718$).

Definition 2.1 The set ℓ_p of *p-summable sequences* for $p \in [1, \infty)$ is the collection of sequences

$$\ell_p = \left\{ (\xi_1, \xi_2, \dots, \xi_n, \dots) : \sum_{n=1}^{\infty} |\xi_n|^p < \infty \right\}.$$

Define the *p-norm* on ℓ_p by

$$\|\xi\|_p = \left(\sum_{n=1}^{\infty} |\xi_n|^p \right)^{1/p}, \quad \xi = (\xi_n)_{n=1}^{\infty} \in \ell_p.$$

The set ℓ_p is a vector space under component-wise addition and scalar multiplication. (This is a nontrivial fact which we will take as given. [See Theorem A.27.]) Furthermore, ℓ_p is a Banach space when given the *p*-norm (for $1 \leq p < \infty$). We leave the proof of this fact to the exercises. (See Exercise 2.7.)

The next lemma will identify the dual space of ℓ_p for $p \in (1, \infty)$.

Lemma 2.2 *For $p \in (1, \infty)$, the space ℓ_p^* can be identified with ℓ_q, where $\frac{1}{p} + \frac{1}{q} = 1$.*

© Springer Science+Business Media, LLC 2014
A. Bowers, N. J. Kalton, *An Introductory Course in Functional Analysis*,
Universitext, DOI 10.1007/978-1-4939-1945-1_2

Proof For simplicity, we start by assuming the scalars are real. (We will consider the complex case at the end of the proof.)

We wish to identify the dual space of ℓ_p for $p \in (1, \infty)$ as the sequence space ℓ_q, where $\frac{1}{p} + \frac{1}{q} = 1$. First, we will demonstrate how elements in ℓ_q determine linear functionals on ℓ_p. Let $\eta = (\eta_n)_{n=1}^{\infty}$ be an element in ℓ_q and define a scalar-valued function ϕ_η on ℓ_p by

$$\phi_\eta(\xi) = \sum_{n=1}^{\infty} \xi_n \eta_n, \tag{2.1}$$

where $\xi = (\xi_n)_{n=1}^{\infty}$ is any sequence in ℓ_p. By Hölder's Inequality (Theorem A.29), this series is absolutely convergent (whence ϕ_η is linear) and

$$|\phi_\eta(\xi)| \leq \left(\sum_{n=1}^{\infty} |\xi_n|^p \right)^{1/p} \left(\sum_{n=1}^{\infty} |\eta_n|^q \right)^{1/q} = \|\xi\|_p \|\eta\|_q.$$

It follows that ϕ_η is a bounded linear functional on ℓ_p and $\|\phi_\eta\| \leq \|\eta\|_q$.

We claim that $\|\phi_\eta\|$ is in fact equal to $\|\eta\|_q$. In order to show this, it will suffice to find a sequence ξ in ℓ_p such that $\|\xi\|_p = 1$ and $\phi_\eta(\xi) = \|\eta\|_q$. We begin by constructing a sequence $\zeta = (\zeta_n)_{n=1}^{\infty}$ so that $\zeta_n = |\eta_n|^{q-1} (\text{sign } \eta_n)$ for each $n \in \mathbb{N}$. Then

$$\sum_{n=1}^{\infty} |\zeta_n|^p = \sum_{n=1}^{\infty} |\eta_n|^{(q-1)p} = \sum_{n=1}^{\infty} |\eta_n|^q,$$

where $(q-1)p = q$ follows from the assumption that $\frac{1}{p} + \frac{1}{q} = 1$. Consequently, the sequence ζ is in ℓ_p and $\|\zeta\|_p = \|\eta\|_q^{q/p}$. Observe that

$$\phi_\eta(\zeta) = \sum_{n=1}^{\infty} \zeta_n \cdot \eta_n = \sum_{n=1}^{\infty} |\eta_n|^{q-1} (\text{sign } \eta_n) \cdot \eta_n = \sum_{n=1}^{\infty} |\eta_n|^q = \|\eta\|_q^q.$$

Let $\xi = \frac{\zeta}{\|\zeta\|_p}$. Then ξ is a sequence in ℓ_p such that $\|\xi\|_p = 1$ and such that

$$\phi_\eta(\xi) = \frac{\phi_\eta(\zeta)}{\|\zeta\|_p} = \frac{\|\eta\|_q^q}{\|\eta\|_q^{q/p}} = \|\eta\|_q^{q - \frac{q}{p}} = \|\eta\|_q.$$

Therefore, for any $\eta = (\eta_n)_{n=1}^{\infty}$ in ℓ_q, there is a linear functional ϕ_η on ℓ_p such that $\|\eta\|_q = \|\phi_\eta\|$ and such that $\phi_\eta(\xi) = \sum_{n=1}^{\infty} \xi_n \eta_n$ for all $\xi = (\xi_n)_{n=1}^{\infty}$ in ℓ_p.

We have demonstrated that any sequence in ℓ_q determines a bounded linear functional on ℓ_p. Next, we will show that all linear functionals on ℓ_p can be obtained in this way. Let $\psi \in \ell_p^*$ and define a sequence $\eta = (\eta_i)_{i=1}^{\infty}$ by letting $\eta_i = \psi(e_i)$ for each $i \in \mathbb{N}$. We will show that the sequence η is an element of ℓ_q such that $\|\eta\|_q = \|\psi\|$ and such that $\psi(\xi) = \sum_{i=1}^{\infty} \xi_i \eta_i$ for all $\xi = (\xi_i)_{i=1}^{\infty}$ in ℓ_p.

First, we will show that η is in fact an element of ℓ_q. For each $i \in \mathbb{N}$, define $\zeta_i = |\eta_i|^{q-1} (\text{sign } \eta_i)$. Then, for any $n \in \mathbb{N}$, we have

$$\sum_{i=1}^{n} |\zeta_i|^p = \sum_{i=1}^{n} |\eta_i|^{(q-1)p} = \sum_{i=1}^{n} |\eta_i|^q.$$

Computing the ℓ_p-norm of the finite sequence $\sum_{i=1}^{n} \zeta_i e_i$, we conclude that

$$\left\| \sum_{i=1}^{n} \zeta_i e_i \right\|_p = \left(\sum_{i=1}^{n} |\zeta_i|^p \right)^{\frac{1}{p}} = \left(\sum_{i=1}^{n} |\eta_i|^q \right)^{\frac{1}{p}}.$$

By assumption, the linear functional ψ is bounded on ℓ_p, and consequently

$$\left| \psi \left(\sum_{i=1}^{n} \zeta_i e_i \right) \right| \leq \|\psi\| \left\| \sum_{i=1}^{n} \zeta_i e_i \right\|_p = \|\psi\| \left(\sum_{i=1}^{n} |\eta_i|^q \right)^{\frac{1}{p}}. \tag{2.2}$$

However, computing directly, we obtain

$$\psi \left(\sum_{i=1}^{n} \zeta_i e_i \right) = \sum_{i=1}^{n} \zeta_i \, \psi(e_i) = \sum_{i=1}^{n} \zeta_i \, \eta_i = \sum_{i=1}^{n} |\eta_i|^q. \tag{2.3}$$

From (2.2) and (2.3), it follows that

$$\sum_{i=1}^{n} |\eta_i|^q \leq \|\psi\| \left(\sum_{i=1}^{n} |\eta_i|^q \right)^{\frac{1}{p}}.$$

Dividing, we see that

$$\|\psi\| \geq \left(\sum_{i=1}^{n} |\eta_i|^q \right)^{1-\frac{1}{p}} = \left(\sum_{i=1}^{n} |\eta_i|^q \right)^{\frac{1}{q}}.$$

This inequality holds for all $n \in \mathbb{N}$, and so $\eta \in \ell_q$ and $\|\psi\| \geq \|\eta\|_q$.

It remains to show that $\|\psi\| \leq \|\eta\|_q$ and that $\psi = \phi_\eta$, where ϕ_η is defined by (2.1). Since we have already demonstrated that $\|\phi_\eta\| \leq \|\eta\|_q$, it suffices to show that $\psi = \phi_\eta$.

Suppose $\xi = (\xi_i)_{i=1}^{\infty} \in \ell_p$ and let $\xi^{(n)} = (\xi_1, \ldots, \xi_n, 0, \ldots)$ for each $n \in \mathbb{N}$. We claim that $\xi^{(n)}$ converges to ξ in the norm on ℓ_p. To see this, observe that $\sum_{i=1}^{\infty} |\xi_i|^p < \infty$, by assumption, and consequently

$$\lim_{n \to \infty} \left\| \xi - \xi^{(n)} \right\|_p = \lim_{n \to \infty} \left(\sum_{i=n+1}^{\infty} |\xi_i|^p \right)^{1/p} = 0. \tag{2.4}$$

Now, observe that

$$\psi\left(\xi^{(n)}\right) = \psi\left(\sum_{i=1}^{n} \xi_i \, e_i\right) = \sum_{i=1}^{n} \xi_i \, \psi(e_i) = \sum_{i=1}^{n} \xi_i \, \eta_i.$$

Therefore, by the continuity of ψ,

$$\psi(\xi) = \lim_{n \to \infty} \psi(\xi^{(n)}) = \sum_{i=1}^{\infty} \xi_i \, \eta_i = \phi_\eta(\xi).$$

It follows that $\psi = \phi_\eta$, which is the desired result.

Now assume the scalar field is $\mathbb{C}$. The proof in this case is essentially the same. However, when defining ζ_n, for $n \in \mathbb{N}$, let $\zeta_n = |\eta_n|^{q-1}\rho_n$, where ρ_n is a complex number such that $|\rho_n| = 1$ and $\eta_n \, \rho_n = |\eta_n|$. The argument proceeds as it did in the real case.   □

The previous theorem identified the dual space of ℓ_p for $p \in (1, \infty)$ as the space ℓ_q, where $\frac{1}{p} + \frac{1}{q} = 1$, via the *dual space action*

$$\eta(\xi) = \sum_{n=1}^{\infty} \xi_n \eta_n,$$

where $\xi = (\xi_n)_{n=1}^{\infty}$ is in ℓ_p and $\eta = (\eta_n)_{n=1}^{\infty}$ is in ℓ_q. In the above equation, we write $\eta(\xi)$ as shorthand for $\phi_\eta(\xi)$, where ϕ_η is the linear functional corresponding to η that appears in (2.1). The object η is a sequence in the space ℓ_q, but we write $\eta(\xi)$ because we are viewing η as a linear functional in ℓ_p^*.

Next, we wish to identify the dual space of ℓ_1. In order to do this, we must introduce a new space of sequences.

Definition 2.3   The set ℓ_∞ of *bounded sequences* is the collection of sequences

$$\ell_\infty = \left\{ (\xi_n)_{n=1}^{\infty} : \sup_{n \in \mathbb{N}} |\xi_n| < \infty \right\}.$$

Define the *supremum norm* on ℓ_∞ by

$$\|\xi\|_\infty = \sup_{n \in \mathbb{N}} |\xi_n|, \quad \xi = (\xi_n)_{n=1}^{\infty} \in \ell_\infty.$$

The set ℓ_∞ is a vector space under component-wise addition and scalar multiplication and is a Banach space when given the supremum norm.

Lemma 2.4   *The space ℓ_1^* can be identified with ℓ_∞.*

Proof   The proof is similar to the proof of Lemma 2.2 and is left to the reader. (See Exercise 2.2.) As in Lemma 2.2, the dual space action is

$$\eta(\xi) = \sum_{n=1}^{\infty} \xi_n \eta_n,$$

where now $\xi = (\xi_n)_{n=1}^{\infty}$ is in ℓ_1 and $\eta = (\eta_n)_{n=1}^{\infty}$ is in ℓ_∞.   □

Since we identify the dual space of ℓ_p with ℓ_q, it is standard to write $\ell_p^* = \ell_q$. When we write this, however, it is understood that we mean there is a way of identifying the linear functionals in ℓ_p^* with the sequences in ℓ_q and the identification is shown explicitly by the dual space action $\eta(\xi) = \sum_{n=1}^{\infty} \xi_n \eta_n$, where $\xi = (\xi_n)_{n=1}^{\infty}$ is in ℓ_p and $\eta = (\eta_n)_{n=1}^{\infty}$ is in ℓ_q.

We summarize our results in the following theorem.

Theorem 2.5 Let $p \in [1, \infty)$ and suppose q is such that $\frac{1}{p} + \frac{1}{q} = 1$, with the convention that $q = \infty$ when $p = 1$. Then $\ell_p^* = \ell_q$.

Proof See Lemmas 2.2 and 2.4.

The relationship between the exponents p and q in Theorem 2.5 motivates the next definition.

Definition 2.6 If $p \in [1, \infty)$ and if q is such that $\frac{1}{p} + \frac{1}{q} = 1$, with the convention that $q = \infty$ when $p = 1$, then p and q are called *conjugate exponents*.

If p and q are conjugate exponents that are both finite, then $\ell_p^* = \ell_q$ and $\ell_q^* = \ell_p$. Since $\ell_1^* = \ell_\infty$, it is natural to ask if ℓ_1 is the dual space of ℓ_∞. While $\ell_1 \subseteq \ell_\infty^*$, the spaces do not coincide. The argument used in the proof of Lemma 2.2 for $p \in (1, \infty)$ fails in the case $p = \infty$ because there exist bounded sequences $(\xi_n)_{n=1}^{\infty}$ that do not satisfy the equation that corresponds to (2.4) for the case $p = \infty$. That is, we can find $\xi \in \ell_\infty$ such that

$$\lim_{n \to \infty} \|\xi - \xi^{(n)}\|_\infty = \lim_{n \to \infty} \left(\sup_{k > n} |\xi_k| \right) \neq 0, \tag{2.5}$$

where $\xi^{(n)} = (\xi_1, \ldots, \xi_n, 0, \ldots)$. As a simple example, let $\xi = e$, the constant sequence having every term equal to 1. We have that $\|e\|_\infty = 1$, and so e is an element of ℓ_∞, but $\|e - e^{(n)}\|_\infty = 1$ for all $n \in \mathbb{N}$.

Let us now consider the space of sequences for which the limit in (2.1.5) is 0.

Definition 2.7 Let c_0 be the space of all sequences converging to 0:

$$c_0 = \left\{ (\xi_n)_{n=1}^{\infty} : \lim_{n \to \infty} \xi_n = 0 \right\}.$$

The space c_0 is a Banach space with the supremum norm

$$\|\xi\|_\infty = \sup_{n \in \mathbb{N}} |\xi_n|, \quad \xi = (\xi_n)_{n=1}^{\infty} \in c_0.$$

Theorem 2.8 $c_0^* = \ell_1$.

The proof is similar to that of Lemma 2.2 and is left as an exercise for the reader. (See Exercise 2.2.) The last sequence space we discuss is the space c.

Definition 2.9 Let c be the space of all convergent sequences:

$$c = \left\{ (\xi_n)_{n=1}^{\infty} : \lim_{n \to \infty} \xi_n \text{ exists} \right\}.$$

The space c is also a Banach space with the supremum norm. Perhaps surprisingly, the dual space of c is also ℓ_1, albeit with a slightly different dual space action. (See

Example 2.22.) It is straightforward to show that c_0 is a closed subspace of c, and in turn c is a closed subspace of ℓ_∞. (See Exercise 2.1.)

2.2 Function Spaces

In this section, let (Ω, Σ, μ) be a measure space, where μ is a positive measure. The theorems in this section are true for positive σ-finite measure spaces, but for simplicity we will assume $\mu(\Omega) < \infty$. As before, $\mathbb{K}$ denotes the underlying scalar field, which is either $\mathbb{R}$ or $\mathbb{C}$. We begin by recalling some definitions from measure theory. (See Appendix A for a more detailed discussion.)

Definition 2.10 If A is a subset of Ω, then the *characteristic function* of A is the function

$$\chi_A(\omega) = \begin{cases} 1 & \text{if } \omega \in A, \\ 0 & \text{if } \omega \notin A. \end{cases}$$

The characteristic function of A is a measurable function if and only if A is a measurable subset of Ω.

Definition 2.11 For $p \in [1, \infty)$, the set of *p-integrable functions* (or L_p-functions) on (Ω, Σ, μ) is the collection

$$L_p(\Omega, \Sigma, \mu) = \left\{ f : \Omega \to \mathbb{K} \text{ a measurable function} : \int_\Omega |f|^p \, d\mu < \infty \right\}.$$

We often write this space as $L_p(\Omega, \mu)$ or $L_p(\mu)$ when there is no risk of confusion.

The set $L_p(\mu)$ is actually a collections of *equivalence classes* of measurable functions. Two functions in $L_p(\mu)$ are considered equivalent if they differ only on a set of μ-measure zero. Despite this, we will usually speak of the elements in $L_p(\mu)$ as functions, rather than equivalence classes of functions. We remark that the set $L_p(\mu)$ is a vector space under pointwise addition and scalar multiplication. (As with ℓ_p in the previous section, this is a nontrivial result. [See Theorem A.27.])

Define the *p-norm* on $L_p(\mu)$ by

$$\|f\|_p = \left(\int_\Omega |f|^p \, d\mu \right)^{\frac{1}{p}}, \quad f \in L_p(\mu).$$

We will show that $L_p(\mu)$ is a Banach space in the p-norm (for $1 \le p < \infty$) in the next section. (See Theorem 2.25.)

As in the case of sequence spaces, we must consider the case $p = \infty$ separately.

Definition 2.12 Let (Ω, Σ, μ) be a measure space. The *essential supremum norm* of a measurable function f is defined to be

$$\|f\|_\infty = \inf \{K : \mu(|f| > K) = 0\} .$$

The set of *essentially bounded functions* (or L_∞-functions) on (Ω, Σ, μ) is the collection

$$L_\infty(\Omega, \Sigma, \mu) = \{f : \Omega \to \mathbb{K} \text{ a measurable function} : \|f\|_\infty < \infty\}.$$

We often write $L_\infty(\Omega, \mu)$ or $L_\infty(\mu)$ when there is no risk of confusion.

As is the case for $L_p(\mu)$ when p is finite, the set $L_\infty(\mu)$ is a collection of equivalence classes of measurable functions and (as before) we consider two functions to be equal in $L_\infty(\mu)$ if they differ only on a set of μ-measure zero.

The set $L_\infty(\mu)$ is a vector space under pointwise addition and scalar multiplication. The essential supremum defines a norm on $L_\infty(\mu)$ and $L_\infty(\mu)$ is a Banach space when given this norm. (See Theorem 2.26.) We use the terminology "essentially bounded" to describe the functions in $L_\infty(\mu)$ and call $\|f\|_\infty$ the "essential supremum" of $|f|$ in $L_\infty(\mu)$, because the quantity $\|f\|_\infty$ is the smallest number K such that $|f| \le K$ a.e.(μ).

We wish to identify the dual space of $L_p(\mu)$, where $p \in [1, \infty)$. We will see that, analogous to the case of sequence spaces, the dual space of $L_p(\mu)$ is $L_q(\mu)$, where p and q are conjugate exponents. In this case, however, the dual action of $L_q(\mu)$ on $L_p(\mu)$ is given by integration. That is, if $f \in L_p(\mu)$ and $g \in L_q(\mu)$, then the dual action of g on f is given by

$$g(f) = \int_\Omega fg d\mu.$$

Notice that g is a function on Ω. Here, however, we write $g(f)$ because we view g as an element of the dual space $L_p(\mu)^*$. As before, we write $L_p(\mu)^* = L_q(\mu)$ to indicate the identification of linear functionals in $L_p(\mu)^*$ with elements of the function space $L_q(\mu)$.

Theorem 2.13 *Let (Ω, Σ, μ) be a positive finite measure space. If p and q are conjugate exponents, where $p \in [1, \infty)$, then $L_p(\mu)^* = L_q(\mu)$.*

Proof Start by assuming the scalars are real. We begin with the case $p \in (1, \infty)$. Let $g \in L_q(\mu)$, where $\frac{1}{p} + \frac{1}{q} = 1$, and define a scalar-valued function ϕ_g on $L_p(\mu)$ by

$$\phi_g(f) = \int_\Omega fg d\mu, \quad f \in L_p(\mu). \tag{2.6}$$

We will show first that ϕ_g is a linear functional on $L_p(\mu)$ such that $\|\phi_g\| = \|g\|_q$, and then we will show that all linear functionals on $L_p(\mu)$ can be achieved in this way.

We note that ϕ_g is linear (by the linearity of the integral) and $|\phi_g(f)| \le \|f\|_p \|g\|_q$ (by Hölder's Inequality). Thus, ϕ_g is a bounded linear functional and $\|\phi_g\| \le \|g\|_q$. In order to show equality of the norms, it suffices to find a function f in $L_p(\mu)$ such that $\|f\|_p = 1$ and such that $\phi_g(f) = \|g\|_q$. First, define a scalar-valued function h on Ω by letting $h(x) = |g(x)|^{q-1} (\text{sign } g(x))$ for each $x \in \Omega$. Then

$$\|h\|_p = \left(\int_\Omega |g|^{(q-1)p}\, d\mu\right)^{\frac{1}{p}} = \left(\int_\Omega |g|^q\, d\mu\right)^{\frac{1}{p}} = \|g\|_q^{q/p},$$

where $(q-1)p = q$ follows from the assumption that $\frac{1}{p} + \frac{1}{q} = 1$. It follows that h is in $L_p(\mu)$. Next, observe that $\phi_g(h) = \|g\|_q^q$. Now, let $f = \frac{h}{\|h\|_p}$. Then $\|f\|_p = 1$ and

$$\phi_g(f) = \frac{\phi_g(h)}{\|h\|_p} = \frac{\|g\|_q^q}{\|g\|_q^{q/p}} = \|g\|_q.$$

Therefore, ϕ_g is a bounded linear functional on $L_p(\mu)$ and $\|\phi_g\| = \|g\|_q$.

We now wish to show that any bounded linear functional in $L_p(\mu)^*$ can be written as in (2.6) for some g in $L_q(\mu)$. To that end, let ψ be a bounded linear functional in the dual space $L_p(\mu)^*$. Define a measure ν on (Ω, Σ) by $\nu(A) = \psi(\chi_A)$ for all $A \in \Sigma$. It is routine to show that ν is finitely additive (by the linearity of ψ), and it is countably additive by the continuity of ψ. We also claim that $\nu \ll \mu$. That is, ν is absolutely continuous with respect to μ. To see this, suppose $A \in \Sigma$ is such that $\mu(A) = 0$. Because ψ is bounded,

$$|\nu(A)| = |\psi(\chi_A)| \le \|\psi\|\, \|\chi_A\|_{L_p(\mu)} = \|\psi\|\, \mu(A)^{1/p} = 0.$$

By the Radon–Nikodým Theorem (Theorem A.24), there exists a measurable function $g \in L_1(\mu)$ such that $\nu(A) = \int_A g\, d\mu$ for all $A \in \Sigma$. Therefore, for every $A \in \Sigma$,

$$\psi(\chi_A) = \nu(A) = \int_A g\,d\mu = \int_\Omega \chi_A g\, d\mu.$$

By linearity, it follows that $\psi(f) = \int_\Omega fg\,d\mu$ whenever f is a simple measurable function. Let $f \in L_\infty(\mu)$ be a real nonnegative essentially bounded measurable function. Since (Ω, Σ, μ) is a finite measure space, it follows that $f \in L_p(\mu)$. Thus, there exists a sequence of simple measurable functions $(f_n)_{n=1}^\infty$ such that $f_n \ge f_{n-1}$ for all $n \in \mathbb{N}$, and such that $\|f - f_n\|_p \to 0$ as $n \to \infty$. By the continuity of ψ,

$$\psi(f) = \lim_{n\to\infty} \psi(f_n) = \lim_{n\to\infty} \int_\Omega f_n g\,d\mu.$$

Therefore, by Lebesgue's Dominated Convergence Theorem (Theorem A.17),

$$\psi(f) = \int_\Omega fg\,d\mu, \quad f \in L_\infty(\mu) \cap L_p(\mu),\ f \ge 0.$$

To extend this to an arbitrary real function in $L_\infty(\mu) \cap L_p(\mu)$, let

$$f^+ = f\, \chi_{\{x:\, f(x) \ge 0\}} \quad \text{and} \quad f^- = -f\, \chi_{\{x:\, f(x) < 0\}},$$

and observe that $f = f^+ - f^-$.

We claim that $g \in L_q(\mu)$. For each $n \in \mathbb{N}$, define a function h_n on Ω by letting $h_n = \chi_{\{|g| \leq n\}} |g|^{q-1} (\text{sign } g)$. Then, for each $n \in \mathbb{N}$, we have $h_n \in L_\infty(\mu) \cap L_p(\mu)$ and

$$\psi(h_n) = \int_{\{|g| \leq n\}} |g|^q \, d\mu.$$

By assumption, the linear functional ψ is bounded on $L_p(\mu)$, and so it follows that $|\psi(h_n)| \leq \|\phi\| \|h_n\|_p$. Computing the L_p-norm of h_n, and once again observing that $(q-1)p = q$, we see that

$$\|h_n\|_p = \left(\int_{\{|g| \leq n\}} |g|^{(q-1)p} \, d\mu \right)^{\frac{1}{p}} = \left(\int_{\{|g| \leq n\}} |g|^q \, d\mu \right)^{\frac{1}{p}}.$$

Therefore,

$$\int_{\{|g| \leq n\}} |g|^q \, d\mu = |\psi(h_n)| \leq \|\psi\| \|h_n\|_p = \|\psi\| \left(\int_{\{|g| \leq n\}} |g|^q \, d\mu \right)^{\frac{1}{p}}.$$

Dividing, we obtain

$$\|\psi\| \geq \left(\int_{\{|g| \leq n\}} |g|^q \, d\mu \right)^{1 - \frac{1}{p}} = \left(\int_{\{|g| \leq n\}} |g|^q \, d\mu \right)^{\frac{1}{q}}.$$

Thus, by Fatou's Lemma (Theorem A.16),

$$\|g\|_q = \left(\int_\Omega \liminf_{n \to \infty} \chi_{\{|g| \leq n\}} |g|^q \, d\mu \right)^{1/q} \leq \left(\liminf_{n \to \infty} \int_\Omega \chi_{\{|g| \leq n\}} |g|^q \, d\mu \right)^{1/q} \leq \|\psi\|.$$

Therefore, g is in $L_q(\mu)$ and $\|g\|_q \leq \|\psi\|$.

It remains to show that $\psi = \phi_g$. If f is a real nonnegative function in $L_p(\mu)$, then we may choose a sequence $(f_n)_{n=1}^\infty$ of simple measurable functions such that f_n increases to f almost everywhere and such that $f_n \to f$ in the L_p-norm as $n \to \infty$. We have already established that $\psi(f_n) = \int_\Omega f_n g \, d\mu$ for all $n \in \mathbb{N}$. We also know that $\psi(f_n) \to \psi(f)$ as $n \to \infty$, because ψ is a continuous linear functional on $L_p(\mu)$. Since $f \in L_p(\mu)$ and $g \in L_q(\mu)$, it follows that $fg \in L_1(\mu)$, by Hölder's Inequality. Thus,

$$\lim_{n \to \infty} \int_\Omega f_n g \, d\mu = \int_\Omega fg \, d\mu,$$

by Lebesgue's Dominated Convergence Theorem. Therefore, $\psi(f) = \phi_g(f)$ for all nonnegative functions f in $L_p(\mu)$. As before, we may extend this to all real functions in $L_p(\mu)$ by writing $f = f^+ - f^-$.

We have now proven the theorem for $p \in (1, \infty)$ when the scalar field is $\mathbb{R}$. In order to extend this result to $\mathbb{C}$, we argue as above, but define the function h by the rule $h = |g|^{q-1} \rho$, where $\rho : \Omega \to \mathbb{C}$ is a function such that $|\rho| = 1$ and $g \rho = |g|$. Similarly, we let $h_n = \chi_{\{|g| \leq n\}} |g|^{q-1} \rho$ for each $n \in \mathbb{N}$. This argument proves that ϕ_g is a bounded linear functional on $L_p(\mu)$ for all $g \in L_q(\mu)$. It also proves that for any bounded linear functional ψ on $L_p(\mu)$, there exists a function $g \in L_q(\mu)$ such that

$\psi(f) = \phi_g(f)$ for all *real* functions f in $L_p(\mu)$. To extend this result to complex functions f in $L_p(\mu)$, write $f = \Re(f) + i\,\Im(f)$, where $\Re(f)$ and $\Im(f)$ are the real and imaginary parts of f, respectively, and use linearity.

For $p = 1$, the proof is similar to the case when $p \in (1, \infty)$ and is left to the reader. (See Exercise 2.3.) $\square$

As is the case with sequence spaces, the dual of $L_\infty(\mu)$ need not be $L_1(\mu)$.

Theorem 2.13 remains true when μ is a positive σ-finite measure. Such a case can be seen in the following example.

Example 2.14 Consider the measure space $(\mathbb{N}, 2^\mathbb{N}, m)$, where $2^\mathbb{N}$ denotes the power set of $\mathbb{N}$ (the collection of all subsets of $\mathbb{N}$), and m is counting measure on $\mathbb{N}$ (i.e., the set function for which $m(A)$ is the cardinality of the set $A \subseteq \mathbb{N}$). Suppose that $f \in L_p(\mathbb{N}, 2^\mathbb{N}, m)$, where $p \in [1, \infty)$. Then,

$$\|f\|_p = \left(\int_\mathbb{N} |f|^p \, dm \right)^{1/p} = \left(\sum_{n=1}^\infty |f(n)|^p \right)^{1/p}.$$

We see that f in $L_p(\mathbb{N}, 2^\mathbb{N}, m)$ corresponds to the sequence $(f(n))_{n=1}^\infty$ in ℓ_p. The same conclusion holds for $p = \infty$. Therefore,

$$L_p(\mathbb{N}, 2^\mathbb{N}, m) = \ell_p, \quad 1 \le p \le \infty.$$

Let us now consider spaces of continuous functions.

Definition 2.15 Let K be a compact metric space. We denote the collection of scalar-valued continuous functions on K by $C(K)$. Define the *supremum norm* on $C(K)$ by

$$\|f\|_\infty = \sup_{t \in K} |f(t)|, \quad f \in C(K).$$

The set $C(K)$ is a vector space under pointwise addition and scalar multiplication and is a Banach space when given the supremum norm. (See Theorem 2.27.) If we wish to emphasize the underlying scalar field, we will write $C_\mathbb{R}(K)$ or $C_\mathbb{C}(K)$.

Observe that, for $f \in C(K)$, the quantity $\|f\|_\infty$ is actually the maximum of $|f|$, since a continuous function attains its supremum on compact sets.

Remark 2.16 We use the notation $\|\cdot\|_\infty$ to represent both the supremum norm on $C(K)$ (for a compact metric space K) and the essential supremum norm on $L_\infty(\mu)$ (for a measure space $(\Omega, \mathcal{A}, \mu)$). If there is any risk of confusion, we will write $\|\cdot\|_{C(K)}$ and $\|\cdot\|_{L_\infty(\mu)}$ to denote the norm on $C(K)$ and $L_\infty(\mu)$, respectively.

We wish to identify the dual space of $C(K)$. To that end, we consider the following example, where $K = [0, 1]$. In this case, we write $C(K) = C[0, 1]$.

Example 2.17 Consider the following linear functionals on $C[0, 1]$:

(a) *(Integration)* $f \to \displaystyle\int_0^1 f(t)\, dt$.

(b) *(Point evaluation)* $f \to f(s)$ for $s \in K$.

(c) *(Integration against L_1 functions)* $f \to \displaystyle\int_0^1 f(t)\, g(t)\, dt$ for $g \in L_1(0, 1)$.

We use $L_1(0, 1)$ to denote the Banach space of L_1-functions on $[0, 1]$ with Lebesgue measure.

Note that point evaluation (Example 2.17(b)) can be thought of as integration, by means of a *Dirac measure*:

$$\delta_s(A) = \begin{cases} 1 & \text{if } s \in A, \\ 0 & \text{if } s \notin A. \end{cases} \tag{2.7}$$

For any $s \in [0, 1]$, the set function δ_s is a measure on $[0, 1]$, and

$$f(s) = \int_{[0,1]} f(t)\,\delta_s(dt).$$

The measure δ_s is also known as the *Dirac mass at s*. A Dirac measure is an example of a *singular measure*, or a measure that is concentrated on a set of Lebesgue measure zero.

Example 2.17(c) provides a bounded linear functional on $C[0, 1]$ because

$$\left| \int_0^1 f(t) g(t)\,dt \right| \leq \sup_{t \in [0,1]} |f(t)| \cdot \int_0^1 |g(t)|\,dt = \|f\|_\infty \|g\|_1,$$

for $f \in C(K)$ and $g \in L_1(0, 1)$. This, too, can be realized as integration against a measure. Let

$$v(A) = \int_A g(t)\,dt, \quad A \in \mathcal{B},$$

where $\mathcal{B}$ denotes the collection of Borel measurable subsets of $[0, 1]$. Then

$$\int_0^1 f(t) g(t)\,dt = \int_{[0,1]} f\,dv.$$

The linear functionals given in Example 2.17 are not all of the linear functionals on $C[0, 1]$ (as will be seen in Theorem 2.20, below), but they do give a hint to the true nature of $(C[0, 1])^*$. Before we identify this space, let us recall several definitions from measure theory.

Definition 2.18 Let $(K, \mathcal{B})$ be a measurable space and let μ be a measure on K. The *total variation* of μ is defined for all $A \in \mathcal{B}$ by

$$|\mu|(A) = \sup \left\{ \sum_{j \in F} |\mu(A_j)| : (A_j)_{j \in F} \text{ is a finite measurable partition of } A \right\}.$$

We remark here that the measure v defined in Example 2.17(c) has total variation

$$|v|(A) = \int_A |g(t)|\,dt, \quad A \subseteq [0, 1] \text{ is a Borel set.}$$

We leave the verification of this as an exercise. (See Exercise 2.4.)

Definition 2.19 Suppose K is a compact metric space. We denote by $M(K)$ the space of all Borel measures on K having finite total variation. This is a Banach space with the *total variation norm*, which is given by $\|\nu\|_M = |\nu|(K)$ for all $\nu \in M(K)$.

The next theorem, called the Riesz Representation Theorem, was proved by F. Riesz in 1909 [31], although not in this generality. It represented a milestone in analysis and identified the dual space of $C(K)$ as a space of measures on K.

Theorem 2.20   (Riesz Representation Theorem) *Suppose K is a compact metric space. The dual of $C(K)$ can be identified with the space $M(K)$. In particular, if $\phi \in C(K)^*$, then there exists a Borel measure ν on K such that*

$$\phi(f) = \int_K f \, d\nu, \quad f \in C(K), \tag{2.8}$$

and $\|\phi\| = |\nu|(K)$. Furthermore, all Borel measures on K determine bounded linear functionals on $C(K)$ according to (2.8).

The Riesz Representation Theorem is a classical result in measure theory and, consequently, we will not prove it here.

If the underlying scaler field is $\mathbb{R}$, then $M(K)$ consists of all bounded signed Borel measures on K. When the scalar field is $\mathbb{C}$, the measures are bounded complex measures, and in this case

$$M(K) = \{\mu + i\nu : \mu, \nu \text{ are signed (real-valued) measures}\}.$$

The Riesz Representation Theorem identifies $(C[0, 1])^*$ as the space $M[0, 1]$ of Borel measures on the unit interval. It is no surprise, then, that each of the bounded linear functionals in Example 2.17 turned out to be given by a measure on $[0, 1]$. We now give another example of a bounded linear functional on $C[0, 1]$.

Example 2.21   (the Cantor function) A *ternary expansion* for $x \in [0, 1]$ is an infinite series $x = \sum_{j=1}^{\infty} \delta_j(x)/3^j$, where $\delta_j(x) \in \{0, 1, 2\}$ for all $j \in \mathbb{N}$. The ternary expansion is said to *terminate* if the sequence $(\delta_j(x))_{j=1}^{\infty}$ has only finitely many nonzero terms; i.e., there exists an $N \in \mathbb{N}$ such that $\delta_j(x) = 0$ for all $j \geq N$. A given number may have two ternary expansions. For example, $1/3 = 1/3 + 0/3^2 + 0/3^3 + \cdots$, but also

$$\frac{1}{3} = \sum_{j=2}^{\infty} \frac{2}{3^j} = \frac{0}{3} + \frac{2}{3^2} + \frac{2}{3^3} + \cdots .$$

(This equality is easily verified with a geometric series argument.) Despite the fact that $x \in (0, 1]$ may have multiple ternary expansions, it can be shown that x has a unique ternary expansion $x = \sum_{j=1}^{\infty} \delta_j(x)/3^j$ that does not terminate.

We will define a map $G : [0, 1] \to [0, 1]$. If $x = 0$, then let $G(x) = 0$. If $x \in (0, 1]$, and x has nonterminating ternary expansion $x = \sum_{j=1}^{\infty} \delta_j(x)/3^j$, then let

$$G(x) = \begin{cases} \displaystyle\sum_{j=1}^{\infty} \frac{\delta_j(x)/2}{2^j} & \text{if } \delta_j(x) \neq 1 \text{ for any } j \in \mathbb{N}, \\[3mm] \displaystyle\sum_{j=1}^{N-1} \frac{\delta_j(x)/2}{2^j} + \frac{1}{2^N} & \text{if } \delta_N(x) = 1 \text{ and } \delta_j(x) \neq 1 \text{ for } j < N. \end{cases} \tag{2.9}$$

An alternate way of describing this procedure is the following: Let $x \in [0,1]$ and express x in base 3. If x contains a 1, then replace each digit after the first 1 by 0. Next, replace each 2 with a 1. Interpret the result as a number in base 2, and this is the value of $G(x)$.

We call G the *Cantor function*. It is also known as *Lebesgue's Devil Staircase*.

Observe that $G(0) = 0$ and $G(1) = 1$. The Cantor function G has the remarkable property that it is continuous (see Exercise 2.17) and nondecreasing on the interval $[0,1]$, but $G'(x) = 0$ for almost every x. Let $\mathcal{C} = \{x : \delta_j(x) \neq 1 \text{ for any } j \in \mathbb{N}\}$. This set, which is precisely the set on which G is not locally constant, is known as the *Cantor set*. It can be shown that $\mathcal{C}$ has Lebesgue measure zero. (See Exercise 2.16.)

Since $G : [0,1] \to [0,1]$ is continuous, positive, and nondecreasing, there exists a measure μ_G on $[0,1]$ such that

$$\mu_G([a,b]) = G(b) - G(a), \quad 0 \leq a < b \leq 1.$$

The measure μ_G, which we call *Cantor measure*, is a Borel probability measure on $[0,1]$, and so the map

$$f \to \int_0^1 f \, d\mu_G, \quad f \in C[0,1],$$

determines a bounded linear functional on $C[0,1]$. The Cantor measure is an example of a *singular measure*, or a measure that is concentrated on a set of Lebesgue measure zero. (See Parts (d) and (e) of Exercise 2.16.)

In fact, the Cantor measure is an example of a *nonatomic* (or *diffuse*) singular measure, which is a singular measure that has no atoms. A measurable set E is called an *atom for a measure* μ if (i) $\mu(E) > 0$ and (ii) $\mu(F) = 0$ for any measurable subset F of E for which $\mu(E) > \mu(F)$.

Example 2.22 The sequence space c can be viewed as a space of continuous functions on a compact metric space. In particular, $c = C(K)$, where $K = \mathbb{N} \cup \{\infty\}$ is the *one-point compactification* of the natural numbers. (See Sect. B.1 for the relevant definitions.) The space K is topologically equivalent to the compact set $\{1/n : n \in \mathbb{N}\} \cup \{0\}$, viewed as a subspace of $\mathbb{R}$. (See Exercise 2.15.) By the Riesz Representation Theorem (Theorem 2.20), the dual space of c is

$$c^* = \ell_1(\mathbb{N} \cup \{\infty\}).$$

In other words, every linear functional on c corresponds to an element of $\ell_1(K)$. We can exhibit this correspondence explicitly. An arbitrary element in $\ell_1(K)$ is of the form $\xi = \big((\xi_n)_{n=1}^{\infty}, \xi_{\infty}\big)$, where $(\xi_n)_{n=1}^{\infty} \in \ell_1(\mathbb{N})$ and $\xi_{\infty} \in \mathbb{K}$. The linear functional that corresponds to ξ is a measure μ_{ξ} on K defined by $\mu_{\xi}(n) = \xi_n$ for $n \in \mathbb{N} \cup \{\infty\}$.

The total variation of μ_{ξ} is $\|\xi\|_{\ell_1(K)}$; that is,

$$|\mu_{\xi}|(K) = \sum_{n=1}^{\infty} |\xi_n| + |\xi_{\infty}|.$$

This quantity is finite, by assumption. The dual action of μ_{ξ} on c is given by integration:

$$\int_K f \, d\mu_{\xi} = \sum_{n=1}^{\infty} f(n)\xi_n + \Big(\lim_{j \to \infty} f(j)\Big)\xi_{\infty}, \quad f \in c = C(K).$$

It is natural to wonder if there are any Banach spaces with trivial dual space. The Hahn–Banach Theorem, which we prove in Sect. 3, guarantees there are no such Banach spaces. In particular, we will prove the following. (See Theorem 3.9.)

Theorem 2.23 (Hahn–Banach Theorem) *Suppose X is a real Banach space and let E be a linear subspace of X. If $\phi \in E^*$, then there exists a bounded linear functional $\psi \in X^*$ such that $\|\psi\| = \|\phi\|$ and $\psi(x) = \phi(x)$ for all $x \in E$.*

The Hahn–Banach Theorem implies there are no Banach spaces X with $X^* = \{0\}$. Any one-dimensional subspace of X will have non-trivial linear functionals, and the Hahn–Banach Theorem states that these can be extended to the entire space X.

2.3 Completeness in Function Spaces

In this section, we will prove that the function spaces of the previous section are complete in their given norms. The sequence spaces are left for the exercises. We begin by proving a very useful lemma.

Lemma 2.24 (Cauchy Summability Criterion) *A normed space X is complete if and only if every series $\sum_{n=1}^{\infty} x_n$ converges in X whenever $\sum_{n=1}^{\infty} \|x_n\| < \infty$.*

Proof First, suppose X is complete in the norm $\|\cdot\|$ and suppose $\sum_{n=1}^{\infty} \|x_n\| < \infty$. For $n \in \mathbb{N}$, let $S_n = x_1 + \cdots + x_n$. Then $\|S_m - S_n\| \leq \sum_{j=n+1}^{m} \|x_k\|$ for each m and n in $\mathbb{N}$, and so $(S_n)_{n=1}^{\infty}$ is a Cauchy sequence. Therefore, since X is complete, the series $\sum_{j=1}^{\infty} x_j$ converges.

Now suppose $\sum_{n=1}^{\infty} x_n$ converges whenever $\sum_{n=1}^{\infty} \|x_n\| < \infty$. Let $(y_n)_{n=1}^{\infty}$ be a Cauchy sequence in X. Pick an increasing sequence $(n_k)_{k \in \mathbb{N}}$ of natural numbers such that

$$\|y_p - y_q\| < \frac{1}{2^k}, \quad \text{whenever } p > q \geq n_k.$$

Let $y_{n_0} = 0$. Then

$$y_{n_k} = \sum_{j=1}^{k} (y_{n_j} - y_{n_{j-1}}), \quad k \in \mathbb{N}.$$

By construction, we have $\|y_{n_j} - y_{n_{j-1}}\| < \frac{1}{2^{j-1}}$ for all $j \geq 2$, and so we see that $\sum_{j=1}^{\infty} \|y_{n_j} - y_{n_{j-1}}\| < \infty$. It follows that the subsequence $(y_{n_k})_{k=1}^{\infty}$ converges to some element in X. Let y be the limit of this subsequence.

Let $\epsilon > 0$ be given. We may choose $K \in \mathbb{N}$ large enough so that $\|y_{n_K} - y\| < \epsilon/2$ and $1/2^K < \epsilon/2$. By definition, if $m > n_K$, then $\|y_m - y_{n_K}\| \leq 1/2^K$. Thus,

$$\|y_m - y\| \leq \|y_m - y_{n_K}\| + \|y_{n_K} - y\| < \frac{1}{2^K} + \frac{\epsilon}{2} < \epsilon,$$

whenever $m > n_K$.

Therefore, $(y_m)_{m=1}^{\infty}$ is a convergent sequence, as required. $\square$

Theorem 2.25 *Let (Ω, Σ, μ) be a positive measure space. If $1 \leq p < \infty$, then $L_p(\Omega, \mu)$ is a Banach space when given the norm*

$$\|f\|_p = \left(\int_{\Omega} |f|^p \, d\mu \right)^{1/p}, \quad f \in L_p(\Omega, \mu).$$

Proof We will make liberal use of the theorems in Appendix A. First, observe that $\| \cdot \|_p$ is a norm on $L_p(\Omega, \mu)$, by Minkowski's Inequality. To show that $\| \cdot \|_p$ is complete, we will use the Cauchy Summability Criterion (Lemma 2.24).

Assume $(f_k)_{k=1}^{\infty}$ is a sequence of functions in $L_p(\Omega, \mu)$ such that $\sum_{k=1}^{\infty} \|f_k\|_p < \infty$. Let $M = \sum_{k=1}^{\infty} \|f_k\|_p$. For each $n \in \mathbb{N}$, define g_n on Ω by $g_n(\omega) = \sum_{k=1}^{n} |f_k(\omega)|$ for all $\omega \in \Omega$. Observe that $(g_n)_{n=1}^{\infty}$ is a sequence of nonnegative measurable functions and that $g_n \leq g_{n+1}$ for each $n \in \mathbb{N}$. By Fatou's Lemma,

$$\int \liminf_{n \to \infty} |g_n|^p \, d\mu \leq \liminf_{n \to \infty} \int |g_n|^p \, d\mu = \liminf_{n \to \infty} \|g_n\|_p^p \leq M^p < \infty.$$

It follows that $\liminf_{n \to \infty} |g_n|^p < \infty$ a.e.(μ), and consequently,

$$\left(\sum_{k=1}^{\infty} |f_k| \right)^p = \lim_{n \to \infty} \left(\sum_{k=1}^{n} |f_k| \right)^p = \liminf_{n \to \infty} |g_n|^p < \infty \quad \text{a.e.}(\mu).$$

Consequently, the function $g = \sum_{k=1}^{\infty} |f_k|$ exists a.e.(μ) and $\|g\|_p \leq M$.

Now let $f = \sum_{k=1}^{\infty} f_k$. It follows that f exists a.e.(μ), because $|f(\omega)| \leq g(\omega)$ for all $\omega \in \Omega$. Observe also that

$$\left| \sum_{k=1}^{n} f_k(\omega) \right| \leq g(\omega), \quad \omega \in \Omega.$$

Therefore, by the Lebesgue Dominated Convergence Theorem, $\sum_{k=1}^{n} f_k$ converges to f in $L_p(\Omega, \mu)$, as required.

Theorem 2.26 *Let (Ω, Σ, μ) be a positive measure space. The space $L_{\infty}(\Omega, \mu)$ is a Banach space when given the essential supremum norm $\| \cdot \|_{\infty}$.*

Proof Again we use Lemma 2.24. Let $(f_k)_{k=1}^{\infty}$ be a sequence of functions in $L_{\infty}(\Omega, \mu)$ and let $M = \sum_{k=1}^{\infty} \|f_k\|_{\infty}$. Suppose $M < \infty$. By assumption, for each $k \in \mathbb{N}$, we have $|f_k| \le \|f_k\|_{\infty}$ a.e.(μ). Therefore, for each $k \in \mathbb{N}$, there exists a measurable set N_k such that $\mu(N_k) = 0$ and $|f_k(\omega)| \le \|f_k\|_{\infty}$ for all $\omega \in \Omega \setminus N_k$. Let $N = \bigcup_{k=1}^{\infty} N_k$. Then $\mu(N) = 0$ (as a countable union of measure-zero sets) and $f_k(\omega) < \infty$ for all $\omega \in \Omega \setminus N$ and all $k \in \mathbb{N}$. Consequently, $f = \sum_{k=1}^{\infty} f_k$ exists a.e.(μ) and

$$|f(\omega)| \le \sum_{k=1}^{\infty} |f_k(\omega)| \le \sum_{k=1}^{\infty} \|f_k\|_{\infty} = M,$$

for all $\omega \in \Omega \setminus N$.

We have determined that $f \in L_{\infty}(\Omega, \mu)$, but now we must show that f is the limit of $\sum_{k=1}^{n} f_k$ in the essential supremum norm. Let $\epsilon > 0$ be given. By assumption, $\sum_{k=1}^{\infty} \|f_k\|_{\infty} < \infty$. Therefore, there exists some $n_0 \in \mathbb{N}$ such that $\sum_{k=m}^{\infty} \|f_k\|_{\infty} < \epsilon$ whenever $m \ge n_0$. Then

$$\|f - \sum_{k=1}^{n} f_k\|_{\infty} \le \sum_{k=n+1}^{\infty} \|f_k\|_{\infty} < \epsilon,$$

provided $n \ge n_0 - 1$. This completes the proof. $\square$

Theorem 2.27 *Let K be a compact metric space. The space $C(K)$ of scalar-valued continuous functions on K is a Banach space when given the supremum norm $\| \cdot \|_{\infty}$.*

Proof The proof is similar to that of Theorem 2.26, with modifications related to continuity, and is left to the reader. (See Exercise 2.5.) $\square$

Theorem 2.27 remains true if we replace K with a locally compact Hausdorff space and consider the space $C_0(K)$ of continuous functions on K vanishing at infinity. (See Sects. 5.1 and A.6 for the relevant definitions.)

When K is a locally compact Hausdorff space, the dual space $C_0(K)^*$ is still $M(K)$, but in this case it denotes the Banach space of *regular* Borel measures on K with the total variation norm. This is not inconsistent notation because, when K is a compact metric space, $C_0(K) = C(K)$ and all finite Borel measures on K are regular.

Note that $M(K)$ is necessarily a Banach space as the dual space of a Banach space. (See Proposition 1.11.)

Exercises

Exercise 2.1 Show that c_0 is a closed subspace of c and that c is a closed subspace of ℓ_{∞}.

Exercise 2.2 Show that the dual of c_0 can be identified with ℓ_1, and that the dual of ℓ_1 can be identified with ℓ_{∞}.

Exercise 2.3 Let (Ω, Σ, μ) be a positive finite measure space. Show that $L_1(\Omega, \mu)^*$ can be identified with $L_\infty(\Omega, \mu)$. (This completes the proof of Theorem 2.13.)

Exercise 2.4 Let $g \in L_1(0, 1)$, the Banach space of L_1-functions on $[0, 1]$ with Lebesgue measure, and define a measure on $[0, 1]$ by $\nu(A) = \int_A g(t)\,dt$, where A is a Borel subset of $[0, 1]$. Show that $|\nu|(A) = \int_A |g(t)|\,dt$ for all Borel subsets A of $[0, 1]$. (See Example 2.17(c) and the comments following it.)

Exercise 2.5 Let K be a compact metric space. Prove that $C(K)$ is a Banach space when given the supremum norm. (That is, prove Theorem 2.27.)

Exercise 2.6 Let δ_0 denote the linear functional on $C[0, 1]$ given by evaluation at 0. That is, $\delta_0(f) = f(0)$ for all $f \in C[0, 1]$. Show that δ_0 is bounded on $C[0, 1]$ when equipped with the $\| \cdot \|_\infty$-norm, but not when equipped with the $\| \cdot \|_1$-norm.

Exercise 2.7 Use the theorems of Sect. 2.3 to prove that ℓ_p is complete in the p-norm for $1 \le p \le \infty$. (You may assume the theorems of Sect. 2.3 remain true for σ-finite measure spaces.)

Exercise 2.8 Verify that any Cauchy sequence in c_0 (equipped with the supremum norm) converges to a limit in c_0. Conclude that c_0 is a Banach space.

Exercise 2.9 Prove that c_0 is not a Banach space in the $\| \cdot \|_2$-norm.

Exercise 2.10 Let $1 \le p < q \le \infty$.

(a) Denote by ℓ_p^n the finite-dimensional vector space $\mathbb{R}^n$ equipped with the norm

$$\|(x_1, \ldots, x_n)\|_p = \left(|x_1|^p + \cdots + |x_n|^p\right)^{1/p}.$$

Show that the norms $\| \cdot \|_p$ and $\| \cdot \|_q$ are equivalent on $\mathbb{R}^n$.
(b) Show that $\ell_p \subseteq \ell_q$, but ℓ_q is not a subset of ℓ_p.

Exercise 2.11 Let $x \in \ell_r$ for some $r < \infty$. Show that $x \in \ell_p$ for all $p \ge r$ and prove that $\|x\|_p \to \|x\|_\infty$ as $p \to \infty$.

Exercise 2.12 Suppose (Ω, μ) is a positive measure space and let $1 \le p < q \le \infty$.

(a) Prove that if $\mu(\Omega) < \infty$, then $\|f\|_p \le C_{p,q} \|f\|_q$ for all measurable functions f, where $C_{p,q}$ is a constant that depends on p and q.
(b) Show that the assumption $\mu(\Omega) < \infty$ cannot be omitted in (a).
(c) Find a real-valued function f on $[0, 1]$ such that $\|f\|_p < \infty$ but $\|f\|_q = \infty$.

Exercise 2.13 Suppose (Ω, μ) is a positive measure space such that $\mu(\Omega) = 1$.

(a) If $1 \le p < q \le \infty$, then show $\|f\|_p \le \|f\|_q$ for all measurable functions f. (See Exercise 2.12.)
(b) Assume that f is an essentially bounded measurable function and prove that $\|f\|_p \to \|f\|_\infty$ as $p \to \infty$.

Exercise 2.14 Let ℓ_p^2 be the finite-dimensional vector space $\mathbb{R}^2$ equipped with the $\| \cdot \|_p$-norm for $1 \le p \le \infty$.

(a) What do the closed unit balls $B_{\ell_1^2}$, $B_{\ell_2^2}$, and $B_{\ell_\infty^2}$ represent geometrically?

(b) Let a and b be nonzero real numbers and define a function on $\mathbb{R}^2$ by

$$\|(x, y)\|_E = \left(\frac{x^2}{a^2} + \frac{y^2}{b^2} \right)^{1/2}, \quad (x, y) \in \mathbb{R}^2.$$

Prove that $\| \cdot \|_E$ is a norm on $\mathbb{R}^2$ and identify geometrically the closed unit ball in $(\mathbb{R}^2, \| \cdot \|_E)$.

Exercise 2.15 Let M be a metric space with subset E. A set V is said to be *open in the subspace topology on E* if there exists a set U that is open in M and $V = U \cap E$.

(a) Show that a closed subset of a complete metric space is a complete metric space.

(b) Show that $\mathbb{N}$, the set of natural numbers, is a locally compact metric space with the metric $d(x, y) = |x - y|$ for all x and y in $\mathbb{N}$. Conclude that the one-point compactification $\mathbb{N} \cup \{\infty\}$ of the natural numbers is a compact metric space. (See Appendix B.1 for the definition of the one-point compactification.)

(c) Show that the one-point compactification of the natural numbers $\mathbb{N} \cup \{\infty\}$ (from part (b)) is homeomorphic to $\{1/n : n \in \mathbb{N}\} \cup \{0\}$, where the latter set is given the subspace topology inherited from $\mathbb{R}$.

(d) Conclude that c is a Banach space. (See Example 2.22.)

Exercise 2.16 Consider the interval $[0, 1]$. From this set, remove the open subinterval $(\frac{1}{3}, \frac{2}{3})$, the so-called *middle third*. This leaves the union of two closed intervals: $[0, \frac{1}{3}] \cup [\frac{2}{3}, 1]$. From each of these, again remove the middle third. What remains is the union of four closed intervals: $[0, \frac{1}{9}] \cup [\frac{2}{9}, \frac{1}{3}] \cup [\frac{2}{3}, \frac{7}{9}] \cup [\frac{8}{9}, 1]$. Once again, from each remaining set remove the middle third. Continue this process indefinitely to create *Cantor's Middle Thirds Set*. This set, which we denote $\mathcal{K}$, can be written explicitly as follows:

$$\mathcal{K} = [0, 1] \setminus \bigcup_{n=1}^{\infty} \bigcup_{k=0}^{3^{n-1}-1} \left(\frac{3k+1}{3^n}, \frac{3k+2}{3^n} \right).$$

(a) Show that $\mathcal{K}$ coincides with the Cantor set $\mathcal{C}$ of Example 2.21. (*Hint:* You may wish to use the fact that every nonzero number has a unique nonterminating ternary expansion.)

(b) Evidently $\mathcal{K}$ is not empty because it contains the endpoints of the middle third sets. Show that $\mathcal{K}$ contains other numbers by showing that $\frac{1}{4} \in \mathcal{K}$. (*Hint:* Consider the geometric series $\sum_{j=1}^{\infty} 3^{-kj}$, where $k \in \mathbb{N}$.)

(c) Show that $\mathcal{K}$ is uncountable. (*Hint:* Use diagonalization and $\mathcal{K} = \mathcal{C}$.)

(d) Let m be Lebesgue measure on $[0, 1]$. Show that $m(\mathcal{K}) = 0$.

(e) Let μ_G be the Cantor measure from Example 2.21. Show that $\mu_G(\mathcal{K}) = 1$.

Exercise 2.17 Let $\mathcal{K}$ be Cantor's Middle Thirds Set from Exercise 2.16. After the first middle third is removed, two closed intervals remain. Call these two sets $E_{1,1}$ and $E_{1,2}$. After the middle thirds are removed from $E_{1,1}$ and $E_{1,2}$, there will remain four closed intervals. Label these sets $E_{2,1}, E_{2,2}, E_{2,3}, E_{2,4}$, ordering them from left to right, as they appear on the unit interval. After the process has been repeated n times, there will remain 2^n closed intervals, each of length $\frac{1}{3^n}$. Label these sets $E_{n,1}, \ldots, E_{n,2^n}$, again from left to right, as they appear on the unit interval, so that $x_1 \in E_{n,k_1}$ and $x_2 \in E_{n,k_2}$ implies that $x_1 < x_2$ whenever $k_1 < k_2$. Observe that

$$\mathcal{K} = \bigcap_{n=1}^{\infty} \bigcup_{k=1}^{2^n} E_{n,k}.$$

For each $n \in \mathbb{N}$ and each $k \in \{1, \ldots, 2^n\}$, let $E_{n,k} = [a_{n,k}, b_{n,k}]$ and define a function $G_n : [0, 1] \to [0, 1]$ as follows:

$$G_n(x) = \begin{cases} \frac{b_{n,k}-x}{3^{-n}}\frac{k-1}{2^n} + \frac{x-a_{n,k}}{3^{-n}}\frac{k}{2^n} & \text{if} \quad a_{n,k} \le x \le b_{n,k} \text{ for } k \in \{1, \ldots, 2^n\}, \\ \frac{k}{2^n} & \text{if} \quad b_{n,k} < x < a_{n,k+1} \text{ for } k \in \{1, \ldots, 2^n - 1\}. \end{cases}$$

show that $(G_n)_{n=1}^{\infty}$ converges uniformly to G, the cantor function (from Example 2.21), and deduce that G is continuous on $[0, 1]$.

Chapter 3
The Hahn–Banach Theorems

Several theorems in functional analysis have been labeled as "the Hahn–Banach Theorem." At the heart of all of them is what we call here the Hahn–Banach Extension Theorem, given in Theorem 3.4, below. This theorem is at the foundation of modern functional analysis, and its use is so pervasive that its importance cannot be overstated.

3.1 The Axiom of Choice

The Zermelo–Fraenkel Axioms (ZF) is a list of accepted statements upon which mathematics can be built. They form what is possibly the most common foundation of mathematics. When the system includes the *Axiom of Choice*, it is often called (ZFC).

Axiom of Choice For any collection $\mathcal{X}$ of nonempty sets, there exists a choice function defined on $\mathcal{X}$.

A *choice function* on a collection $\mathcal{X}$ of nonempty sets is a function c, defined on $\mathcal{X}$, such that $c(A) \in A$ for every set A in $\mathcal{X}$.

The Axiom of Choice was formulated by Ernst Zermelo in 1904, and is usually accepted by mathematicians. Indeed, it is often used without realizing it. For example, consider the sequential characterization of continuity: Suppose f is a continuous function. From a countable collection of open sets $(U_n)_{n=1}^{\infty}$ that decrease to a point x_0, we choose a sequence $(x_n)_{n=1}^{\infty}$ such that $x_n \in U_n$, and so $x_n \to x_0$ as $n \to \infty$. By the choice of x_n, and the continuity of f, it follows that $f(x_n) \to f(x_0)$ as $n \to \infty$, and so on. We are using the Axiom of Choice when we choose $x_n \in U_n$ for each $n \in \mathbb{N}$; but the Axiom of Choice is even stronger, allowing us to choose uncountably many points simultaneously, each one from a different set.

While the Axiom of Choice is easy to believe, accepting it results in some unexpected consequences.

© Springer Science+Business Media, LLC 2014

A. Bowers, N. J. Kalton, *An Introductory Course in Functional Analysis*,
Universitext, DOI 10.1007/978-1-4939-1945-1_3

Banach–Tarski Paradox A three-dimensional solid sphere of radius 1 can be split into a finite number of disjoint pieces, and those pieces can be reassembled to form two solid spheres of radius 1.

The disjoint pieces will by necessity be *non-measurable*, and so difficult to imagine. The Banach–Tarski Paradox violates standard geometric intuition, and as a result has led some to question the Axiom of Choice. Despite these reservations, we will take the Axiom of Choice for granted.

Of interest to us is an alternate (but equivalent) formulation of the Axiom of Choice known as Zorn's Lemma.

Zorn's Lemma Suppose $(P, \leq)$ is a partially ordered set. If every chain in P has an upper bound, then P contains a maximal element.

Let us recall the relevant definitions.

Definition 3.1 Let P be a set. A relation $\leq$ on P is said to be a *partial order* if for every a, b, and c in P:

  (i) $(a \leq b$ and $b \leq c) \Rightarrow a \leq c$,
 (ii) $(a \leq b$ and $b \leq a) \Rightarrow a = b$,
(iii) $a \leq a$.

A subset C of P is a *chain* in P if $(C, \leq)$ is totally ordered:

$$\{a, b\} \subseteq C \quad \Longrightarrow \quad (a \leq b \text{ or } b \leq a).$$

If $A \subseteq P$, then b is called an *upper bound* for A if $a \leq b$ for all $a \in A$. An element $a \in P$ is called *maximal* if $a \leq b$ implies $b = a$.

Naturally, we can define a reverse partial order $\geq$ in the obvious way, and reinterpret Zorn's Lemma in terms of lower bounds and minimal elements. Another (equivalent) formulation of the Axiom of Choice is the Hausdorff Maximality Principle.

Hausdorff Maximality Principle In a partially ordered set, every chain is contained in a maximal chain.

The Hausdorff Maximality Principle was formulated by Felix Hausdorff in 1914. Zorn's lemma was proposed later, in 1935, by Max Zorn. Both are equivalent to the Axiom of Choice. A detailed discussion of the history of the Axiom of Choice and its equivalent formulations can be found in [26]. The formulation we will usually use is Zorn's Lemma.

3.2   Sublinear Functionals and the Extension Theorem

We briefly turn our attention from linear functionals to sublinear functionals.

Definition 3.2 Suppose V is a vector space. A map $p : V \to \mathbb{R}$ is called a *sublinear functional* if

(i) $p(\alpha x) = \alpha\, p(x)$ for all $\alpha \geq 0$ and $x \in V$ *(positive homogeneity)*, and

(ii) $p(x + y) \leq p(x) + p(y)$ for all x and y in V *(subadditivity)*.

Note that condition *(i)* implies that $p(0) = 0$. Condition *(ii)* is also called the *triangle inequality*.

Example 3.3 The following are examples of sublinear functionals:

(a) Any linear functional is a sublinear functional.

(b) If E is a normed space, then the norm function defined by $p(x) = \|x\|$ for $x \in E$ is a sublinear functional.

(c) If $C(K)$ is the Banach space of real-valued continuous functions on a compact metric space K, then

$$p(f) = \max_{s \in K} f(s), \quad f \in C(K),$$

defines a sublinear functional.

(d) If ℓ_∞ is the Banach space of bounded sequences of real numbers, then

$$p(\xi) = \sup_{n \in \mathbb{N}} \xi_n, \quad \xi = (\xi_n)_{n=1}^\infty \in \ell_\infty,$$

defines a sublinear functional.

Theorem 3.4 (Hahn–Banach Extension Theorem) *Let E be a real vector space and let p be a sublinear functional on E. If V is a subspace of E and if $f : V \to \mathbb{R}$ is a linear functional such that*

$$f(x) \leq p(x), \quad x \in V,$$

then there is an extension $\hat{f} : E \to \mathbb{R}$ of f that is linear and satisfies

$$\hat{f}(x) \leq p(x), \quad x \in E.$$

Before proving the Hahn–Banach Extension Theorem, we will prove several preliminary lemmas.

Lemma 3.5 *A sublinear functional p on a real vector space E is linear if and only if $p(-x) = -p(x)$ for all $x \in E$.*

Proof Certainly, if p is linear, then the conclusion follows. Suppose now that $p(-x) = -p(x)$ for all $x \in E$. Then, by the triangle inequality,

$$p(x + y) = -p(-x - y) \geq -(p(-x) + p(-y)) = p(x) + p(y).$$

The reverse inequality is simply subadditivity of p.   □

Let $\mathfrak{P}_E$ be the collection of all sublinear functionals on a real vector space E. We define an order on the set $\mathfrak{P}_E$ by saying $p \leq q$ whenever $p(x) \leq q(x)$ for all $x \in E$.

Lemma 3.6 *Suppose $p \in \mathfrak{P}_E$ and V is a linear subspace of a real vector space E. If q is a sublinear functional on V such that $q \leq p|_V$, then there exists a sublinear functional $r \in \mathfrak{P}_E$ such that $r|_V = q$ and $r \leq p$.*

Proof Define

$$r(x) = \inf\{q(v) + p(x - v) : v \in V\}, \quad x \in E.$$

To show that r is well defined, we show the set $\{q(v) + p(x - v) : v \in V\}$ has a lower bound for all $x \in E$. Fix $x \in E$ and let $v \in V$. By assumption, $q(-v) \leq p(-v)$, and so, by the subadditivity of q,

$$0 \leq q(v) + q(-v) \leq q(v) + p(-v).$$

It follows that $-p(-v) \leq q(v)$. Next, using the subadditivity of p, we have

$$p(-v) \leq p(-x) + p(x - v).$$

Rearranging this inequality, and also using the fact that $-p(-v) \leq q(v)$, we have

$$-p(-x) \leq -p(-v) + p(x - v) \leq q(v) + p(x - v).$$

This is true for all $v \in V$, and so the set $\{q(v) + p(x - v) : v \in V\}$ has a lower bound. Thus, the quantity $r(x)$ is well-defined for each $x \in E$.

It is clear that r is positively homogeneous and that $r(x) \leq p(x)$ for all $x \in E$ (by taking $v = 0$). We claim that also $r(x) = q(x)$ for each $x \in V$. To see this, suppose that $x \in V$. Then for all $v \in V$,

$$q(v) + p(x - v) \geq q(v) + q(x - v) \geq q(v + (x - v)) = q(x).$$

Thus,

$$r(x) = \inf\{q(v) + p(x - v) : v \in V\} \geq q(x).$$

This infimum is achieved (when $v = x$), and so $r(x) = q(x)$.

It remains only to show that r is subadditive on E. Let x and y be in E and suppose $\epsilon > 0$. Pick v and w in V so that

$$q(v) + p(x - v) < r(x) + \frac{\epsilon}{2} \quad \text{and} \quad q(w) + p(y - w) < r(y) + \frac{\epsilon}{2}.$$

By the sublinearity of p and q,

$$q(v + w) + p(x + y - (v + w)) < r(x) + r(y) + \epsilon.$$

Taking the infimum over elements in V, we have $r(x + y) < r(x) + r(y) + \epsilon$. Since $\epsilon > 0$ was arbitrary, r is subadditive and the proof is complete. $\square$

Lemma 3.7 *If $p \in \mathfrak{P}_E$, then there exists a minimal $q \in \mathfrak{P}_E$ such that $q \leq p$.*

Proof Let $P = \{r \in \mathfrak{P}_E : r \leq p\}$ and let $C = (r_i)_{i \in I}$ be a chain in P. For $x \in E$, let

$$r(x) = \inf_{i \in I} r_i(x).$$

We claim that r is a sublinear functional on E. First we must show that r is well-defined. Let $i \in I$ and $x \in E$. By subadditivity, $0 \leq r_i(x) + r_i(-x)$. Then, since $r_i \in P$,

$$r_i(x) \geq -r_i(-x) \geq -p(-x).$$

Thus, the set $\{r_i(x) : i \in I\}$ has a lower bound, and thus r is well-defined. Positive homogeneity is clear. Now we show subaditivity. Let x and y be elements of E. For any $\epsilon > 0$, there exist indices i and j in I such that

$$r_i(x) < r(x) + \frac{\epsilon}{2} \quad \text{and} \quad r_j(y) < r(y) + \frac{\epsilon}{2}.$$

Since C is a chain, r_i and r_j are comparable. Without loss of generality, assume $r_j \leq r_i$. Then

$$r(x + y) \leq r_j(x + y) \leq r_j(x) + r_j(y) \leq r_i(x) + r_j(y).$$

Therefore

$$r(x + y) < r(x) + r(y) + \epsilon.$$

This is true for all $\epsilon > 0$, and so r is subadditive. By construction, r is a lower bound of the chain C. Thus, by Zorn's Lemma, P contains a minimal element q.

We claim that q is actually minimal in $\mathfrak{P}_E$. Suppose q_0 is an element of $\mathfrak{P}_E$ such that $q_0 \leq q$. Then $q_0 \leq p$, and so $q_0 \in P$. It follows that q_0 is an element of P such that $q_0 \leq q$, and so $q_0 = q$ by the minimality of q in P. This completes the proof. □

Lemma 3.8 *If $q \in \mathfrak{P}_E$ is minimal, then q is linear.*

Proof By Lemma 3.5, it suffices to show $q(-x) = -q(x)$ for all $x \in E$.

Fix an $x \in E$ and let $V = \{\alpha x : \alpha \in \mathbb{R}\}$. Define a linear functional on V by

$$f(\alpha x) = -\alpha \, q(-x), \quad \alpha \in \mathbb{R}.$$

If $\alpha < 0$, then $f(\alpha x) = q(\alpha x)$, by the positive homogeneity of q. Suppose $\alpha \geq 0$. By the subadditivity of q, we have that $0 \leq q(\alpha x) + q(-\alpha x)$, and so

$$f(\alpha x) = -\alpha \, q(-x) = -q(-\alpha x) \leq q(\alpha x).$$

It follows that $f \leq q$ on the subspace V. Thus, by Lemma 3.6, there exists a sublinear functional r on E such that $r \leq q$ and $r|_V = f$. Then $r = q$ by the minimality of q, and so $f = q|_V$. Therefore (taking $\alpha = 1$), we have

$$q(x) = f(x) = -q(-x).$$

The choice of x was arbitrary, and so we have the desired result. □

We are now ready to prove the Hahn–Banach Extension Theorem (Theorem 3.4).

Proof of the Hahn–Banach Extension Theorem By Lemma 3.6, there exists a sublinear functional q on E such that $q|_V = f$ and $q \leq p$. By Lemma 3.7, there exists a minimal sublinear functional q_0 on E such that $q_0 \leq q$. By Lemma 3.8, the map q_0 is linear on E. Let $\hat{f} = q_0$. Then $\hat{f}$ is linear and $\hat{f} \leq p$. We must show $\hat{f}|_V = f$. Since $\hat{f} \leq q$ and $q|_V = f$, we have $\hat{f} \leq f$ on V. Let $x \in V$. Then $\hat{f}(x) \leq f(x)$. Furthermore, $\hat{f}(-x) \leq f(-x)$. By linearity, we then have $\hat{f}(x) \geq f(x)$. It follows that $\hat{f}(x) = f(x)$ for all $x \in V$, as required. $\square$

Theorem 3.9 (Hahn–Banach Theorem for real normed spaces) *Suppose X is a real normed vector space and let V be a linear subspace of X. If $f \in V^*$, then there exists an extension $\hat{f} \in X^*$ such that $\hat{f}|_V = f$ and $\|\hat{f}\| = \|f\|$.*

Proof Define $p(x) = \|f\| \, \|x\|$ for $x \in X$. Then $f \leq p$ on V. (Note that f is defined only on V.) Therefore, by the Hahn–Banach Extension Theorem (Theorem 3.4), there is a linear functional $\hat{f} \in X^*$ such that $\hat{f} \leq p$ and $\hat{f}|_V = f$. It follows directly that $\|\hat{f}\| = \|f\|$. $\square$

Example 3.10 Consider the space c of real convergent sequences. The norm on this Banach space is $\|\xi\| = \sup_{n \in \mathbb{N}} |\xi_n|$, where $\xi = (\xi_n)_{n=1}^\infty$. We know that c is a subspace of ℓ_∞, the space of all real bounded sequences. (See Exercise 2.1.) Define a linear functional $f : c \to \mathbb{R}$ by

$$f(\xi) = \lim_{n \to \infty} \xi_n, \quad \xi = (\xi_n)_{n=1}^\infty \in c. \tag{3.1}$$

The map f is bounded and $\|f\| = 1$. By Theorem 3.9 (the Hahn–Banach Theorem for real normed spaces), there exists a linear extension $\hat{f} : \ell_\infty \to \mathbb{R}$ such that $\hat{f}|_c = f$ and $\|\hat{f}\| = 1$.

By construction, $\hat{f} \in (\ell_\infty)^*$. We will now show that $\hat{f} \notin \ell_1$. (Note that, among other things, this shows that ℓ_1 is not reflexive.) Suppose to the contrary that $\hat{f} \in \ell_1$. Then there exists a sequence of scalars $(\alpha_n)_{n=1}^\infty$ such that $\sum_{n=1}^\infty |\alpha_n| = 1$ and such that

$$\hat{f}(\xi) = \sum_{n=1}^\infty \alpha_n \xi_n, \quad \xi = (\xi_n)_{n=1}^\infty \in \ell_\infty. \tag{3.2}$$

If x is a scalar-valued sequence, then denote the nth coordinate of x by $x(n)$. Let e_m be the sequence with a 1 in the mth coordinate and zeros elsewhere, so that $e_m(m) = 1$ and $e_m(n) = 0$ if $m \neq n$. Certainly, we have $e_m \in c$ for every $m \in \mathbb{N}$. Thus, by (3.1),

$$\hat{f}(e_m) = f(e_m) = \lim_{n \to \infty} e_m(n) = 0.$$

On the other hand, by (3.2),

$$\hat{f}(e_m) = \sum_{n=1}^\infty \alpha_n e_m(n) = \alpha_m.$$

Consequently, $\alpha_m = 0$ for all $m \in \mathbb{N}$. This implies that $\hat{f} = 0$, which is a contradiction (because $\|\hat{f}\| = 1$). We conclude that $\hat{f} \notin \ell_1$.

Example 3.11 Consider again the space c of real convergent sequences, this time with the sublinear functional $p(\xi) = \sup_{n \in \mathbb{N}} \xi_n$, where $\xi = (\xi_n)_{n=1}^{\infty}$. As in Example 3.10, let

$$f(\xi) = \lim_{n \to \infty} \xi_n, \quad \xi = (\xi_n)_{n=1}^{\infty} \in c.$$

Then f is a norm one linear functional on c. It is clear that $f \leq p$ on c. Thus, by Theorem 3.4 (the Hahn–Banach Extension Theorem), there exists a linear functional $\tilde{f} : \ell_{\infty} \to \mathbb{R}$ such that $\tilde{f}|_c = f$ and $\tilde{f} \leq p$. As was the case in Example 3.10, it can be shown that $\tilde{f} \notin \ell_1$.

It is worth noting that $\tilde{f}$ has an additional property: If ξ is a bounded sequence of nonnegative real numbers, then $\tilde{f}(\xi) \geq 0$. To see this, suppose that $\xi = (\xi_n)_{n=1}^{\infty}$ is a bounded sequence such that $\xi_n \geq 0$ for all $n \in \mathbb{N}$. Then

$$p(-\xi) = \sup_{n \in \mathbb{N}} (-\xi_n) \leq 0.$$

By construction, $\tilde{f} \leq p$ on ℓ_{∞}, and so $\tilde{f}(-\xi) \leq 0$. Therefore, $\tilde{f}(\xi) \geq 0$, by the linearity of $\tilde{f}$.

As a simple extension, observe that, for sequences $\xi = (\xi_n)_{n=1}^{\infty}$ and $\eta = (\eta_n)_{n=1}^{\infty}$ in ℓ_{∞}, we have $\tilde{f}(\xi) \geq \tilde{f}(\eta)$ whenever $\xi_n \geq \eta_n$ for all $n \in \mathbb{N}$.

Remark 3.12 The existence of $\hat{f}$ and $\tilde{f}$ in the previous examples is guaranteed by the Hahn–Banach Theorem, but this relies on the Axiom of Choice. No formula exists for constructing $\hat{f}$ or $\tilde{f}$ and, in fact, no formula can exist. If the Axiom of Choice is replaced by a weaker assumption, then $\ell_{\infty}^* = \ell_1$. (See [36].) This means, for one thing, that any linear functional on ℓ_{∞} which can be written explicitly must belong to ℓ_1.

When an extension of a bounded linear functional is found using the Hahn–Banach Theorem, it is sometimes called a *Hahn–Banach extension* of the functional. The extensions $\hat{f}$ and $\tilde{f}$ in Examples 3.10 and 3.11 (respectively) are both Hahn–Banach extensions of the same bounded linear functional f. Hahn–Banach extensions are generally not unique, as the following example illustrates.

Example 3.13 Let c be the space of convergent sequences and let $f(\xi) = \lim_{n \to \infty} \xi_n$ for all $\xi = (\xi_n)_{n=1}^{\infty}$ in c. Then f is a linear functional on c with $\|f\| = 1$.

Now let $1_E = (0, 1, 0, 1, \dots)$ be the sequence having 0 in each odd coordinate and 1 in each even coordinate. Let X denote the subspace of ℓ_{∞} generated by c and 1_E. (That is, let X be the smallest subspace of ℓ_{∞} that contains all sequences in c and the sequence 1_E.) We will extend the linear functional f to X in two ways:

$$f_E(x) = \lim_{n \to \infty} x_{2n} \quad \text{and} \quad f_O(x) = \lim_{n \to \infty} x_{2n+1},$$

where $x = (x_n)_{n=1}^{\infty}$ is a sequence in X. Observe that $f_E(\xi) = f(\xi) = f_O(\xi)$ for all sequences $\xi \in c$.

Each of the linear functionals f_E and f_O are bounded on X and have norm one. By the Hahn–Banach Theorem, these bounded linear functionals can be extended to norm one linear functionals $\widehat{f_E}$ and $\widehat{f_O}$ (respectively) on ℓ_∞. Because

$$\widehat{f_E}|_c = \widehat{f_O}|_c = f,$$

and both $\widehat{f_E}$ and $\widehat{f_O}$ have norm 1, they are both Hahn–Banach extensions of f to ℓ_∞. It is easily seen, however, that $\widehat{f_E}$ and $\widehat{f_O}$ are distinct linear functionals on ℓ_∞, because $\widehat{f_E}(1_E) = 1$ and $\widehat{f_O}(1_E) = 0$.

In the preceding example, we found two distinct Hahn–Banach extensions for the bounded linear functional f by partitioning $\mathbb{N}$ into two sets, namely the set of even numbers and the set of odd numbers. We can find further distinct Hahn–Banach extensions for f by repeating the same argument using different partitions of $\mathbb{N}$.

When $p \in [1, \infty)$, we can actually write down an explicit formula for the linear functionals on ℓ_p. Ultimately, this is because ℓ_p is a separable space when $p \in [1, \infty)$. The space ℓ_∞, however, is vast, and consequently ℓ_∞^* is a "monster." To get some perspective on the size of ℓ_∞, consider the set S of all sequences of zeros and ones, which is certainly contained in ℓ_∞. Any two distinct elements ξ and η in S must differ in at least one coordinate, and so $\|\xi - \eta\|_\infty = 1$. The size of S is $|S| = 2^{\aleph_0}$, the size of the continuum (also denoted $\mathfrak{c}$), and so ℓ_∞ is far from separable.

Remark 3.14 (Separability and classical spaces) The comments above show that the space ℓ_∞ of bounded sequences is not a separable space. On the other hand, the space ℓ_p of p-summable sequences is separable when $p \in [1, \infty)$, because the set $\{e_k : k \in \mathbb{N}\}$ is a countable dense subset of ℓ_p, where e_k is the sequence with 1 in the kth coordinate and zero in every other coordinate.

The space $C[0, 1]$ of continuous functions on $[0, 1]$ is separable, because the set $\{t^k : k \in \mathbb{N} \cup \{0\}\}$ is dense in $C[0, 1]$, by the Weierstrass Approximation Theorem. Consequently, the space $L_p(0, 1)$ of (equivalence classes of) p-integrable measurable functions on $[0, 1]$, where $p \in [1, \infty)$, is also separable, because $C[0, 1]$ is dense in this space, by Lusin's Theorem (Theorem A.36). The space $L_\infty(0, 1)$ of essentially bounded measurable functions on $[0, 1]$ is not separable, however. Observe that $\{\chi_{[0,x]} : 0 < x \leq 1\}$ is an uncountable collection of functions in $L_\infty(0, 1)$ and $\|\chi_{[0,s]} - \chi_{[0,t]}\|_\infty = 1$ whenever $s \neq t$.

Theorem 3.15 (Hahn–Banach Theorem for complex normed spaces) *Suppose X is a complex normed vector space and let V be a linear subspace of X. If $f \in V^*$, then there exists an extension $\hat{f} \in X^*$ such that $\hat{f}|_V = f$ and $\|\hat{f}\| = \|f\|$.*

Proof Let $X_\mathbb{R}$ be the underlying real Banach space (i.e., forget you can use complex scalars). Define $f_0 : V \to \mathbb{R}$ by $f_0(v) = \Re(f(v))$ for all $v \in V$. Then f_0 is a real linear functional on V. Furthermore, for all $v \in V$,

$$|f_0(v)| \leq |f(v)| \leq \|f\| \|v\|,$$

and so $\|f_0\| \leq \|f\|$. By Theorem 3.9 (the Hahn–Banach Theorem for real normed spaces), the linear functional f_0 has an extension $\hat{f_0} : X_\mathbb{R} \to \mathbb{R}$ that is linear and such that $\|\hat{f_0}\| = \|f_0\| \leq \|f\|$.

Define

$$\hat{f}(x) = \hat{f_0}(x) - i\, \hat{f_0}(ix), \quad x \in X. \tag{3.3}$$

Observe that if $v \in V$, then (since f is complex-linear)

$$f_0(iv) = \Re\left(f(iv)\right) = \Re\left(if(v)\right) = -\Im\left(f(v)\right).$$

It follows that, for all $v \in V$,

$$\hat{f}(v) = f_0(v) - if_0(iv) = \Re\left(f(v)\right) + i\,\Im\left(f(v)\right) = f(v).$$

Therefore, $\hat{f}$ is an extension of f. By construction, $\hat{f}$ is $\mathbb{R}$-linear.

To see that it is also $\mathbb{C}$-linear, put ix into (3.2.3): For $x \in X$,

$$\hat{f}(ix) = \hat{f}_0(ix) + i\,\hat{f}_0(x) = i\,\hat{f}(x).$$

It remains to show that $\|\hat{f}\| = \|f\|$. We know $\|\hat{f}\| \geq \|f\|$, since $\hat{f}$ is an extension of f. Suppose $x \in X$ with $\|x\| \leq 1$. We know that $\hat{f}(x) \in \mathbb{C}$. Fix $\theta \in \mathbb{R}$ so that $e^{i\theta}\hat{f}(x) \in \mathbb{R}$. Then, by linearity, $\hat{f}(e^{i\theta}x) \in \mathbb{R}$. Therefore,

$$\hat{f}(e^{i\theta}x) = \hat{f}_0(e^{i\theta}x),$$

and so

$$|\hat{f}(x)| = |e^{i\theta}\hat{f}(x)| = |\hat{f}_0(e^{i\theta}x)| \leq \|\hat{f}_0\|\,\|e^{i\theta}x\| \leq \|f\|\,\|x\|.$$

Consequently, $\|\hat{f}\| \leq \|f\|$, and the proof is complete.   □

3.3   Banach Limits

In Example 3.11, we showed the existence of a bounded linear functional $L : \ell_\infty \to \mathbb{R}$ on the real Banach space ℓ_∞ such that $\|L\| = 1$ and $L\left((\xi_n)_{n=1}^\infty\right) \leq \sup_{n \in \mathbb{N}} \xi_n$, and such that

$$L\left((\xi_n)_{n=1}^\infty\right) = \lim_{n\to\infty} \xi_n, \qquad (3.4)$$

whenever this limit exists. (We called the linear functional $\hat{f}$ in Example 3.11.) In this section, we will show the existence of bounded linear functionals L that satisfy an additional property, called *shift-invariance*:

$$L\left((\xi_n)_{n=1}^\infty\right) = L\left((\xi_{n+1})_{n=1}^\infty\right). \qquad (3.5)$$

Shift-invariance, together with the other properties mentioned above, leads to the following inequality:

$$\liminf_{n\to\infty} \xi_n \;\leq\; L\left((\xi_n)_{n=1}^\infty\right) \;\leq\; \limsup_{n\to\infty} \xi_n, \qquad (3.6)$$

for all $(\xi_n)_{n=1}^\infty$ in ℓ_∞. Notice that (3.6) implies (3.4). Linear functionals that satisfy (3.6) are of interest because they generalize the notion of limits.

We define the *shift operator* $T : \ell_\infty \to \ell_\infty$ by

$$T\left((\xi_n)_{n=1}^\infty\right) = (\xi_{n+1})_{n=1}^\infty, \quad (\xi_n)_{n=1}^\infty \in \ell_\infty.$$

Given this definition, we can restate the shift-invariance property in terms of the shift operator T:

$$L(T\xi) = L(\xi), \quad \xi \in \ell_\infty. \tag{3.7}$$

When L satisfies this equation, we say that L is *invariant under the shift operator*, or is *shift-invariant*.

We now prove a form of the Hahn–Banach Theorem for sublinear functionals which are invariant under some collection of linear maps.

Theorem 3.16 (Invariant Hahn–Banach Theorem) *Let V be a real vector space and suppose that $\mathcal{T}$ is a commutative collection of linear maps on V (i.e., $ST = TS$ for all S and T in $\mathcal{T}$). If p is a sublinear functional on V such that*

$$p(Tx) \le p(x), \quad x \in V, \ T \in \mathcal{T},$$

then there exists a linear functional f on V such that $f \le p$ and

$$f(Tx) = f(x), \quad x \in V, \ T \in \mathcal{T}.$$

Before proving Theorem 3.16, let us consider some situations where it can be used.

Example 3.17 In many cases, we will have a set $\mathcal{T}$ containing only one map, and so the commutativity assumption will be satisfied trivially. For example, consider the real Banach space ℓ_∞ and let $T : \ell_\infty \to \ell_\infty$ be the shift operator defined in (3.3).

The following functions are sublinear functionals on ℓ_∞ that satisfy the hypotheses of Theorem 3.16 with $\mathcal{T} = \{T\}$:

(i) $p(\xi) = \sup_{n \in \mathbb{N}} \xi_n$,

(ii) $p(\xi) = \limsup_{n \to \infty} \xi_n$, and

(iii) $p(\xi) = \sup_{n \in \mathbb{N}} |\xi_n|$,

where $\xi = (\xi_n)_{n=1}^\infty \in \ell_\infty$. By Theorem 3.16, each one of these sublinear functionals will lead to a shift-invariant bounded linear functional on ℓ_∞.

Proof of Theorem 3.16 Consider the collection $\mathcal{C}$ of sublinear functionals q such that $q \le p$ and such that $q(Tx) \le q(x)$ for all $x \in V$ and $T \in \mathcal{T}$. We will use Zorn's Lemma to show there exists a minimal element $q \in \mathcal{C}$.

Suppose $(q_i)_{i \in I}$ is a chain in $\mathcal{C}$ and let $q = \inf_{i \in I} q_i$. Then q is a sublinear functional on V. (See the proof of Lemma 3.7.) Let $T \in \mathcal{T}$ and $x \in V$. Then, by the definition of q,

$$q(Tx) \le q_i(Tx), \quad i \in I.$$

We assumed $q_i \in \mathcal{C}$, and hence $q_i(Tx) \le q_i(x)$, for each $i \in I$. It therefore follows that $q(Tx) \le q_i(x)$ for all $i \in I$. Consequently,

$$q(Tx) \le \inf_{i \in I} q_i(x) = q(x), \quad x \in V, \; T \in \mathcal{T}.$$

We conclude that $q \in \mathcal{C}$ and q is a lower bound for the chain $(q_i)_{i \in I}$. Thus, any chain in $\mathcal{C}$ has a lower bound. Therefore, $\mathcal{C}$ contains a minimal element, by Zorn's Lemma.

Now let q be a minimal element of $\mathcal{C}$ and let $T \in \mathcal{T}$. Let $n \in \mathbb{N}$ and define the nth *Cesàro mean* by

$$q_n(x) = q\left(\frac{x + Tx + \cdots + T^{n-1}x}{n}\right), \quad x \in V.$$

Note that q_n is a sublinear functional, because q is sublinear and T is linear. We wish to show that $q_n \in \mathcal{C}$. Suppose $S \in \mathcal{T}$. By assumption, S and T commute, and so

$$q_n(Sx) = q\left(\frac{Sx + TSx + \cdots + T^{n-1}Sx}{n}\right) = q\left(S\left(\frac{x + Tx + \cdots + T^{n-1}x}{n}\right)\right)$$

$$\le q\left(\frac{x + Tx + \cdots + T^{n-1}x}{n}\right) = q_n(x).$$

Thus, $q_n \in \mathcal{C}$ for each $n \in \mathbb{N}$.

Observe that, since $q \in \mathcal{C}$, we have

$$q(T^{n-1}x) \le q(T^{n-2}x) \le \cdots \le q(T^2x) \le q(Tx) \le q(x),$$

for all $x \in V$. Consequently,

$$q_n(x) = q\left(\frac{x + Tx + \cdots + T^{n-1}x}{n}\right) \le q\left(\frac{x}{n}\right) + \cdots + q\left(\frac{x}{n}\right) = q(x),$$

for all $x \in V$. By the minimality of q in $\mathcal{C}$, it follows that $q_n = q$ for all $n \in \mathbb{N}$, and hence

$$q(x) = q\left(\frac{x + Tx + \cdots + T^{n-1}x}{n}\right),$$

for all $x \in V$, $T \in \mathcal{T}$, and $n \in \mathbb{N}$.

Now, let $x \in V$ and $T \in \mathcal{T}$. For all $n \in \mathbb{N}$,

$$q(x - Tx) = q\left(\frac{(x - Tx) + T(x - Tx) + \cdots + T^{n-1}(x - Tx)}{n}\right)$$

$$= q \left(\frac{x - T^n x}{n} \right) \le \frac{1}{n} q(x) + \frac{1}{n} q(-T^n x)$$

$$\le \frac{1}{n} q(x) + \frac{1}{n} q(-x).$$

Since this is true for all $n \in \mathbb{N}$, we conclude that $q(x - Tx) \le 0$ for all $x \in V$ and $T \in \mathcal{T}$. By a similar argument, we also deduce that $q(-x - T(-x)) \le 0$ for all $x \in V$ and $T \in \mathcal{T}$.

By Lemmas 3.7 and 3.8, there exists a linear functional f such that $f \le q$. It follows that both $f(x - Tx) \le 0$ and $f(-x - T(-x)) \le 0$ for all $x \in V$ and $T \in \mathcal{T}$. Therefore, by the linearity of f, we have $f(x) = f(Tx)$ for all $x \in V$ and $T \in \mathcal{T}$. This is the desired result, and so the proof is complete.   □

Example 3.17 (revisited). Let $T : \ell_\infty \to \ell_\infty$ be the shift operator on the real Banach space ℓ_∞ and consider the sublinear functional p on ℓ_∞ defined by $p(\xi) = \sup_{n \in \mathbb{N}} \xi_n$ for all $\xi = (\xi_n)_{n=1}^\infty$ in ℓ_∞. Observe that $p(T\xi) \le p(\xi)$ for all $\xi \in \ell_\infty$. We now invoke Theorem 3.16 with $\mathcal{T} = \{T\}$ to conclude that there exists a linear functional L on ℓ_∞ such that $L \le p$ that is shift-invariant; that is, $L(\xi) = L(T\xi)$ for all $\xi \in \ell_\infty$. We claim that the map L has the following properties, where $\xi = (\xi_n)_{n=1}^\infty \in \ell_\infty$:

(i) $\liminf\limits_{n \to \infty} \xi_n \le L(\xi) \le \limsup\limits_{n \to \infty} \xi_n$,

(ii) $L(\xi) = \lim\limits_{n \to \infty} \xi_n$ whenever the limit exists, and

(iii) $L(\xi) \ge 0$ whenever $\xi_n \ge 0$ for all $n \in \mathbb{N}$.

Observe that *(i)* implies both *(ii)* and *(iii)*. To show that *(i)* is true, we use the invariance of L under the shift operator. If $k \in \mathbb{N}$, then (by shifting k times)

$$L(\xi) = L\left((\xi_{n+k})_{n=1}^\infty \right) \le \sup_{n \in \mathbb{N}} \xi_{n+k} = \sup_{n \ge k} \xi_n, \quad \xi \in \ell_\infty.$$

This is true for all $k \in \mathbb{N}$, and so we conclude that $L(\xi) \le \limsup\limits_{n \to \infty} \xi_n$. A similar argument shows $\liminf\limits_{n \to \infty} \xi_n \le L(\xi)$, which proves *(i)*.

Motivated by the preceding example, we make a definition.

Definition 3.18 A linear functional L on ℓ_∞ is called a *Banach limit* if, for any sequence $\xi = (\xi_n)_{n=1}^\infty$ in ℓ_∞,

(i) $L(T\xi) = L(\xi)$, where T is the shift operator,

(ii) $L(\xi) = \lim\limits_{n \to \infty} \xi_n$ whenever the limit exists, and

(iii) $L(\xi) \ge 0$ whenever $\xi_n \ge 0$ for all $n \in \mathbb{N}$.

Shortly, we will present an application of a Banach limit. First, we recall some definitions.

Definition 3.19 A real vector space H is called a *real inner product space* if there is a map $(\cdot, \cdot) : H \times H \to \mathbb{R}$, called an *inner product*, that satisfies the following properties:

(i) $(x,x) \geq 0$ for all $x \in H$, and $(x,x) = 0$ if and only if $x = 0$,

(ii) $(x,y) = (y,x)$, and

(iii) $(\alpha x + \beta x', y) = \alpha(x,y) + \beta(x',y)$ and $(x, \alpha y + \beta y') = \alpha(x,y) + \beta(x,y')$,

where $\{x, x', y, y'\} \subseteq H$ and $\{\alpha, \beta\} \subseteq \mathbb{R}$.

When a map possesses the three properties listed in Definition 3.19, it is called *(i) positive definite*, *(ii) symmetric*, and *(iii) bilinear*, respectively. The concept of an inner product space exists also when the underlying scalar field is $\mathbb{C}$, but the properties defining an inner product must be modified in this setting. (See Definition 7.1.)

An inner product $(\cdot, \cdot)$ on a vector space H can always be used to define a norm by the formula

$$\|x\| = \sqrt{(x,x)}, \quad x \in H.$$

This norm on H is said to be *induced* by the inner product. (We will verify that this formula defines a norm in Section 7.1.)

Definition 3.20 A real inner product space H is called a *real Hilbert space* if it is a complete normed space when given the norm induced by the inner product.

If H is an inner product space with induced norm $\| \cdot \|$, then H is a Hilbert space precisely when $(H, \| \cdot \|)$ is a Banach space. We will study the topic of Hilbert spaces in greater depth in Chapter 7.

Suppose H is a real Hilbert space with inner product $(\cdot, \cdot)$. A bounded linear map $S : H \to H$ is said to be an *orthogonal operator* if $(Sx, Sy) = (x, y)$ for all x and y in H. Note that S is an orthogonal operator if and only if $\|S^n x\| = \|x\|$ for all $n \in \mathbb{N}$ and $x \in H$. (See Exercise 3.3.) In particular, if $n = 1$, then $\|Sx\| = \|x\|$, and so S is necessarily bounded.

A bounded linear map $S : H \to H$ is said to be *similar to an orthogonal operator* if it is orthogonal with respect to an equivalent inner product on H. (Two inner products on H are *equivalent* if they induce equivalent norms.)

We will make use of the following significant fact: If H is an inner product space, then $|(x, y)| \leq \|x\| \|y\|$ for all x and y in H. This inequality, known as the *Cauchy–Schwarz Inequality*, is a fundamental tool. We will make use of it now, but will not prove it until later. (See Theorem 7.2.)

Proposition 3.21 *Let H be a real Hilbert space and suppose $\| \cdot \|$ is the complete norm induced by the inner product on H. If $S : H \to H$ is an invertible bounded linear map and there exist positive constants c and C such that*

$$c\|x\| \leq \|S^n x\| \leq C\|x\|, \tag{3.8}$$

for all $n \in \mathbb{N}$ and $x \in H$, then S is similar to an orthogonal operator.

Proof Denote the inner product on H by $(\cdot, \cdot)$. By the Cauchy–Schwarz Inequality,

$$|(S^n x, S^n y)| \leq \|S^n x\| \|S^n y\| \leq C^2 \|x\| \|y\|,$$

for $n \in \mathbb{N}$ and $\{x, y\} \subseteq H$. It follows that $((S^n x, S^n y))_{n \in \mathbb{N}}$ is a sequence in ℓ_∞ for each x and y in H. Let L be a Banach limit on ℓ_∞. Define a new inner product on

H by

$$\langle x, y \rangle = L \left(\left((S^n x, S^n y) \right)_{n \in \mathbb{N}} \right), \quad \{x, y\} \subseteq H.$$

(See Exercise 3.15 to show that this defines a real inner product on H.) If $x \in H$ and $|||x||| = \sqrt{\langle x, x \rangle}$, then $||| \cdot |||$ is a norm on H and

$$|||x|||^2 = \langle x, x \rangle = L \left(\left(\| S^n x \|^2 \right)_{n \in \mathbb{N}} \right).$$

By (3.8) and property *(iii)* of a Banach limit,

$$c\|x\| \leq |||x||| \leq C\|x\|, \quad x \in H.$$

Consequently, $||| \cdot |||$ and $\| \cdot \|$ are equivalent norms on H, and so $\langle \cdot, \cdot \rangle$ and $(\cdot, \cdot)$ are equivalent inner products on H.

By the shift-invariance of a Banach limit (property *(i)*),

$$\langle Sx, Sy \rangle = L \left(\left((S^{n+1}x, S^{n+1}y) \right)_{n \in \mathbb{N}} \right) = \langle x, y \rangle, \quad \{x, y\} \subseteq H.$$

Therefore, S is orthogonal with respect to $\langle \cdot, \cdot \rangle$, where $\langle \cdot, \cdot \rangle$ is an inner product equivalent to the original. Hence, S is similar to an orthogonal operator. $\square$

3.4 Haar Measure for Compact Abelian Groups

In this section, we will apply the Hahn–Banach Theorem to the setting of compact groups. A *group* is a pair $(G, \cdot)$, where G is a set and $\cdot$ is a binary operation on G, called *multiplication*, that satisfies the following properties:

(i) *(closure)* $x \cdot y \in G$ for all $\{x, y\} \subseteq G$.
(ii) *(associativity)* $(x \cdot y) \cdot z = x \cdot (y \cdot z)$ for all $\{x, y, z\} \subseteq G$.
(iii) *(identity)* There exists an element $e \in G$ such that $x \cdot e = x = e \cdot x$ for all $x \in G$.
(iv) *(inverses)* For $x \in G$, there exists $x^{-1} \in G$ such that $x \cdot x^{-1} = e = x^{-1} \cdot x$.

Properties *(i)–(iv)* are known as the *group axioms*. When the multiplication is understood, the group $(G, \cdot)$ is often abbreviated to G. Frequently, group multiplication is denoted by juxtaposition, so that $x \cdot y$ is written xy. We will adopt this convention when there is no risk of confusion.

A simple calculation shows that, for a given $x \in G$, the inverse x^{-1} is necessarily unique. Therefore, the map $x \mapsto x^{-1}$ is a well-defined operation on G (called *inversion*).

We call G a *metric group* if it is both a group and a metric space, and if the group operations of multiplication and inversion are continuous on G; that is, if both the maps $(x, y) \mapsto xy$ and $x \mapsto x^{-1}$ are continuous for x and y in G. If a metric group G is also a compact topological space, then G is called a *compact metric group*.

Example 3.22 The following are examples of compact metric groups:

(i) The unit circle in $\mathbb{C}$, written $\mathbb{T} = \{e^{i\theta} : 0 \le \theta < 2\pi\}$, is a compact group, often called the *circle group* or the *torus*. Multiplication in $\mathbb{T}$ is taken from $\mathbb{C}$, and so is the metric. Consequently, the identity in $\mathbb{T}$ is 1 and the inverse of $e^{i\theta}$ is $e^{-i\theta}$. The *punctured complex plane* $\mathbb{C}\backslash\{0\}$ with the standard multiplication and metric is itself a metric group, but it is not compact. (Nor is it complete, because $\mathbb{C}\backslash\{0\}$ is an open subset of $\mathbb{C}$.)

(ii) Let $\mathcal{O}_n$ denote the collection of $n \times n$ orthogonal matrices $(n < \infty)$. Then $\mathcal{O}_n$ is a compact metric group, called the *orthogonal group*. The group operations are given by matrix multiplication and matrix inversion. The metric on $\mathcal{O}_n$ is induced by the operator norm $||| \cdot |||$ on $\mathcal{O}_n$. That is, $d(X, Y) = |||Y - X|||$ for all X and Y in $\mathcal{O}_n$.

(iii) The *Cantor group* is the countable product $\prod_{n=1}^{\infty} \mathbb{Z}_2 = \{0, 1\}^{\mathbb{N}}$. Elements of the Cantor group are sequences of zeros and ones, and the group operation is given by component-wise addition (mod 2). Note that the Cantor group is a compact space by Tychonoff's Theorem. (See Theorem B.4 in the appendix.) The Cantor group can be given a metric using the formula

$$d(x, y) = \sum_{k=1}^{\infty} \frac{1}{2^k} \frac{|x_k - y_k|}{1 + |x_k - y_k|},$$

where $x = (x_k)_{k=1}^{\infty}$ and $y = (y_k)_{k=1}^{\infty}$ are elements of the set $\{0, 1\}^{\mathbb{N}}$.

Of particular interest in this section are abelian groups. A group G is called *abelian* if $xy = yx$ for all x and y in G. (In other words, if the group multiplication is commutative.) Both the torus and the Cantor group are abelian, but the orthogonal group is not if $n \ge 2$. When a group is abelian, the group multiplication is often denoted by addition $(+)$ and the inverse x^{-1} is then written as $-x$ (provided this causes no confusion).

Definition 3.23 Let $(G, +)$ be an abelian metric group. A Borel measure λ on G is *translation-invariant* if $\lambda(B) = \lambda(x + B)$ for all $x \in G$ and Borel subsets $B \subseteq G$.

Theorem 3.24 *If $(G, +)$ is a compact abelian metric group, then there is a unique translation-invariant Borel probability measure on G.*

Proof. Let $C(G)$ be the space of real-valued continuous functions on G equipped with the norm

$$\|f\|_{\infty} = \max_{x \in G} |f(x)|, \quad f \in C(G).$$

Note that this maximum is attained because f is continuous on the compact set G.

For each $x \in G$, define an operator $T_x : C(G) \to C(G)$ by $T_x f(y) = f(x + y)$ for all $f \in C(G)$ and $y \in G$. Let $\mathcal{T} = \{T_x : x \in G\}$. The set $\mathcal{T}$ is a commuting family of operators on $C(G)$ because G is abelian. We call the elements of $\mathcal{T}$ *rotations*.

Define a sublinear functional p on $C(G)$ by

$$p(f) = \max_{x \in G} f(x), \quad f \in C(G).$$

Certainly, $p(T_x f) = p(f)$ for all $x \in G$ and $f \in C(G)$, and so p is invariant under rotations. Thus, by Theorem 3.16 (the Invariant Hahn–Banach Theorem), there exists a linear functional ϕ on $C(G)$ such that $\phi \leq p$ and $\phi(T_x f) = \phi(f)$ for all $x \in G$ and $f \in C(G)$.

Let $f \in C(G)$. By construction,

$$\phi(f) \leq \max_{x \in G} f(x) \quad \text{and} \quad \phi(-f) \leq \max_{x \in G} (-f(x)) = -\min_{x \in G} f(x).$$

Consequently,

$$\min_{x \in G} f(x) \leq \phi(f) \leq \max_{x \in G} f(x). \tag{3.9}$$

In particular, we see $|\phi(f)| \leq \|f\|_\infty$ for all $f \in C(G)$. It follows that ϕ is a bounded linear functional on $C(G)$. Thus, by Theorem 2.20 (the Riesz Representation Theorem), there exists a Borel measure λ on G such that

$$\phi(f) = \int_G f \, d\lambda, \quad f \in C(G).$$

Since $\phi(f) \geq 0$ whenever $f \geq 0$, we have that λ is a positive measure. Furthermore, because of (3.4.1), we have that $\phi(1) = 1$. It follows that $\lambda(G) = 1$, and consequently λ is a probability measure on G.

We now show λ is translation-invariant. Let $x \in G$ and define a measure λ_x on G by $\lambda_x(B) = \lambda(x + B)$ for all Borel sets B in G. Our goal is to show that $\lambda = \lambda_x$. To that end, we make the following claim:

$$\int_G f \, d\lambda_x = \int_G T_{-x} f \, d\lambda, \quad f \in C(G). \tag{3.10}$$

To prove this, let B be a Borel set in G and let $y \in G$. By the definition of the map T_{-x}, we have that $T_{-x} \chi_B(y) = \chi_B(y - x)$. Thus,

$$\int_G T_{-x} \chi_B(y) \, \lambda(dy) = \int_G \chi_B(y-x) \, \lambda(dy) = \lambda(B+x) = \lambda_x(B) = \int_G \chi_B(y) \, \lambda_x(dy).$$

Therefore, (3.4.2) holds for $f = \chi_B$, where B is a Borel subset of G. By linearity, (3.4.2) holds for simple functions, and by the density of simple functions in $C(G)$, (3.4.2) holds for continuous functions, as well.

The linear functional ϕ was chosen (via the Invariant Hahn–Banach Theorem) so that $\phi(T_{-x} f) = \phi(f)$ for all $f \in C(G)$. Therefore, using (3.10),

$$\int_G f \, d\lambda_x = \int_G T_{-x} f \, d\lambda = \phi(T_{-x} f) = \phi(f) = \int_G f \, d\lambda,$$

for all $f \in C(G)$. It follows that $\lambda_x = \lambda$. This is true for all $x \in G$, and so λ is translation-invariant.

It remains to show that λ is unique. Assume μ is a translation-invariant probability measure on G. Let $f \in C(G)$. By Fubini's Theorem,

$$\int_G \left(\int_G f(x+y)\,\lambda(dx) \right) \mu(dy) = \int_G \left(\int_G f(x+y)\,\mu(dy) \right) \lambda(dx).$$

By translation-invariance,

$$\int_G f(x+y)\,\lambda(dx) = \int_G f(x)\,\lambda(dx), \quad y \in G,$$

and

$$\int_G f(x+y)\,\mu(dy) = \int_G f(y)\,\mu(dy), \quad x \in G.$$

Therefore, since λ and μ are probability measures (and so $\lambda(G) = \mu(G) = 1$),

$$\int_G f(x)\,\lambda(dx) = \int_G f(y)\,\mu(dy), \quad f \in C(G).$$

It follows that $\lambda = \mu$, and so the proof is complete. $\qquad\square$

Definition 3.25 The unique translation-invariant Borel probability measure on a compact abelian metric group is called *Haar measure*.

Remark 3.26 In the proof of Theorem 3.24, we do not actually use the metric on G. Indeed, our proof requires only that G is a compact abelian topological group. At this time, however, we have restricted our attention to topologies arising from a metric. We will consider more general topological spaces in Chapter 5 and we will revisit the topic of Haar measure at that time. (See Section 5.9.)

Example 3.27 Let $\mathbb{T} = \{e^{i\theta} : 0 \le \theta < 2\pi\}$ be the torus from Example 3.22 *(i)*. Haar measure on $\mathbb{T}$ is given by $m/(2\pi)$, where m is Lebesgue measure on $[0, 2\pi)$. To be more precise, if $f \in C(\mathbb{T})$ and λ is Haar measure on $\mathbb{T}$, then

$$\int_{\mathbb{T}} f\,d\lambda = \frac{1}{2\pi} \int_0^{2\pi} f\left(e^{i\theta}\right) m(d\theta).$$

For this reason, some authors write $\mathbb{T} = [0, 2\pi)$ and $\lambda = m/(2\pi)$.

(The factor 2π is needed in the denominator so that λ is a probability measure.)

Corollary 3.28 *Let G be a compact abelian metric group with Haar measure λ. If B is any Borel subset of G, and if $-B$ is the set $\{-x : x \in B\}$ of inverses of elements in B, then $\lambda(-B) = \lambda(B)$.*

Proof Define a probability measure μ on G by $\mu(B) = \lambda(-B)$ for all Borel subsets B in G. By the translation invariance of λ,

$$\mu(x + B) = \lambda(-x + (-B)) = \lambda(-B) = \mu(B).$$

Thus, μ is a translation invariant Borel probability measure on G. Since Haar measure is the unique measure with these properties, we conclude that $\mu = \lambda$. $\qquad\square$

3.5 Duals, Biduals, and More

Let X be a Banach space. Recall that the dual space X^* of X is the space of bounded linear functionals on X. When X^* is equipped with the operator norm $\|x^*\| = \sup\{|x^*(x)| : x \in B_X\}$, where B_X is the closed unit ball of X, the space X^* is a Banach space. (See Proposition 1.11.)

Proposition 3.29 *Let X be a Banach space with dual space X^*. If $x \in X$, then*

$$\|x\| = \sup\big\{|x^*(x)| : x^* \in B_{X^*}\big\},$$

and there exists an $x^* \in X^*$ such that $\|x^*\| = 1$ and $x^*(x) = \|x\|$.

Proof Let $\mathbb{K}$ denote the scalar field. For $x \in X$, define a closed linear subspace E_x of X by $E_x = \{\alpha x : \alpha \in \mathbb{K}\}$. Define a map $f : E_x \to \mathbb{K}$ by $f(\alpha x) = \alpha \|x\|$, for all $\alpha \in \mathbb{K}$. Then f is linear and $\|f\| = 1$.

By the Hahn–Banach Theorem for normed spaces (Theorem 3.9 for real spaces, Theorem 3.15 for complex spaces), there exists a bounded linear functional $x_f^* \in X^*$ that extends f. That is, the map x_f^* is a bounded linear functional on X such that $x_f^*(x) = \|x\|$ and $\|x_f^*\| = 1$.

Observe that $|x^*(x)| \leq \|x\|$ for all $x^* \in B_{X^*}$. Therefore,

$$\|x\| \geq \sup\big\{|x^*(x)| : x^* \in B_{X^*}\big\}.$$

On the other hand, the linear functional x_f^* is an element in B_{X^*} with the property that $x_f^*(x) = \|x\|$. The result follows. □

For any x in a Banach space X, Proposition 3.29 guarantees the existence of a so-called *norming element* in the dual space X^*; that is, an element x^* of norm 1 such that $x^*(x) = \|x\|$. This element may or may not be unique.

Example 3.30 We consider the existence of norming elements in a few real sequence spaces.

(i) Let $X = \ell_1$, and so $X^* = \ell_\infty$. Consider the summable sequence $\xi = (1, \frac{1}{4}, \frac{1}{9}, \frac{1}{16}, \ldots, \frac{1}{n^2}, \ldots)$. Then $\xi \in \ell_1$ and $\|\xi\| = \sum_{n=1}^\infty \frac{1}{n^2} = \frac{\pi^2}{6}$. In this case, there is a unique norming element in ℓ_∞, and that element is the constant sequence $(1, 1, 1, \ldots)$.

(ii) Once again, let $X = \ell_1$. This time, let $\xi = (1, 0, 0, 0, \ldots)$. In this case, $\xi \in \ell_1$ and $\|\xi\| = 1$. There are many norming functionals in this case. Indeed, any element in ℓ_∞ of the form $(1, a_2, a_3, a_4, \ldots)$ with $|a_j| \leq 1$ for all $j \geq 2$ will determine a norming functional for ξ.

(iii) In ℓ_2, the norming functional is always unique. Let $\xi = (\xi_1, \xi_2, \ldots)$ be an element of ℓ_2. Then $\|\xi\| = \left(\sum_{n=1}^\infty \xi_n^2\right)^{1/2}$ and the norming element is $\xi/\|\xi\|$. (Recall that $\ell_2^* = \ell_2$ (Theorem 2.5).)

(iv) Consider the space ℓ_∞ and let $\xi = (1, 1, 1, \ldots)$. Any linear functional ϕ on ℓ_∞ will be a norming element for ξ provided both $\phi(\xi) = 1$ and $\|\phi\| = 1$. Any Banach limit will satisfy these criteria, as well as other linear functionals.

Definition 3.31 Let X be a Banach space. The *bidual* of X is the space $X^{**} = (X^*)^*$.

Let X be a Banach space. Define a map $j : X \to X^{**}$ by letting $j(x) \in X^{**}$ be the linear functional on X^* defined by

$$j(x)(x^*) = x^*(x), \quad x^* \in X^*, \tag{3.11}$$

for all $x \in X$. The equation in (3.11) is sometimes written $\langle x^*, j(x) \rangle = \langle x, x^* \rangle$. We call j the *natural embedding of X into its bidual* and $\langle \cdot, \cdot \rangle$ the *dual space action* between a Banach space (written on the left) and its dual (written on the right).

Theorem 3.32 *The natural embedding of a Banach space into its bidual is an isometric isomorphism onto a closed subspace of the bidual.*

Proof Let X be a Banach space and suppose j is the natural embedding of X into its bidual X^{**}. By direct computation, one can show that j is a linear injection onto its image. (See Exercise 3.5.)

We now show that j is an isometry. If $x \in X$, then $j(x) \in X^{**}$. Thus, for all $x \in X$,

$$\| j(x) \| = \sup_{\|x^*\| \le 1} |j(x)(x^*)| = \sup_{\|x^*\| \le 1} |x^*(x)| = \|x\|.$$

(The last equality follows from Proposition 3.29.) It follows that j is an isometry. From this, we can conclude that $j(X)$ is closed in X^{**} and that j is an isomorphism onto $j(X)$. (See Exercise 3.5.)   □

Theorem 3.32 suggests that an exact copy of X sits inside of X^{**}. In light of this, it is common to suppress the map j and simply view X as a closed subspace of X^{**}.

Definition 3.33 A Banach space X is called *reflexive* if the natural embedding j of X into its bidual is a surjection; that is, if $j(X) = X^{**}$.

Example 3.34

(i) The sequence spaces ℓ_p are reflexive whenever $1 < p < \infty$. The spaces ℓ_1 and ℓ_∞ are not reflexive, having vast and mysterious biduals.

(ii) Let $(\Omega, \mathcal{B}, \mu)$ be a measure space. The function spaces $L_p(\Omega, \mathcal{B}, \mu)$ are reflexive whenever $1 < p < \infty$. The spaces $L_1(\mu)$ and $L_\infty(\mu)$ will not be reflexive unless the support of μ is a finite set.

(iii) The sequence space c_0 is not reflexive. In fact, $c_0^{**} = \ell_1^* = \ell_\infty$. This is possibly the only case where we can explicitly see the proper inclusion of a Banach space in its bidual; i.e., $c_0 \subset \ell_\infty$.

Proposition 3.35 *A Banach space X is reflexive if and only if its dual X^* is reflexive.*

Proof First, assume X is reflexive. Then $(X^*)^* = X$. By definition, the bidual of X^* is

$$(X^*)^{**} = \left((X^*)^* \right)^* = X^*.$$

The last equality follows from the assumption that X is reflexive. Therefore, X^* is reflexive.

Now assume X^* is reflexive. We wish to show X is reflexive. Let $j : X \to X^{**}$ be the natural embedding of X into its bidual. Assume that j is not a surjection. Then there exists an $x^{**} \in X^{**}$ such that $x^{**} \notin j(X)$. Let

$$\delta = d(x^{**}, j(X)) = \inf\{\|x^{**} - j(x)\| : x \in X\}.$$

Then $\delta > 0$, by assumption, because $j(X)$ is closed in X^{**}. (See Theorem 3.32.)

Let $E = \operatorname{span}\{x^{**}, j(x) : x \in X\}$. Define a linear functional $\phi : E \to \mathbb{K}$ (where $\mathbb{K}$ denotes the scalar field) by

$$\phi\left(\lambda x^{**} + j(x)\right) = \lambda, \quad \lambda \in \mathbb{K}, \ x \in X.$$

This map is well-defined because j is an injection onto $j(X)$. For $\lambda \in \mathbb{K}$ and $x \in X$,

$$\|\lambda x^{**} + j(x)\| = |\lambda| \left\|x^{**} - j(-\lambda^{-1}x)\right\| \geq |\lambda| \, d(x^{**}, j(X)) = \delta \, |\lambda|.$$

It follows that
$$|\phi(\lambda x^{**} + j(x))| = |\lambda| \leq \frac{1}{\delta} \|\lambda x^{**} + j(x)\|.$$

We conclude that ϕ is bounded on $E \subseteq X^{**}$ and $\|\phi\| \leq 1/\delta$. Therefore, by the Hahn–Banach Extension Theorem (Theorem 3.4), there exists an element of X^{***} that extends ϕ. In particular, there exists $x^{***} \in X^{***}$ such that $x^{***}(x^{**}) = 1$ and $x^{***}(j(x)) = 0$ for all $x \in X$. By assumption, X^* is reflexive, and so x^{***} corresponds to some $x^* \in X^*$. Consequently, there exists an element $x^* \in X^*$ such that $x^{**}(x^*) = 1$ and $x^*|_X = 0$. This implies both $\|x^*\| > 0$ and $x^* = 0$, a contradiction. Therefore, X is reflexive.   □

3.6   The Adjoint of an Operator

Suppose n is a natural number. If $A = (a_{jk})_{j,k=1}^n$ is an $n \times n$ complex matrix, then the *matrix adjoint* (or *conjugate transpose*) of A is the $n \times n$ matrix $A^* = (b_{jk})_{j,k=1}^n$ with entries $b_{jk} = \overline{a_{kj}}$ for each j and k in the set $\{1, \dots, n\}$. One of the important properties of the matrix adjoint is

$$(Ax, y) = (x, A^*y), \quad \{x, y\} \subseteq \mathbb{C}^n,$$

where $(\cdot, \cdot)$ denotes the inner product on $\mathbb{C}^n$. In this section we will generalize the notion of a matrix adjoint to infinite-dimensional Banach spaces.

Definition 3.36 Let X and Y be Banach spaces and let $T : X \to Y$ be a bounded linear operator. The map $T^* : Y^* \to X^*$ defined by

$$(T^*y^*)(x) = (y^* \circ T)(x), \quad x \in X, \ y^* \in Y^*,$$

is called the *adjoint* of T.

Owing to the abundance of Banach spaces, we will sometimes find it convenient to denote the norm on a Banach space X by $\|\cdot\|_X$. The proof of the following proposition will afford one such occasion.

Proposition 3.37 *Let X and Y be two Banach spaces. If $T : X \to Y$ is a bounded linear operator, then the adjoint T^* is a bounded linear operator and $\|T^*\| = \|T\|$.*

Proof It is not hard to show T^* is linear. To show T^* is bounded, let $y^* \in Y^*$. Then

$$\|T^*y^*\|_{X^*} = \sup_{\|x\| \leq 1} |(T^*y^*)(x)| = \sup_{\|x\| \leq 1} |y^*(Tx)|.$$

Therefore,

$$\|T^*y^*\|_{X^*} \leq \sup_{\|x\| \leq 1} \|y^*\|_{Y^*} \|Tx\|_Y = \|T\| \, \|y^*\|_{Y^*}.$$

Hence, T is bounded and $\|T^*\| \leq \|T\|$.

To prove the reverse inequality, we begin by letting $\epsilon > 0$. There exists $x \in X$ such that $\|x\|_X \leq 1$ and $\|Tx\|_Y > \|T\| - \epsilon$. By Proposition 3.29, there exists $y^* \in Y^*$ such that $\|y^*\|_{Y^*} = 1$ and $y^*(Tx) = \|Tx\|_Y$; whence,

$$\|Tx\|_Y = (T^*y^*)(x) \leq \|T^*y^*\|_{X^*} \|x\|_X \leq \|T^*y^*\|_{X^*} \leq \|T^*\|.$$

Consequently, $\|T\| < \|T^*\| + \epsilon$. Since the choice of ϵ was arbitrary, we conclude that $\|T\| \leq \|T^*\|$. $\square$

Corollary 3.38 *Let X and Y be Banach spaces and let $\mathcal{L}(X, Y)$ denote the Banach space of bounded linear operators from X to Y. The map taking T to T^* is a linear isometry from $\mathcal{L}(X, Y)$ to $\mathcal{L}(Y^*, X^*)$.*

Proof This follows from Proposition 3.37. $\square$

For an operator T between Banach spaces X and Y, the adjoint $T^* : Y^* \to X^*$ also has an adjoint $T^{**} : X^{**} \to Y^{**}$. In the following proposition, we think of X as a subspace of its bidual X^{**}.

Proposition 3.39 *Let X and Y be Banach spaces. If $T : X \to Y$ is a bounded linear operator, then $T^{**}|_X = T$. That is, $T^{**}(x) = T(x)$ for all $x \in X$.*

Proof Let j be the natural embedding of X into its dual X^{**} (which we think of as the inclusion map). Suppose $x \in X$. By definition, $T^{**}x \in Y^{**}$. Let the action of Y^{**} on Y^* be represented by $\langle \cdot, \cdot \rangle$. Then for any $y^* \in Y^*$,

$$(T^{**}x)(y^*) = \langle y^*, T^{**}x \rangle = \langle T^*y^*, j(x) \rangle.$$

We have $T^*y^* \in X^*$, and so, by (3.11) and Definition 3.36,

$$\langle T^*y^*, j(x) \rangle = \langle x, T^*y^* \rangle = \langle Tx, y^* \rangle.$$

It follows that $\langle y^*, T^{**}x \rangle = \langle Tx, y^* \rangle$ for all $y^* \in Y^*$, and hence the result. $\square$

Proposition 3.40 *Let X, Y, and Z be Banach spaces and suppose both $S : X \to Y$ and $T : Y \to Z$ are bounded linear operators. If $TS = T \circ S : X \to Z$, then the adjoint map $(TS)^* : Z^* \to X^*$ is given by $(TS)^* = S^*T^*$.*

Proof The proof is left to the reader. (See Exercise 3.8.)   □

Example 3.41 (The Volterra operator). Let p and q be conjugate exponents (so that $1/p + 1/q = 1$), where $1 < p < \infty$. Let $L_p(0, 1)$ denote the real Banach space of p-integrable real-valued functions on $[0, 1]$ with respect to Lebesgue measure. Define a map $V : L_p(0, 1) \to L_p(0, 1)$ by

$$Vf(t) = \int_0^t f(s)\,ds, \quad f \in L_p(0,1),\ t \in [0,1].$$

We call V the *Volterra operator* on $L_p(0, 1)$. We must show that V is well-defined.

By Hölder's Inequality, for all $t \in [0, 1]$,

$$|Vf(t)| \leq \left(\int_0^t 1^q\,ds \right)^{1/q} \left(\int_0^t |f(s)|^p\,ds \right)^{1/p} \leq t^{1/q}\,\|f\|_p.$$

Therefore,

$$\|Vf\|_p \leq \left(\int_0^1 (t^{1/q})^p\,dt \right)^{1/p} \|f\|_p = \left(\frac{1}{p} \right)^{1/p} \|f\|_p.$$

It follows that V is bounded and $\|V\| \leq (1/p)^{1/p}$.

We now compute the adjoint of the operator V. Observe that the adjoint operator $V^* : L_q(0, 1) \to L_q(0, 1)$ satisfies the equation

$$\int_0^1 f(s)\,V^*g(s)\,ds = \int_0^1 Vf(s)\,g(s)\,ds = \int_0^1 \left(\int_0^s f(t)\,dt \right) g(s)\,ds,$$

for all $f \in L_p(0, 1)$ and $g \in L_q(0, 1)$. By Fubini's Theorem,

$$\int_0^1 \left(\int_0^s f(t)\,dt \right) g(s)\,ds = \int_0^1 \left(\int_t^1 g(s)\,ds \right) f(t)\,dt,$$

for all $f \in L_p(0, 1)$ and $g \in L_q(0, 1)$. We therefore conclude that

$$V^*g(t) = \int_t^1 g(s)\,ds, \quad g \in L_q(0, 1).$$

We can define the Volterra operator V for $p = 1$, as well. A similar argument will yield the same adjoint V^*. If $p = \infty$, however, then V^* is a map from $(L_\infty(0, 1))^*$ to $(L_\infty(0, 1))^*$, and this map is not so easy to compute.

The Volterra operator defined in Example 3.41 is a special case of the next example.

Example 3.42 Suppose $K \in L_\infty([0, 1] \times [0, 1])$. For $p \in [1, \infty)$, define a linear map $T_K : L_p(0, 1) \rightarrow L_p(0, 1)$ by

$$T_K f(s) = \int_0^1 K(s, t) f(t) \, dt, \quad f \in L_p(0, 1).$$

Then $|T_K f(s)| \le \|K\|_\infty \|f\|_p$ for all $s \in [0, 1]$, and so $\|T_K f\|_p \le \|K\|_\infty \|f\|_p$. Thus, $\|T_K\| \le \|K\|_\infty$. A calculation similar to that in Example 3.41 reveals

$$T_K^* g(s) = \int_0^1 K(t, s) g(t) \, dt, \quad g \in L_q(0, 1).$$

This example can be thought of as a "continuous" analog of a matrix adjoint. This demonstrates the original goal of functional analysis: To generalize linear algebra to an infinite-dimensional setting.

3.7 New Banach Spaces From Old

In this section, we will show two common ways to construct new Banach spaces from given ones. The first method is essentially a means of summing two spaces, while the second is comparable to subtraction.

Definition 3.43 Let X an Y be Banach spaces. The *direct sum* of X and Y is the set $X \times Y$ equipped with component-wise addition and scalar multiplication:

• $(x_1, y_1) + (x_2, y_2) = (x_1 + x_2, y_1 + y_2)$, and
• $\lambda \cdot (x, y) = (\lambda x, \lambda y)$,

where (x, y), (x_1, y_1), and (x_2, y_2) are in $X \times Y$ and λ is a scalar. When given this vector space structure, the direct sum is denoted by $X \oplus Y$.

Proposition 3.44 *Let X and Y be Banach spaces. The direct sum $X \oplus Y$ is a Banach space under the norm*

$$\|(x, y)\| = \|x\|_X + \|y\|_Y, \quad (x, y) \in X \times Y.$$

Proof The proof that this norm is complete follows directly from the fact that $\| \cdot \|_X$ and $\| \cdot \|_Y$ are complete norms. □

In some cases, a Banach space can be decomposed into a direct sum of closed subspaces.

Proposition 3.45 *Let X be a Banach space. Suppose V and W are closed subspaces of X. If $X = V + W$ and $V \cap W = \{0\}$, then X is isomorphic (as a vector space) to $V \oplus W$. (In this case, we write $X = V \oplus W$.)*

Proof We wish to establish a vector space isomorphism between the spaces X and $V \oplus W$. That is, we wish to find a linear bijection (which need not be a homeomorphism). By assumption, $X = \{v+w : v \in V, w \in W\}$. Define a map $\phi : X \to V \oplus W$ by

$$\phi(v + w) = (v, w), \quad (v, w) \in V \times W.$$

A priori, it may not be clear that this map is well-defined. Suppose that $x \in X$ can be written in two ways as the sum of elements from V and W; i.e., suppose that $x = v + w$ and $x = v' + w'$, where (v, w) and (v', w') are in $V \times W$. Then $v + w = v' + w'$, and consequently

$$v - v' = w' - w \in V \cap W = \{0\}.$$

Therefore, $v = v'$ and $w = w'$. It follows that each $x \in X$ has a unique representation of the form $x = v + w$, where $v \in V$ and $w \in W$, and so ϕ is well-defined.

By construction, ϕ is onto. Furthermore, ϕ is one-to-one, because $x = v + w$ if $\phi(x) = (v, w)$. Next, let (v, w) and (v', w') be elements of $V \times W$. If $x = v + w$ and $x' = v' + w'$, then

$$\phi(x + x') = \phi\big((v + v') + (w + w')\big) = (v + v', w + w') = (v, w) + (v', w').$$

Consequently, $\phi(x + x') = \phi(x) + \phi(x')$. Furthermore, if λ is a scalar, then

$$\phi(\lambda x) = \phi(\lambda v + \lambda w) = (\lambda v, \lambda w) = \lambda \phi(x).$$

Therefore, ϕ is linear, and hence a vector space isomorphism. $\qquad\square$

It is perhaps worth mentioning that the map ϕ in the proof of Proposition 3.45 need not be an isometry between the given norm on X and the norm on $V \oplus W$ (as given in Proposition 3.44). These two norms will always be equivalent, however, because ϕ is, in fact, a homeomorphism (a continuous bijection with continuous inverse). This will follow from the Bounded Inverse Theorem (Corollary 4.30), which we shall meet in Section 4.3 as a consequence of the Open Mapping Theorem.

We now consider a second operation used to create new Banach spaces.

Definition 3.46 Let X be a Banach space and let Y be a closed subspace of X. The *quotient space* X/Y is the set of all *cosets* of Y in X. That is, $X/Y = \{x + Y : x \in X\}$. The map $Q : X \to X/Y$ defined by $Qx = x + Y$ for $x \in X$ is called the *quotient map*.

The quotient space X/Y is a vector space with addition and scalar multiplication given by $(x + Y) + (x' + Y) = (x + x') + Y$ and $\alpha(x + Y) = (\alpha x) + Y$, respectively, where x and x' are in X and α is a scalar. We leave it to the reader to verify this fact. Note that the zero vector in X/Y is $Y = 0 + Y$.

Proposition 3.47 *Let X be a Banach space with closed subspace Y. For each $x \in X$, let $\|x + Y\| = \inf_{y \in Y} \|x + y\|$. Then $\| \cdot \|$ defines a complete norm on X/Y (called the quotient norm).*

Proof We first show that $\| \cdot \|$ is a norm on the quotient X/Y. It is clear that $\| \cdot \|$ is nonnegative and $\|0 + Y\| = 0$. Suppose $\|x + Y\| = 0$. We will show that $x + Y = Y$. By the definition of the norm, for all $n \in \mathbb{N}$, there exists $y_n \in Y$ such that $\|x + y_n\| < 2^{-n}$. Therefore, $\|x - (-y_n)\| < 2^{-n}$, and so the sequence $(-y_n)_{n=1}^{\infty}$ converges to x. Since Y is a closed subspace of X, it follows that $x \in Y$, and consequently $x + Y = Y$.

Now let α be a nonzero scalar. If $x \in X$, then

$$\|\alpha x + Y\| = \inf_{y \in Y} \|\alpha x + y\| = \inf_{z \in Y} \|\alpha x + \alpha z\| = |\alpha| \inf_{z \in Y} \|x + z\|.$$

(Observe that $z = y/\alpha$.) Therefore, we conclude $\|\alpha(x + Y)\| = |\alpha| \|x + Y\|$, and so $\| \cdot \|$ is homogeneous.

Next, we show the triangle inequality. Let x_1 and x_2 be in X and suppose $\epsilon > 0$. There exist elements y_1 and y_2 in Y such that

$$\|x_1 + y_1\| < \|x_1 + Y\| + \epsilon/2 \ \text{ and } \ \|x_2 + y_2\| < \|x_2 + Y\| + \epsilon/2.$$

Then

$$\|x_1 + x_2 + y_1 + y_2\| \le \|x_1 + y_1\| + \|x_2 + y_2\| < \|x_1 + Y\| + \|x_2 + Y\| + \epsilon.$$

Since the choice of ϵ was arbitrary, we conclude that $\| \cdot \|$ satisfies the triangle inequality, and hence is a norm on X/Y.

It remains to show the norm $\| \cdot \|$ is complete on X/Y. For this we use the Cauchy Summability Criterion (Lemma 2.24). Suppose $(x_n)_{n=1}^{\infty}$ is a sequence in X such that $\sum_{n=1}^{\infty} \|x_n + Y\| < \infty$. For each $n \in \mathbb{N}$, pick $x_n' \in x_n + Y$ such that $\|x_n'\| \le 2\|x_n + Y\|$. Then $\sum_{n=1}^{\infty} \|x_n'\| < \infty$, and hence $\sum_{n=1}^{\infty} x_n'$ converges in X (by completeness of the norm on X). Suppose $\sum_{n=1}^{\infty} x_n'$ converges to $x \in X$. Then, for any $\epsilon > 0$, there exists an $N \in \mathbb{N}$ such that $\|x - \sum_{k=1}^{n} x_k'\| < \epsilon$ for all $n \ge N$. Therefore,

$$\left\| (x + Y) - \sum_{k=1}^{n} (x_k' + Y) \right\| \le \left\| x - \sum_{k=1}^{n} x_k' \right\| < \epsilon, \quad n \ge N. \tag{3.12}$$

Consequently, $\sum_{n=1}^{\infty} (x_n + Y)$ converges to $x + Y$ in X/Y. It follows that the quotient norm is complete, as required. $\square$

In (3.15), we used the fact that $\|x + Y\| \le \|x\|$ for each $x \in X$. This follows from the definition of the norm on X/Y, because $0 \in Y$. Equivalently, the quotient map $Q : X \to X/Y$ of Definition 3.46 has norm 1.

The quotient map $Q : X \to X/Y$ has the additional property that it is also an open map. A map $T : X \to Z$ is called an *open map* if $T(U)$ is an open set in Z whenever U is an open set in X. If X and Z are Banach spaces and T is linear, then in order to show that T is an open map, it suffices to show that the open unit ball of X is mapped to an open set in Z. (See Exercise 3.7.)

Let U_X and $U_{X/Y}$ be the open unit balls in X and X/Y, respectively. (Recall Definition 1.5.) We will prove the quotient map $Q : X \to X/Y$ is an open map by showing that $Q(U_X) = U_{X/Y}$.

We already know that $\|Q\| \leq 1$, and so $Q(U_X) \subseteq U_{X/Y}$. Now let $z + Y$ be any element of $U_{X/Y}$. By assumption,

$$\|z + Y\| = \inf\{\|z + y\| : y \in Y\} < 1.$$

Thus, there is some $y' \in Y$ such that $\|z + y'\| < 1$. If $z' = z + y'$, then $\|z'\| < 1$ and $Qz' = z + Y$. This proves that $U_{X/Y} \subseteq Q(U_X)$. We have established that the image of U_X is $U_{X/Y}$, and hence Q is an open map, as claimed.

Definition 3.48 Let $T : X \to Y$ be a bounded linear operator between Banach spaces. The set $\{x : Tx = 0\}$ is called the *kernel* of T and is denoted $\ker(T)$.

We now derive an important proposition which is analogous to a well-known fact of linear algebra. This result will prove valuable to us on several occasions.

Proposition 3.49 *Let X and Y be Banach spaces. If $T : X \to Y$ is a bounded linear operator, then $\ker(T)$ is a closed subspace of X and there exists an injective linear operator $T_0 : X/\ker(T) \to Y$ such that $\|T_0\| = \|T\|$. Furthermore, T_0 makes the following diagram commute, where $Q : X \to X/\ker(T)$ is the quotient map.*

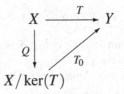

In particular, if T is a surjection, then T_0 is a continuous linear bijection.

Proof Let $E = \ker(T)$. Then $E = T^{-1}(\{0\})$ is a closed set in X because T is continuous and $\{0\}$ is a closed set in Y. A simple calculation shows that E is a linear subspace of X.

Define the map $T_0 : X/E \to Y$ by $T_0(x + E) = Tx$ for all $x \in X$. We must verify that the map T_0 is well-defined. To that end, let $x + E = x' + E$ for x and x' in X. It follows that $x - x' \in E$, and so $T(x - x') = 0$, because E is the kernel of T. Therefore, we have that $Tx - Tx' = T(x - x') = 0$, and hence $Tx = Tx'$. Consequently, the map T_0 is well-defined, as required.

For any $x \in X$, we have $\|Tx\| = \|T_0(x + E)\| \leq \|T_0\|\,\|x + E\| \leq \|T_0\|\,\|x\|$, and so $\|T\| \leq \|T_0\|$. To show the reverse inequality, let $\epsilon > 0$ and choose $x' \in x + E$ such that $\|x'\| < \|x + E\| + \epsilon$. Then $\|Tx\| = \|Tx'\|$ and

$$\|T_0(x + E)\| = \|Tx'\| \leq \|T\|\,\|x'\| \leq \|T\|\,\|x + E\| + \epsilon\|T\|.$$

Since the choice of ϵ was arbitrary, it follows that $\|T_0\| \leq \|T\|$. We therefore conclude that $\|T_0\| = \|T\|$. The rest of the proposition follows directly.   □

There is a close connection between direct sums and quotient spaces. In Section 4.4, we will show that $X = V \oplus W$ if and only if there exists a continuous projection $P : X \to V$ such that $W = \ker(P)$ and $V = P(X)$. (Thus, by Proposition 3.49, we may identify X/W with V.)

3.8 Duals of Quotients and Subspaces

Definition 3.50 Let X be a Banach space. If E is a closed subspace of X, then the *annihilator* of E is the set

$$E^\perp = \{x^* \in X^* : x^*(x) = 0 \text{ for all } x \in E\} \subseteq X^*.$$

If F is a closed subspace of X^*, then the *pre-annihilator* of F is the set

$$F_\perp = \{x \in X : x^*(x) = 0 \text{ for all } x^* \in F\} \subseteq X.$$

It is easy to check that $E^\perp$ and $F_\perp$ are always closed in X^* and X, respectively.

Proposition 3.51 *If X is a Banach space and E is a closed subspace of X, then:*

 (i) E^* can be naturally identified with $X^*/E^\perp$, and
 (ii) $(X/E)^*$ can be naturally identified with $E^\perp$.

Proof (i) Define a map $\rho : X^*/E^\perp \to E^*$ by

$$\rho(x^* + E^\perp) = x^*|_E, \quad x^* \in X^*.$$

We claim this map is well-defined. To see this, suppose x_1^* and x_2^* are two elements of X^* such that $x_1^* + E^\perp = x_2^* + E^\perp$ (i.e., they are in the same coset). It follows that $x_1^* - x_2^* \in E^\perp$, and consequently $(x_1^* - x_2^*)(e) = 0$ for all $e \in E$. Thus $x_1^*|_E = x_2^*|_E$, and so $\rho(x_1^* + E^\perp) = \rho(x_2^* + E^\perp)$. We have established that the definition of $\rho(x^* + E^\perp)$ does not depend on the choice of representative in the coset $x^* + E^\perp$, and hence the map ρ is well-defined. We can show by direct computation that ρ is linear. We also have that ρ is injective, because $x^*|_E = 0$ if and only if $x^* \in E^\perp$.

We now define a map $\psi : E^* \to X^*/E^\perp$. For any $\phi \in E^*$, let

$$\psi(\phi) = x_\phi^* + E^\perp,$$

where x_ϕ^* is any Hahn–Banach extension of ϕ to an element of X^*. In order to show this map is well-defined, we must demonstrate that $\psi(\phi)$ is independent of choice of Hahn–Banach extension of ϕ. To that end, let $\phi \in E^*$ and suppose ϕ has Hahn–Banach extensions x_1^* and x_2^* in X^*. Since these are both extensions of ϕ, it follows that

$$x_1^*|_E = x_2^*|_E = \phi.$$

Thus, $(x_1^* - x_2^*)|_E = 0$, and so $x_1^* - x_2^* \in E^\perp$. We conclude that x_1^* and x_2^* are in the same coset, and therefore $x_1^* + E^\perp = x_2^* + E^\perp$.

Once again, it is easy to see that ψ is an injective linear map. It is also easy to see that $\psi \circ \rho = \text{Id}_{X^*/E^\perp}$ and $\rho \circ \psi = \text{Id}_{E^*}$. (Here, we use $\text{Id}_{X^*/E^\perp}$ and Id_{E^*} to denote the identity maps on $X^*/E^\perp$ and E^*, respectively.) Consequently, ρ is a linear bijection with inverse ψ.

We now show ρ is an isometry by showing that $\|\phi\| = \|x_\phi^* + E^\perp\|$, where $\phi \in E^*$ and x_ϕ^* is any Hahn–Banach extension of ϕ. Certainly, $\|x_\phi^* + E^\perp\| \leq \|x_\phi^*\| = \|\phi\|$. Suppose $\|x_\phi^* + E^\perp\| < \|x_\phi^*\|$. Then there exists some $z^* \in x_\phi^* + E^\perp$ with $\|z^*\| < \|x_\phi^*\|$. Since z^* and x_ϕ^* are in the same coset, it must be the case that $z^*|_E = x_\phi^*|_E = \phi$, but $\|z^*|_E\| \leq \|z^*\| < \|\phi\|$, a contradiction. Therefore, $\|x_\phi^* + E^\perp\| = \|\phi\|$.

(ii) Denote the quotient map by $Q : X \to X/E$. Let $\phi \in (X/E)^*$. For any $e \in E$, $Qe = E$. Consequently, $\phi \circ Q(e) = \phi(E) = 0$, and so ϕ determines an element of $E^\perp$ through the identification $\phi \mapsto \phi \circ Q$. This identification preserves norms because $Q(U_X) = U_{X/E}$ (see the comments following the proof of Proposition 3.47), and so $\phi \circ Q(U_X) = \phi(U_{X/E})$.

To see that an element of $E^\perp$ determines an element of $(X/E)^*$, let $x^* \in E^\perp$. Define $\phi \in (X/E)^*$ by $\phi(x + E) = x^*(x)$ for all $x \in X$. We must show ϕ is well-defined. Let x and x' be in the same coset, so that $x - x' \in E$. Then, $x^*(x - x') = 0$, and hence $x^*(x) = x^*(x')$. Thus, $\phi(x + E) = \phi(x' + E)$, and so ϕ is well-defined. To complete the proof, observe that $\phi \circ Q = x^*$, and refer to the previous paragraph. $\square$

The identifications in the preceding proposition lead to a remarkable corollary.

Corollary 3.52 *Let X be a Banach space. For any closed set E in X, we have the identification $E^{**} = E^{\perp\perp}$.*

Proof First apply (i), and then (ii), of Proposition 3.51. $\square$

Exercises

Exercise 3.1 Let X be a real inner product space with inner product $(\cdot, \cdot)$ and associated norm $\|\cdot\|$. Prove the *Parallelogram Law:* If x and y are elements of X, then

$$\|x + y\|^2 + \|x - y\|^2 = 2(\|x\|^2 + \|y\|^2).$$

Exercise 3.2 Let X be a real inner product space with inner product $(\cdot, \cdot)$ and associated norm $\|\cdot\|$. Verify the *polarization formula:* If x and y are in X, then

$$(x, y) = \frac{1}{4}\left(\|x + y\|^2 - \|x - y\|^2\right).$$

Exercise 3.3 Let H be a Hilbert space with norm $\|\cdot\|$. Show that a bounded linear map $S : H \to H$ is an orthogonal operator if and only if $\|S^n x\| = \|x\|$ for all $n \in \mathbb{N}$ and $x \in H$. (*Hint:* Use Exercise 3.2.)

Exercise 3.4 Let X and Y be normed vector spaces. If x is a nonzero vector in X, and $y \in Y$, show there exists a bounded linear map $T : X \to Y$ such that $T(x) = y$.

Exercise 3.5 In this exercise, we complete the proof of Theorem 3.32. Let X be a Banach space with bidual X^{**} and let $j : X \to X^{**}$ be the natural embedding.

(a) Show that j is an injective bounded linear map.

(b) Show that $j(X)$ is a closed subset of X^{**}. (You may wish to use Exercise 1.10.)

Exercise 3.6 Let X and Y be normed spaces. Show that if $\mathcal{L}(X, Y)$ is a Banach space, then Y must be a Banach space. (This is the converse to Proposition 1.11.)

Exercise 3.7 Let X and Z be Banach spaces and let $T \in \mathcal{L}(X, Z)$. Show that T is an open map if and only if T maps the open unit ball of X to an open set in Z.

Exercise 3.8 Prove Proposition 3.40.

Exercise 3.9 Let p be a sublinear functional on a real vector space V. Show that

$$p(x) = \max\{f(x) : f \le p, \ f \text{ linear}\}, \quad x \in V.$$

Conversely, show that a functional q of the form

$$q(x) = \sup_{f \in A} f(x), \quad x \in V,$$

where A is some collection of linear functionals, is necessarily sublinear.

Exercise 3.10 (Fekete's Lemma [10]) Let $(a_n)_{n=1}^{\infty}$ be a sequence of real numbers such that

$$a_{m+n} \le a_m + a_n, \quad \{m, n\} \subseteq \mathbb{N}.$$

Show that if the sequence $(a_n/n)_{n=1}^{\infty}$ is bounded below, then

$$\lim_{n \to \infty} \frac{a_n}{n} = \inf_{n \in \mathbb{N}} \frac{a_n}{n}.$$

(*Hint:* For any $m \in \mathbb{N}$, show that $\limsup\limits_{n \to \infty} \dfrac{a_n}{n} \le \dfrac{a_m}{m}$ by writing $n = km + r$, where $\{k, r\} \subseteq \mathbb{N}$ and $0 \le r \le m - 1$.)

Exercise 3.11 Let V be a real vector space and suppose p is a sublinear functional on V. Suppose $T : V \to V$ is a linear map such that $p(Tx) = p(x)$. Show, using Exercise 3.10, that

$$q(x) = \lim_{n \to \infty} \frac{1}{n} p(x + Tx + \cdots + T^{n-1}x), \quad x \in V,$$

defines a sublinear functional with $q \le p$. Show further that if f is a linear functional with $f \le p$, then f is T-*invariant* (i.e., $f(Tx) = f(x)$ for all $x \in V$) if and only if $f \le q$.

Exercise 3.12. Show that a linear functional L on ℓ_∞ is a Banach limit if and only if

$$L(\xi) \le \lim_{n \to \infty} \sup_{k \in \mathbb{N}} \frac{\xi_{k+1} + \cdots + \xi_{k+n}}{n}, \quad \xi = (\xi_j)_{j=1}^{\infty} \in \ell_\infty.$$

Exercise 3.13 A sequence $\xi \in \ell_\infty$ is called *almost convergent* to α if $L(\xi) = \alpha$ for all Banach limits L.

(a) Show that ξ is almost convergent to α if and only if

$$\lim_{n \to \infty} \sup_{k \in \mathbb{N}} \left| \frac{\xi_{k+1} + \cdots + \xi_{k+n}}{n} - \alpha \right| = 0.$$

(b) Show that for any θ, the sequence $(\sin(n\theta))_{n=1}^{\infty}$ is almost convergent to 0.

Exercise 3.14 If $x = (x_j)_{j=1}^{\infty}$ and $y = (y_j)_{j=1}^{\infty}$ are sequences in ℓ_{∞}, let xy be the sequence $(x_j \, y_j)_{j=1}^{\infty}$. Show for any Banach limit L, there are sequences x and y in ℓ_{∞} such that $L(xy) \neq L(x)L(y)$. (Notice that $L(xy) = L(x)L(y)$ if x and y are in c.)

Exercise 3.15 Prove that $\langle \cdot, \cdot \rangle$ is an inner product on H in the proof of Proposition 3.21.

Exercise 3.16 Show that a bounded linear functional $f : c_0 \to \mathbb{R}$ has a *unique* Hahn–Banach extension $\tilde{f} : \ell_{\infty} \to \mathbb{R}$.

Exercise 3.17 Let $E = \{(x_n)_{n=1}^{\infty} \in \ell_1 : x_{2k-1} = 0 \text{ for all } k \in \mathbb{N}\}$. Show that E is a closed subspace of ℓ_1. Prove that any nonzero bounded linear functional on E has more than one Hahn–Banach extension to ℓ_1.

Exercise 3.18 Let $E = \left\{ f \in L_2(0,1) : \int_0^1 xf(x)\,dx = 0 \right\}$. Define a bounded linear functional Λ on E by

$$\Lambda(f) = \int_0^1 x^2 f(x)\,dx, \quad f \in E.$$

Find the (unique) Hahn–Banach extension of Λ to $L_2(0,1)$ and determine $\|\Lambda\|$. (*Hint:* Use the fact that $\int_0^1 x^2 f(x)\,dx = \int_0^1 (x^2 + ax)f(x)\,dx$ on E, for all $a \in \mathbb{R}$.)

Chapter 4
Consequences of Completeness

The space $C[0, 1]$ of continuous functions on the interval $[0, 1]$ can be equipped with many metrics. Two important examples are the metrics arising from the norms

$$\|f\|_\infty = \max_{s \in [0,1]} |f(s)| \quad \text{and} \quad \|f\|_2 = \left(\int_0^1 |f(s)|^2 \, ds \right)^{1/2},$$

where $f \in C[0, 1]$. The metric arising from the first norm is *complete*, whereas the metric induced by the second norm is not (i.e., there exist Cauchy sequences that fail to converge). Completeness of a metric is a very profitable property, as we shall see in this chapter. The first theorem we shall meet is a classical result about metric spaces called the Baire Category Theorem. It originated in Baire's 1899 doctoral thesis, although metric spaces were not formally defined until later.

4.1 The Baire Category Theorem

In this section, we will state and prove the Baire Category Theorem and see some of its applications. A notion that will prove fruitful is that of a G_δ-set. We remind the reader that a G_δ-*set* is a countable intersection of open sets. Certainly, all open sets are G_δ-sets, but not all G_δ-sets are open. Correspondingly, an F_σ-*set* is the countable union of closed sets. Naturally, all closed sets are F_σ-sets, but not all F_σ-sets are closed. (See Exercise 4.1.)

Theorem 4.1 (Baire Category Theorem) *Suppose (M, d) is a complete metric space. If $(U_n)_{n=1}^\infty$ is a sequence of dense open subsets of M, then $\bigcap_{n=1}^\infty U_n$ is dense in M.*

Recall that a set D is *dense* in a topological space M if and only if $D \cap U \neq \emptyset$ for all nonempty open sets U in M.

Observe that, while the conclusion of the Baire Category Theorem is topological in nature, the hypothesis is not. The two spaces $\mathbb{R}$ and $(-\frac{\pi}{2}, \frac{\pi}{2})$ are homeomorphic (via the mapping $x \mapsto \arctan x$); however, $\mathbb{R}$ is complete, while $(-\frac{\pi}{2}, \frac{\pi}{2})$ is not. Therefore, Theorem 4.1 applies directly to $\mathbb{R}$, but not to $(-\frac{\pi}{2}, \frac{\pi}{2})$. However, the

© Springer Science+Business Media, LLC 2014
A. Bowers, N. J. Kalton, *An Introductory Course in Functional Analysis,*
Universitext, DOI 10.1007/978-1-4939-1945-1_4

conclusion of Theorem 4.1 holds for every metric space which is homeomorphic to a complete metric space.

Before proving Theorem 4.1, let us consider some consequences of it.

Corollary 4.2 *If* $(G_n)_{n=1}^{\infty}$ *is a sequence of dense G_δ-sets, then $\bigcap_{n=1}^{\infty} G_n$ is also a dense G_δ-set.*

Proof By assumption, for each $n \in \mathbb{N}$, the set G_n can be written as an intersection of countably many open sets, say $G_n = \bigcap_{m=1}^{\infty} U_{mn}$. Since G_n is assumed to be dense, it must be the case that U_{mn} is dense for each m and n in $\mathbb{N}$. By Theorem 4.1 (The Baire Category Theorem), the set

$$\bigcap_{n=1}^{\infty} G_n = \bigcap_{n=1}^{\infty} \bigcap_{m=1}^{\infty} U_{mn}$$

is a dense set. Since the sequence of open sets $(U_{mn})_{m,n=1}^{\infty}$ is countable, the set $\bigcap_{n=1}^{\infty} G_n$ is a G_δ-set. $\qquad\square$

Proposition 4.3 *If $f : \mathbb{R} \to \mathbb{R}$ is a function, then the set of points of continuity of f is a G_δ-set.*

Proof For any open interval I in $\mathbb{R}$, let the *oscillation* of f over I be given by

$$\operatorname{osc}_I f = \sup\{|f(x) - f(y)| : \{x, y\} \subseteq I\}.$$

Let $(\epsilon_n)_{n=1}^{\infty}$ be a sequence of positive real numbers such that $\epsilon_n \to 0$ as $n \to \infty$. For each $n \in \mathbb{N}$, let

$$U_n = \{x \in \mathbb{R} : \exists \text{ an open interval } I \text{ in } \mathbb{R} \text{ such that } x \in I \text{ and } \operatorname{osc}_I f < \epsilon_n\}.$$

Suppose $x \in U_n$. By definition, there exists an open interval I such that $x \in I$ and $\operatorname{osc}_I f < \epsilon_n$. The same statement holds for all $x' \in I$, and so $I \subseteq U_n$. Consequently, U_n is an open set. Therefore, the intersection $\bigcap_{n=1}^{\infty} U_n$ is a G_δ-set. The set $\bigcap_{n=1}^{\infty} U_n$ is precisely the set of all points of continuity of f, and hence the result. $\qquad\square$

Example 4.4 Define a function $f : \mathbb{R} \to \mathbb{R}$ by $f(x) = 0$ for all $x \notin \mathbb{Q}$, $f(0) = 1$, and $f(x) = 1/q$ for all $x \in \mathbb{Q}\backslash\{0\}$, where $x = p/q$ is written in lowest terms. The function f is continuous precisely on the set $\mathbb{R}\backslash\mathbb{Q}$, and consequently (by Proposition 4.3), the set of irrational numbers is a G_δ-set.

The preceding example shows that the set of irrational numbers $\mathbb{R}\backslash\mathbb{Q}$ is a G_δ-set, but we could have shown that directly, without the aid of Proposition 4.3. Note that

$$\mathbb{R}\backslash\mathbb{Q} = \bigcap_{q \in \mathbb{Q}} \mathbb{R}\backslash\{q\}.$$

Thus, $\mathbb{R}\backslash\mathbb{Q}$ is the intersection of countably many sets, each of which is open in $\mathbb{R}$.

Example 4.4 does lead one to ask a natural question: Does there exist a function $g : \mathbb{R} \to \mathbb{R}$ which is continuous precisely on $\mathbb{Q}$? By Proposition 4.3, this can happen only if $\mathbb{Q}$ is a G_δ-set. Both $\mathbb{Q}$ and $\mathbb{R}\backslash\mathbb{Q}$ are dense in $\mathbb{R}$, and $\mathbb{R}\backslash\mathbb{Q}$ is a G_δ-set. If $\mathbb{Q}$ is also a G_δ-set, then the intersection $\mathbb{Q}\cap(\mathbb{R}\backslash\mathbb{Q}) = \emptyset$ would have to be dense, by Corollary 4.2 (a consequence of the Baire Category Theorem). This is a clear contradiction. We conclude there exists no function that has $\mathbb{Q}$ as the set of points of continuity.

Proof of Theorem 4.1 (the Baire Category Theorem) Let (M,d) be a complete metric space and suppose $(U_n)_{n=1}^\infty$ is a sequence of dense open subsets of M. Our goal is to show that $\bigcap_{n=1}^\infty U_n$ is also dense in M.

For any $z \in M$ and $\epsilon > 0$, let $B(z,\epsilon) = \{x \in M : d(x,z) < \epsilon\}$ be the open ball about z of radius ϵ.

Let V be a nonempty open set in M. We will show that $\bigcap_{n=1}^\infty U_n$ and V have nonempty intersection. Pick any $x_0 \in V$ and $\epsilon_0 \in (0,1)$ such that $B(x_0,\epsilon_0) \subseteq V$. By assumption, the set U_1 is dense in M, and so $B\left(x_0,\frac{\epsilon_0}{2}\right) \cap U_1 \neq \emptyset$. Thus, there exists a point $x_1 \in U_1$ and an $\epsilon_1 \leq \frac{\epsilon_0}{2}$ such that $B(x_1,\epsilon_1) \subseteq B\left(x_0,\frac{\epsilon_0}{2}\right) \cap U_1 \subseteq V \cap U_1$. By assumption, as before, the set U_2 is dense in M, and so $B\left(x_1,\frac{\epsilon_1}{2}\right) \cap U_2 \neq \emptyset$. Again arguing as before, there exists a point $x_2 \in U_2$ and an $\epsilon_2 \leq \frac{\epsilon_1}{2}$ such that $B(x_2,\epsilon_2) \subseteq B\left(x_1,\frac{\epsilon_1}{2}\right) \cap U_2 \subseteq V \cap U_1 \cap U_2$. Continuing inductively, we construct sequences $(x_n)_{n=0}^\infty$ and $(\epsilon_n)_{n=0}^\infty$ such that $\epsilon_n \leq \frac{\epsilon_{n-1}}{2}$ and $x_n \in V \cap (U_1 \cap \cdots \cap U_n)$, and with the further property that

$$B(x_n,\epsilon_n) \subseteq B\left(x_{n-1},\frac{\epsilon_{n-1}}{2}\right) \cap U_n \subseteq V \cap (U_1 \cap \cdots \cap U_n),$$

for all $n \in \mathbb{N}$.

Observe that $\epsilon_n < \frac{1}{2^n}$ for all $n \in \mathbb{N}$. Suppose $m > n$. By the triangle inequality,

$$d(x_m,x_n) \leq d(x_m,x_{m-1}) + \cdots + d(x_{n+1},x_n).$$

The sequence $(x_n)_{n=0}^\infty$ was chosen so that $d(x_{n+1},x_n) \leq \frac{\epsilon_n}{2}$. Consequently,

$$d(x_m,x_n) \leq \frac{\epsilon_{m-1}}{2} + \cdots + \frac{\epsilon_n}{2} < \frac{1}{2^m} + \cdots + \frac{1}{2^{n+1}} = \frac{1}{2^n} - \frac{1}{2^m} < \frac{1}{2^n}.$$

It follows that $(x_n)_{n=0}^\infty$ is a Cauchy sequence. Hence, by completeness, there exists a point $x \in M$ such that $x = \lim_{n\to\infty} x_n$. If $m > n$, then $x_m \in B(x_n,\epsilon_n)$, by construction. We conclude that $d(x_m,x_n) < \epsilon_n$ for all $m > n$, and hence $d(x,x_n) \leq \epsilon_n$ for all $n \in \mathbb{N}$. Thus,

$$x \in \overline{B(x_n,\epsilon_n)} \subseteq \overline{B\left(x_{n-1},\frac{\epsilon_{n-1}}{2}\right)} \subseteq B(x_{n-1},\epsilon_{n-1}),$$

and so $x \in V \cap (U_1 \cap \cdots \cap U_{n-1})$ for all $n \in \mathbb{N}$. It follows that, $x \in V \cap \left(\bigcap_{n=1}^\infty U_n\right)$.

We have shown that the intersection of $\bigcap_{n=1}^\infty U_n$ and any open set V in M is nonempty. Therefore, the set $\bigcap_{n=1}^\infty U_n$ is dense in M. □

The name *Baire Category Theorem* is derived from a complementary formulation of Theorem 4.1. If U is a dense open set in M, then its complement $M\backslash U$ is closed with empty interior. This fact motivates the following definitions.

Definition 4.5   Let M be a topological space. A set $E \subseteq M$ is said to be *nowhere dense* if the closure of E in M has empty interior; that is to say, if $\operatorname{int}(\overline{E}) = \emptyset$. A set $G \subseteq M$ is called *first category* (also known as *meager*) if there exists a sequence $(E_n)_{n=1}^{\infty}$ of nowhere dense sets such that $G \subseteq \bigcup_{n=1}^{\infty} E_n$. A set is called *second category* if it is not first category. A second category set is also known as *non-meager*. The complement of a meager set is called a *residual* set.

Proposition 4.6   *A countable union of first category sets is first category.*

Proof   A countable union of countably many sets is a countable union of sets.   □

Theorem 4.7 (Baire Category Theorem, complementary version) *In a complete metric space, any dense G_{δ}-set is second category.*

Proof   Let G be a dense G_{δ}-set. Suppose that G is first category. Then there exists a sequence $(E_n)_{n=1}^{\infty}$ of nowhere dense sets such that $G \subseteq \bigcup_{n=1}^{\infty} E_n$. Without loss of generality, we may assume E_n is closed for each $n \in \mathbb{N}$. Now, for each $n \in \mathbb{N}$, let V_n be the complement of E_n. Then V_n is both open and dense for each $n \in \mathbb{N}$. Therefore, by Theorem 4.1, the intersection $\bigcap_{n=1}^{\infty} V_n$ is a dense G_{δ}-set.

Since G is a subset of $\bigcup_{n=1}^{\infty} E_n$, it is disjoint from $\bigcap_{n=1}^{\infty} V_n$. Thus, we have disjoint dense G_{δ}-sets. This contradicts Corollary 4.2, and so G is not first category.   □

The intuition behind the proof of Theorem 4.7 is that a countable union of *small* sets is still *small*. In this case, by *small* we mean *meager*. This notion of size applies to any metric space.

Example 4.8   Consider the real line $\mathbb{R}$. We have now for the set of real numbers two natural notions of *smallness*: first category and zero measure. It is natural to wonder if there is any relationship between the two. Consider the set of rational numbers $\mathbb{Q} \subseteq \mathbb{R}$. If λ represents Lebesgue measure on $\mathbb{R}$, then $\lambda(\mathbb{Q}) = 0$ (because the set of rational numbers is countable). Therefore, there exists a sequence $(U_n)_{n=1}^{\infty}$ of open sets such that $\mathbb{Q} \subseteq U_n$ and $\lambda(U_n) < \frac{1}{n}$ for each $n \in \mathbb{N}$. Let $G = \bigcap_{n=1}^{\infty} U_n$. Then $\lambda(G) = 0$, but G cannot be first category, since $\mathbb{Q} \subseteq G$ and $\mathbb{Q}$ is dense in $\mathbb{R}$. On the other hand, the complement of G is first category, but must be of infinite measure.

For further reading on the analogies between topological spaces and measure spaces, the curious reader might consider *Measure and Category* by John Oxtoby [29].

4.2   Applications of Category

In this section, we will investigate some implications of category in Banach spaces. In particular, we will learn the Uniform Boundedness Principle and see the important role it plays in the study of Fourier series.

Theorem 4.9 (Uniform Boundedness Principle) *Let X and Y be Banach spaces and let $T_i : X \to Y$ be a bounded linear operator for each $i \in I$, where I is an index* *set. In order that there exists a uniform bound M such that $\|T_i\| \le M$ for all $i \in I$, it is necessary and sufficient that for each $x \in X$, there exists a constant $C_x > 0$ such that $\|T_i(x)\|_Y \le C_x$ for all $i \in I$.*

Proof   Necessity is immediate. To show sufficiency, assume that for each $x \in X$ there exists a constant $C_x > 0$ such that $\|T_i(x)\|_Y \le C_x$ for all $i \in I$. For each $n \in \mathbb{N}$, let

$$A_n = \{x : \|T_i(x)\|_Y \le n \ \text{ for all } \ i \in I\} = \bigcap_{i \in I}\{x : \|T_i(x)\|_Y \le n\} = \bigcap_{i \in I} T_i^{-1}\left(n B_Y\right).$$

The set $n B_Y$ is the closed ball centered at $0 \in Y$ with radius n. Since T_i is continuous for each $i \in I$, it follows that A_n is a closed set.

By assumption, every $x \in X$ is contained in A_n for some $n \in \mathbb{N}$, and therefore we must have $X = \bigcup_{n=1}^{\infty} A_n$. By Proposition 4.6, not every A_n can be nowhere dense, and so there exists some $n_0 \in \mathbb{N}$ such that the closed set A_{n_0} has nonempty interior. Thus, there exists some $x_0 \in A_{n_0}$ and $\delta > 0$ such that

$$x_0 + \delta B_X = \{x : \|x - x_0\|_X \le \delta\} \subseteq A_{n_0}.$$

Now let $u \in B_X$. Then $x_0 + \delta u \in A_{n_0}$, and so it follows that $\|T_i(x_0 + \delta u)\|_Y \le n_0$ for all $i \in I$. We chose x_0 to be in A_{n_0}, and thus $\|T_i(x_0)\|_Y \le n_0$ for all $i \in I$. Consequently, it must be that $\|T_i(\delta u)\|_Y \le 2n_0$ for all $i \in I$. By linearity, since $\delta > 0$ is a constant, we conclude that $\|T_i(u)\|_Y \le 2n_0/\delta$ for all $u \in B_X$. Therefore, $\|T_i\| \le 2n_0/\delta$ for each $i \in I$, and the proof is complete. $\square$

The following theorem is a more concise statement of Theorem 4.9.

Theorem 4.10   (Uniform Boundedness Principle) *Let X and Y be Banach spaces. If $\{T_i : i \in I\}$ is a collection of bounded linear operators from X to Y, then $\sup_{i \in I} \|T_i(x)\|_Y < \infty$ for each $x \in X$ if and only if $\sup_{i \in I} \|T_i\| < \infty$.*

We now provide some definitions that lead to a straightforward application of the Uniform Boundedness Principle.

Definition 4.11   Let X be a Banach space and let A be a subset of X. The set A is said to be *bounded* if $\sup_{a \in A} \|a\| < \infty$. The set A is called *weakly bounded* if $\sup_{a \in A} |x^*(a)| < \infty$ for all $x^* \in X^*$.

Theorem 4.12   *Let X be a Banach space. A subset of X is bounded if and only if it is weakly bounded.*

Proof   Let A be a subset of X. Certainly, if A is bounded, then it is weakly bounded. Assume now that A is weakly bounded. For each $a \in A$, define a scalar-valued function ϕ_a on X^* by

$$\phi_a(x^*) = x^*(a), \quad x^* \in X^*.$$

For each $a \in A$, the function ϕ_a is bounded and linear. Indeed, for a given $a \in A$,

$$\|\phi_a\| = \sup_{x^* \in B_{X^*}} |\phi_a(x^*)| = \sup_{x^* \in B_{X^*}} |x^*(a)| = \|a\|.$$

By the weakly bounded assumption on A, for each $x^* \in X^*$,

$$\sup_{a \in A} |\phi_a(x^*)| = \sup_{a \in A} |x^*(a)| < \infty.$$

Thus, by Theorem 4.10 (the Uniform Boundedness Principle), we conclude that $\sup_{a \in A} \|\phi_a\| < \infty$. Since $\|\phi_a\| = \|a\|$, we have shown that A is bounded. □

Definition 4.13 Let X be a Banach space and let A be a subset of X^*. The set A is called *weak**-bounded* if $\sup_{a^* \in A} |a^*(x)| < \infty$ for all $x \in X$.

Theorem 4.14 *Let X be a Banach space. A subset of X^* is bounded if and only if it is weak*-bounded.*

Proof The argument parallels the proof of Theorem 4.12 and is left to the reader. □

The following significant theorem is a consequence of the Uniform Boundedness Principle.

Theorem 4.15 (Banach-Steinhaus Theorem) *Let X and Y be Banach spaces. Suppose $(S_n)_{n=1}^{\infty}$ is a sequence of bounded linear operators from X to Y. If $\lim_{n \to \infty} S_n x$ exists for each $x \in X$, then*

(i) $\sup_{n \in \mathbb{N}} \|S_n\| < \infty$, *and*

(ii) *if $Tx = \lim_{n \to \infty} S_n x$ for all $x \in X$, then T is a bounded linear operator.*

Proof (i) If $(S_n x)_{n=1}^{\infty}$ converges, then $\sup_{n \in \mathbb{N}} \|S_n x\| < \infty$. The Uniform Boundedness Principle then implies that $\sup_{n \in \mathbb{N}} \|S_n\| < \infty$.

(ii) A simple check reveals that T is linear. To show that T is bounded, observe that

$$\|Tx\| \le \sup_{n \in \mathbb{N}} \|S_n x\| \le \left(\sup_{n \in \mathbb{N}} \|S_n\| \right) \|x\|.$$

Taking the supremum over $x \in B_X$ provides the desired result. □

The next result is in some sense a converse to Theorem 4.15.

Theorem 4.16 *Let X and Y be Banach spaces. If $(S_n)_{n=1}^{\infty}$ is a sequence of uniformly bounded linear operators from X to Y, then the set $E = \{x : \lim_{n \to \infty} S_n x \text{ exists}\}$ is a closed linear subspace of X.*

Proof We need only show that E is closed. Suppose that $x \in \overline{E}$, the closure of E. By assumption, there exists some $M > 0$ such that $\|S_n\| \le M$ for all $n \in \mathbb{N}$. Let $\epsilon > 0$. Since x is in the closure of E, there exists some $y \in E$ such that $\|x - y\| < \epsilon/(3M)$.

By assumption, the sequence $(S_n y)_{n=1}^{\infty}$ is a Cauchy sequence in Y, and so there exists some natural number N such that $\|S_m y - S_n y\| < \epsilon/3$ whenever $m \ge N$ and $n \ge N$. Let m and n be natural numbers such that $m \ge N$ and $n \ge N$. Then,

$$\|S_m x - S_n x\| \le \|S_m x - S_m y\| + \|S_m y - S_n y\| + \|S_n y - S_n x\|$$

$$< M\left(\frac{\epsilon}{3M}\right) + \frac{\epsilon}{3} + M\left(\frac{\epsilon}{3M}\right) = \epsilon.$$

Therefore, the sequence $(S_n x)_{n=1}^{\infty}$ is a Cauchy sequence in Y. Since Y is complete, $\lim_{n\to\infty} S_n x$ exists, and so $x \in E$, as required.   □

Example 4.17 (Convergence of Fourier Series) Let $\mathbb{T} = \{e^{i\theta} : 0 \le \theta < 2\pi\}$ be the one-dimensional *torus*, a subset of the complex plane $\mathbb{C}$. We will consider the space $C(\mathbb{T})$ of complex-valued continuous functions on the compact Hausdorf space $\mathbb{T}$. Let $f \in C(\mathbb{T})$. By an abuse of notation, we write $f(\theta) = f(e^{i\theta})$ for all $\theta \in [0, 2\pi)$.

For each $n \in \mathbb{Z}$, define the nth *Fourier coefficient* by

$$\hat{f}(n) = \int_0^{2\pi} f(\theta) e^{-in\theta} \frac{d\theta}{2\pi}. \tag{4.1}$$

We wish to determine if it is possible to reconstruct f from its Fourier coefficients. To that end, we define the *Fourier series* of f by

$$\sum_{n\in\mathbb{Z}} \hat{f}(n) e^{in\theta}. \tag{4.2}$$

If f is a trigonometric polynomial, then there exists a sequence of scalars $(a_k)_{k\in\mathbb{Z}}$ with only finitely many nonzero terms such that $f(\theta) = \sum_{k\in\mathbb{Z}} a_k e^{ik\theta}$ for all $\theta \in \mathbb{T}$. In this case, it is easy to see that $\hat{f}(n) = a_n$ for each $n \in \mathbb{Z}$. (See Exercise 4.9.)

Since a trigonometric polynomial is equal to its Fourier series, it is natural to ask the following: Does the Fourier series of a general continuous function f converge to f? To make this question more precise, define for each $N \in \mathbb{N}$ the Nth *partial sum operator* $S_N : C(\mathbb{T}) \to C(\mathbb{T})$ as follows:

$$S_N f(\theta) = \sum_{n=-N}^{N} \hat{f}(n) e^{in\theta}, \quad \theta \in [0, 2\pi). \tag{4.3}$$

The question now becomes: Is it the case that $\|S_N f - f\|_\infty \to 0$ as $N \to \infty$ for all f in $C(\mathbb{T})$? (In other words, does $S_N f$ always converge uniformly to f?) If the answer to this question is "yes," then the partial sum operators S_N must be uniformly bounded, by the Uniform Boundedness Principle. We will therefore show the answer is "no" by computing the operator norm $\|S_N\|$ for each $N \in \mathbb{N}$, and then showing that these norms are not uniformly bounded.

Fix $N \in \mathbb{N}$ and $f \in C(\mathbb{T})$. Computing directly, by substituting (4.1) into (4.3), we have for each $\theta \in [0, 2\pi)$,

$$S_N f(\theta) = \sum_{n=-N}^{N} \left(\int_0^{2\pi} f(\phi) e^{-in\phi} \frac{d\phi}{2\pi} \right) e^{in\theta} = \int_0^{2\pi} \left(f(\phi) \sum_{n=-N}^{N} e^{in(\theta-\phi)} \right) \frac{d\phi}{2\pi}.$$

The sum appearing in the rightmost integral above is a geometric series with constant ratio $e^{i(\theta-\phi)}$. We calculate the sum of this geometric series as follows:

$$\sum_{n=-N}^{N} e^{in(\theta-\phi)} = e^{-iN(\theta-\phi)} \sum_{n=0}^{2N} e^{in(\theta-\phi)} = e^{-iN(\theta-\phi)} \frac{1 - e^{i(2N+1)(\theta-\phi)}}{1 - e^{i(\theta-\phi)}}$$

$$= \frac{e^{-iN(\theta-\phi)} - e^{i(N+1)(\theta-\phi)}}{1 - e^{i(\theta-\phi)}} = \frac{e^{i(N+1)(\theta-\phi)} - e^{-iN(\theta-\phi)}}{e^{i(\theta-\phi)} - 1}.$$

We reduce the final fraction by dividing the numerator and denominator by $e^{\frac{i}{2}(\theta-\phi)}$, and so the above equation becomes

$$\sum_{n=-N}^{N} e^{in(\theta-\phi)} = \frac{e^{i(N+\frac{1}{2})(\theta-\phi)} - e^{-i(N+\frac{1}{2})(\theta-\phi)}}{e^{\frac{i}{2}(\theta-\phi)} - e^{-\frac{i}{2}(\theta-\phi)}} = \frac{\sin\left((N+\frac{1}{2})(\theta-\phi)\right)}{\sin\left(\frac{\theta-\phi}{2}\right)}.$$

Substituting this into the formula for $S_N f(\theta)$, we conclude

$$S_N f(\theta) = \int_0^{2\pi} f(\phi) \frac{\sin\left((N+\frac{1}{2})(\theta-\phi)\right)}{\sin\left(\frac{\theta-\phi}{2}\right)} \frac{d\phi}{2\pi}. \tag{4.4}$$

Assume there exists a constant $M > 0$ such that $\|S_N\| \le M$ for all $N \in \mathbb{N}$. By Proposition 3.37, each adjoint operator $S_N^* : M(\mathbb{T}) \to M(\mathbb{T})$ must also be bounded by M; that is, $\|S_N^*\| \le M$ for all $N \in \mathbb{N}$. Let δ_0 denote the Dirac measure at 0 (which was defined in (2.8)). It is easy to see that $\|\delta_0\|_M = 1$, and as a consequence,

$$\|S_N^* \delta_0\|_M \le \|S_N^*\| \le M. \tag{4.5}$$

Let $f \in C(\mathbb{T})$. Applying the definitions, and using (4.4), we see

$$\langle f, S_N^* \delta_0 \rangle = S_N f(0) = \int_0^{2\pi} f(\phi) \frac{\sin\left((N+\frac{1}{2})\phi\right)}{\sin(\phi/2)} \frac{d\phi}{2\pi}.$$

This equality is valid for all $f \in C(\mathbb{T})$. Therefore, because $S_N^* \delta_0$ is in $M(\mathbb{T})$,

$$\|S_N^* \delta_0\|_M = \int_0^{2\pi} \left| \frac{\sin\left((N+\frac{1}{2})\phi\right)}{\sin(\phi/2)} \right| \frac{d\phi}{2\pi}.$$

We make a change of variables: Let $\psi = (N+\frac{1}{2})\phi$, and so $d\psi = (N+\frac{1}{2})d\phi$. Then

$$\|S_N^* \delta_0\|_M = \int_0^{(2N+1)\pi} \left| \frac{\sin \psi}{\sin \frac{\psi}{2N+1}} \right| \frac{d\psi}{(2N+1)\pi}.$$

Combining this with (4.5), we conclude that

$$\frac{1}{\pi} \int_0^{\infty} \chi_{(0,(2N+1)\pi)} \frac{1}{2N+1} \left| \frac{\sin \psi}{\sin \frac{\psi}{2N+1}} \right| d\psi \le M,$$

for all $N \in \mathbb{N}$. A quick application of l'Hôpital's Rule reveals

$$\lim_{N \to \infty} \frac{1}{2N+1} \frac{\sin \psi}{\sin \frac{\psi}{2N+1}} = \frac{\sin \psi}{\psi}.$$

Consequently, by Fatou's Lemma,

$$\frac{1}{\pi} \int_0^\infty \left| \frac{\sin \psi}{\psi} \right| d\psi \leq M.$$

This is, however, a contradiction, since $\int_0^\infty |\frac{\sin \psi}{\psi}| d\psi = \infty$. To see this, observe that

$$\int_0^\infty \left| \frac{\sin \psi}{\psi} \right| d\psi = \sum_{k=0}^\infty \int_{k\pi}^{(k+1)\pi} \left| \frac{\sin \psi}{\psi} \right| d\psi$$

$$\geq \sum_{k=0}^\infty \frac{1}{(k+1)\pi} \int_0^\pi \sin \psi \, d\psi = \frac{2}{\pi} \sum_{k=0}^\infty \frac{1}{k+1} = \infty.$$

Therefore, it cannot be true that $\sup_{N \in \mathbb{N}} \|S_N\| \leq M$ for some $M \geq 0$, and so it is not possible that the Fourier series of f converges uniformly to f for all $f \in C(\mathbb{T})$.

Remark 4.18 Before the advent of functional analysis, and in particular the Uniform Boundedness Principle, the only way to demonstrate that $S_N f$ did not converge uniformly to f for all $f \in C(\mathbb{T})$ was to construct an explicit function for which the desired convergence failed.

In fact, in Example 4.17, we actually proved that $(S_N f(0))_{N=1}^\infty$ fails to converge to $f(0)$ for all f in a dense G_δ-subset of $C(\mathbb{T})$. Thus, there are many functions for which the Fourier series of f does not even converge pointwise to f.

Definition 4.19 For any natural number N, the *Dirichlet kernel of degree N* is the function

$$D_N(\alpha) = \sum_{n=-N}^N e^{in\alpha} = \frac{\sin \left((N + \frac{1}{2}) \alpha \right)}{\sin (\alpha/2)}, \quad \alpha \in \mathbb{R}.$$

Observe that

$$S_N f(\theta) = \int_0^{2\pi} f(\phi) D_N(\theta - \phi) \frac{d\phi}{2\pi} = (D_N * f)(\theta), \tag{4.6}$$

where S_N is the Nth partial sum operator.

We know from Example 4.17 that $D_N * f$ does not converge uniformly to f for each $f \in C(\mathbb{T})$; however, we can find a related kernel for which uniform convergence does hold for all continuous functions on $\mathbb{T}$.

Definition 4.20 For any natural number N, the *Fejér kernel of degree N* is the function

$$K_N(t) = \frac{1}{N} \sum_{k=0}^{N-1} D_k(t), \quad t \in \mathbb{R}.$$

The Nth *Cesàro mean* of $f \in C(\mathbb{T})$ is the function given by the formula

$$T_N f = \frac{1}{N}(S_0 f + \cdots + S_{N-1} f).$$

Using (4.6), we can derive a relationship between K_N and T_N. For any $N \in \mathbb{N}$ and $\theta \in [0, 2\pi)$,

$$T_N f(\theta) = \int_0^{2\pi} f(t) \, K_N(\theta - t) \, \frac{dt}{2\pi} = (K_N * f)(\theta).$$

We now find a closed-form formula for the Fejér kernel of degree N.

Lemma 4.21 *If K_N is the Fejér kernel of degree N, then*

$$K_N(t) = \frac{1}{2N} \frac{1 - \cos(Nt)}{\sin^2(t/2)} = \frac{\sin^2(Nt/2)}{N \sin^2(t/2)}.$$

Proof Recall the identity $\sin A \sin B = \frac{1}{2}(\cos(A - B) - \cos(A + B))$, where A and B are real numbers. For $t \in \mathbb{R}$,

$$K_N(t) = \frac{1}{N} \sum_{k=0}^{N-1} \frac{\sin\left((k + \frac{1}{2})t\right)}{\sin(t/2)} = \frac{1}{N} \sum_{k=0}^{N-1} \frac{\sin\left((k + \frac{1}{2})t\right) \sin(t/2)}{\sin^2(t/2)}$$

$$= \frac{1}{2N} \sum_{k=0}^{N-1} \frac{\cos(kt) - \cos((k+1)t)}{\sin^2(t/2)}.$$

The first equality in the conclusion of the lemma follows directly from the above series, because it is a telescoping series. The second equality in the conclusion of the lemma follows from the application of a half-angle identity.   □

Lemma 4.22 *If T_N is the Nth Cesàro mean, then $\|T_N\| = 1$.*

Proof Let $f \in C(\mathbb{T})$. By a straightforward change of variables, and using the translation-invariance of Lebesgue measure, coupled with the periodicity of functions on $\mathbb{T}$, we have

$$T_N f(\theta) = \int_0^{2\pi} f(\theta - t) \, K_N(t) \, \frac{dt}{2\pi}, \quad \theta \in [0, 2\pi).$$

It follows that $\|T_N f\|_{C(\mathbb{T})} \leq \|f\|_{C(\mathbb{T})} \|K_N\|_{L^1(\mathbb{T})}$.

By Lemma 4.21, we know that $K_N \geq 0$, and so $\|K_N\|_{L^1(\mathbb{T})} = \frac{1}{2\pi} \int_0^{2\pi} K_N(t)\, dt$. To find the value of $\|K_N\|_{L^1(\mathbb{T})}$, we compute it directly:

$$\frac{1}{2\pi} \int_0^{2\pi} K_N(t)\, dt = \frac{1}{2\pi} \int_0^{2\pi} \frac{1}{N} \sum_{k=0}^{N-1} D_k(t)\, dt = \frac{1}{2\pi N} \sum_{k=0}^{N-1} \left(\int_0^{2\pi} \sum_{n=-k}^{k} e^{int}\, dt \right).$$

Observe that

$$\int_0^{2\pi} e^{int}\, dt = \begin{cases} 0 & \text{if } n \neq 0, \\ 2\pi & \text{if } n = 0. \end{cases}$$

Consequently,

$$\frac{1}{2\pi} \int_0^{2\pi} K_N(t)\, dt = \frac{1}{2\pi N} \sum_{k=0}^{N-1} \sum_{n=-k}^{k} \left(\int_0^{2\pi} e^{int}\, dt \right) = \frac{1}{2\pi N} \sum_{k=0}^{N-1} 2\pi = 1.$$

It follows that $\|T_N\| \leq \|K_N\|_{L^1(\mathbb{T})} = 1$. Since $T_N 1 = \frac{1}{2\pi} \int_0^{2\pi} K_N(t)\, dt = 1$, we conclude that $\|T_N\| = 1$. $\square$

The next proposition provides a good example of how the Uniform Boundedness Principle is actually used in practice (in the guise of the Banach-Steinhaus Theorem).

Proposition 4.23   *If* $f \in C(\mathbb{T})$, *then* $K_N * f \to f$ *uniformly.*

Proof If f is a trigonometric polynomial, then it is easy to see that $T_N f \to f$ uniformly. By the Weierstrass Approximation Theorem, the trigonometric polynomials are dense in $C(\mathbb{T})$. By Theorem 4.16, the set of functions for which this limit exists is closed. It follows that $T_N f \to f$ in $C(\mathbb{T})$ as $N \to \infty$ for all $f \in C(\mathbb{T})$. Since $T_N f = K_N * f$, the result follows. $\square$

4.3   The Open Mapping and Closed Graph Theorems

We begin this section with a topological definition.

Definition 4.24   Let X and Y be topological spaces. A map $T : X \to Y$ is called *open* (or an *open map*) if $T(U)$ is open in Y whenever U is open in X.

Before stating the next proposition, we recall that $B_X = \{x : \|x\| \leq 1\}$ and $\text{int}(B_X) = \{x : \|x\| < 1\}$, where X is a normed space. Naturally, B_X is a closed set and $\text{int}(B_X)$ is an open set (in X).

Proposition 4.25   *Let* X *and* Y *be Banach spaces and suppose* $T : X \to Y$ *is a bounded linear operator. The map* T *is an open map if and only if there exists a* $\delta > 0$ *such that* $\delta B_Y \subseteq T(B_X)$.

Proof Assume T is an open map. The set $\text{int} B_X$ is open in X, and so $T(\text{int} B_X)$ is open in Y. By the linearity of T, we have $0 \in T(\text{int} B_X)$. It follows that $\text{int} B_X$

contains a basic neighborhood of 0. Therefore, there exists a $\delta > 0$ such that

$$\delta B_Y \subseteq T(\text{int} B_X) \subseteq T(B_X).$$

Conversely, assume there exists a $\delta > 0$ such that $\delta B_Y \subseteq T(B_X)$. We wish to show that $T(U)$ is open in Y whenever U is an open set in X. To that end, let U be open in X and let $y \in T(U)$. There exists some $x \in U$ such that $Tx = y$. Since $x \in U$, and U is open in X, there is some $\nu > 0$ such that $x + \nu B_X \subseteq U$. It follows that $y + \nu T(B_X) \subseteq T(U)$. By our assumption, this implies that $y + \nu \delta B_Y \subseteq T(U)$. Consequently, $y \in \text{int} T(U)$, and so the set $T(U)$ is open in Y. Therefore, T is an open map, as required. $\qquad \square$

Corollary 4.26 *Let X and Y be Banach spaces and suppose $T : X \to Y$ is a bounded linear operator.*

(i) If T is open, then T maps X onto Y.
(ii) If T is one-to-one and open, then T is invertible and T^{-1} is continuous.

Proof (i) Suppose T is an open map. By Proposition 4.25, there exists a $\delta > 0$ such that $\delta B_Y \subseteq T(B_X)$. Let $y \in Y$. It must be that $y \in \|y\| B_Y$, and so

$$y \in \frac{\|y\|}{\delta} \cdot \delta B_Y \subseteq \frac{\|y\|}{\delta} T(B_X).$$

It follows that $y = T\left(\frac{\|y\|}{\delta} x\right)$ for some $x \in B_X$, and so T is onto.

(ii) If T is open, then T is onto, by *(i)*. Since T is also assumed to be one-to-one, the inverse function T^{-1} is well-defined. It remains to show that $T^{-1} : Y \to X$ is continuous. If U is an open set in X, then $(T^{-1})^{-1}(U) = T(U)$ is open in Y (since T is an open map). The preimage of an open set is open, and so we conclude that T^{-1} is continuous. $\qquad \square$

Definition 4.27 Let X and Y be Banach spaces and suppose $T : X \to Y$ is a bounded linear operator. The map T is said to be *almost open* if there exists a $\delta > 0$ such that $\delta B_Y \subseteq \overline{T(B_X)}$.

Proposition 4.28 *Let X and Y be Banach spaces and suppose $T : X \to Y$ is a bounded linear operator. If T is almost open, then T is open.*

Proof By assumption, T is almost open, and therefore there exists a $\delta \in (0, 1)$ such that $\delta B_Y \subseteq \overline{T(B_X)}$. If $y \in B_Y$, then $\delta y \in \delta B_Y \subseteq \overline{T(B_X)}$, and so for any $\nu > 0$, there exists some $x \in B_X$ such that $\|\delta y - Tx\| < \nu$. Consequently, for any $y \in B_Y$ and $\nu > 0$, there exists an $x \in B_X$ such that

$$\|y - T(x/\delta)\| < \nu/\delta.$$

If $\hat{y}$ is an element of Y such that $0 < \|\hat{y}\| < \delta$, then $\hat{y}/\|\hat{y}\| \in B_Y$. Therefore, for any $\nu > 0$, there exists an element $x \in B_X$ such that

$$\left\| \frac{\hat{y}}{\|\hat{y}\|} - T\left(\frac{x}{\delta}\right) \right\| < \frac{\nu}{\delta}. \tag{4.7}$$

Consequently, for any $\hat{y} \in \text{int}(\delta B_Y)$ and $\nu > 0$, we can find an $\hat{x} \in X$ such that

$$\|\hat{y} - T(\hat{x})\| \le \left(\frac{\nu}{\delta}\right) \|\hat{y}\|, \quad \|\hat{x}\| \le \frac{\|\hat{y}\|}{\delta}. \tag{4.8}$$

If $\hat{y} = 0$, we choose $\hat{x} = 0$. If $\hat{y} \neq 0$, then we choose $\hat{x} = \frac{x\|\hat{y}\|}{\delta}$, where x is the element of B_X chosen to satisfy (4.7).

Our goal is to show that T is an open map. By Proposition 4.25, it will suffice to show that $\text{int}(\delta B_Y) \subseteq T(B_X)$. To that end, let $y \in \text{int}(\delta B_Y)$. Then $\|y\| < \delta$. Let β be any real number such that $\|y\| < \beta < \delta$ and choose a real number ν such that $0 < \nu < \delta - \beta$.

We assumed $\|y\| < \delta$, and so, by (4.8), there exists an $x_1 \in X$ such that

$$\|y - T(x_1)\| \le \left(\frac{\nu}{\delta}\right)\beta, \quad \|x_1\| \le \frac{\beta}{\delta}.$$

Now observe that $\nu < \delta$. This implies that $y - T(x_1)$ is an element of Y with the property that $\|y - T(x_1)\| < \beta < \delta$. Again we use (2.24), only this time with the element $y - T(x_1)$, and we find an $x_2 \in X$ such that

$$\|y - T(x_1) - T(x_2)\| \le \left(\frac{\nu}{\delta}\right)^2 \beta, \quad \|x_2\| \le \frac{\nu}{\delta^2}\beta.$$

Continuing inductively, we construct a sequence $(x_n)_{n=1}^{\infty}$ such that

$$\left\|y - \sum_{k=1}^{n} T(x_k)\right\| \le \left(\frac{\nu}{\delta}\right)^n \beta, \quad \|x_n\| \le \frac{\nu^{n-1}}{\delta^n}\beta, \tag{4.9}$$

for each $n \in \mathbb{N}$.

Since $\nu < \delta - \beta$, we have $\beta < \delta - \nu$, and so

$$\sum_{k=1}^{\infty} \|x_k\| \le \sum_{k=1}^{\infty} \frac{\beta}{\delta} \left(\frac{\nu}{\delta}\right)^{n-1} = \frac{\beta}{\delta} \frac{1}{1 - \frac{\nu}{\delta}} = \frac{\beta}{\delta - \nu} < 1.$$

Since X is a Banach space, $\sum_{k=1}^{\infty} x_k$ converges to an element in X, by the Cauchy Summability Criterion (Lemma cauchy-criterion). Denote this limit in X by x. By the triangle inequality, $\|x\| \le \sum_{k=1}^{\infty} \|x_k\| < 1$, and so $x \in B_X$.

By (4.9) and the continuity of T,

$$y = \lim_{n \to \infty} \sum_{k=1}^{n} T(x_k) = T(x).$$

Consequently, we have $y \in T(B_X)$. Therefore, $\text{int}(\delta B_Y) \subseteq T(B_X)$, as required, and so T is an open map. □

The next theorem is one of the cornerstones of functional analysis.

Theorem 4.29 (Open Mapping Theorem) *Suppose X and Y are Banach spaces. If $T : X \to Y$ is a bounded surjective operator, then T is an open map.*

Proof Observe that $Y = T(X) = \bigcup_{n=1}^{\infty} n\,\overline{T(B_X)}$. Therefore, by Theorem 4.7, the set $\overline{T(B_X)}$ has non-empty interior. Consequently, there exists an element $y \in Y$ and a number $\delta > 0$ such that

$$y + \delta B_Y \subseteq \overline{T(B_X)}.$$

A simple calculation reveals that $-y + \delta B_Y \subseteq \overline{T(B_X)}$, as well, and so it must be the case that $\delta B_Y \subseteq \overline{T(B_X)}$. Therefore, T is almost open. It follows that T is an open map, by Proposition 4.28.   □

Corollary 4.30 (Bounded Inverse Theorem) *Let X and Y be Banach spaces. If $T : X \to Y$ is a bounded linear bijection, then T^{-1} is a bounded linear operator. Consequently, any continuous linear bijection between Banach spaces is an isomorphism.*

Proof By assumption, the map T is a continuous bijection. Since T is surjective, it is an open map, by Theorem 4.29. Because T is an injective open map, it follows that the inverse T^{-1} is bounded, by Corollary 4.26. Therefore, T is a continuous bijection with continuous inverse, and so is an isomorphism between the Banach spaces X and Y.   □

We have stated the Bounded Inverse Theorem as a corollary to the Open Mapping Theorem, but they are in fact equivalent. (See Exercise 4.21.)

Example 4.31 By Theorem 4.29, any quotient map is an open map. Up to an equivalence, the converse is also true. Suppose X and Y are Banach spaces and let $T : X \to Y$ be a bounded linear operator that is an open mapping. Consider the following commuting diagram:

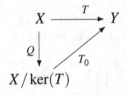

In this diagram, $T = T_0 \circ Q$, where Q is the quotient map onto $X/\ker(T)$. By assumption, the bounded linear operator T is an open map. Thus, by Corollary 4.26, we know that T is a surjection. Consequently, by Proposition 3.49, the map T_0 is a continuous linear bijection. Therefore, T_0 is an isomorphism, by the Bounded Inverse Theorem (Corollary 4.30).

This demonstrates that the open map T can be written as $T_0 \circ Q$, where Q is a quotient map and T_0 is an isomorphism. This means that any open map is a quotient map, up to an isomorphism. (Note that the norms of T and Q might not be equal, because T_0 need not be an isometry.)

Example 4.32 Banach-Mazur Characterization of Separable Spaces) A rather remarkable result (dating back to 1933) is that every *separable* Banach space can be realized as a quotient of ℓ_1. To see this, let X be a separable Banach space. The closed unit ball B_X has a countable dense subset, say $(x_n)_{n=1}^{\infty}$. Define a bounded linear operator $T : \ell_1 \to X$ by

$$T(\xi) = \sum_{n=1}^{\infty} \xi_n x_n, \quad \xi = (\xi_n)_{n=1}^{\infty} \in \ell_1.$$

The linearity of T follows from the summability of the terms in the sequence $\xi \in \ell_1$. To show that T is bounded, observe:

$$\|T(\xi)\| = \Big\| \sum_{n=1}^{\infty} \xi_n x_n \Big\| \leq \sum_{n=1}^{\infty} |\xi_n| \, \|x_n\| \leq \sum_{n=1}^{\infty} |\xi_n| = \|\xi\|.$$

(Recall $x_n \in B_X$ for all $n \in \mathbb{N}$.) Consequently, $\|T\| \leq 1$, and so $T(B_{\ell_1}) \subseteq B_X$.

Recall that e_n denotes the sequence with 1 in the nth coordinate and 0 elsewhere, so that $e_n = (0, \dots, 0, 1, 0, \dots)$. Certainly $e_n \in B_{\ell_1}$ for each $n \in \mathbb{N}$. Since $T(e_n) = x_n$ for each $n \in \mathbb{N}$, we conclude that $(x_n)_{n=1}^{\infty} \subseteq T(B_{\ell_1})$. By assumption, $(x_n)_{n=1}^{\infty}$ is dense in B_X, and so $B_X = \overline{T(B_{\ell_1})}$.

The above argument shows that T is an almost open map (with $\delta = 1$). By Proposition 4.28, an almost open map is an open map. Therefore, by Corollary 4.26, the map T is a surjection.

Consider the diagram from Example 4.31, but in our current context:

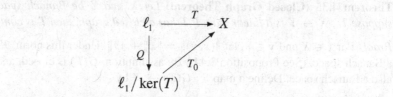

As we saw in Example 4.31, the map T_0 is an isomorphism. Furthermore, T_0 is an isometry because $B_X = \overline{T(B_{\ell_1})}$. Thus, the separable Banach space X is isometrically isomorphic to $\ell_1/\ker(T)$.

The following factorization result is a consequence of the Bounded Inverse Theorem, and one we will use on several occasions.

Lemma 4.33 *Let X and Y be Banach spaces and suppose $T : X \to Y$ is a surjective bounded linear operator. If $x^* \in (\ker T)^{\perp}$, then there exists some $f \in Y^*$ such that $x^* = f \circ T$.*

Proof Let $Q : X \to X/\ker(T)$ be the quotient map. By Proposition 3.49 (noting that T is a surjection), there is a continuous linear bijection $T_0 : X/\ker(T) \to Y$ such that $T = T_0 \circ Q$. By the Bounded Inverse Theorem (Corollary 4.30), we conclude that T_0 is an isomorphism of Banach spaces.

It was assumed that $x^* \in (\ker T)^{\perp}$. Hence, by Proposition 3.51, the linear functional x^* determines an element $\hat{f}$ of $(X/\ker T)^*$ via the identification

$$\hat{f}(x + \ker T) = x^*(x),$$

for all $x \in X$. Since $X/\ker(T)$ is isomorphic to Y, there exists a bounded linear functional $f : Y \to \mathbb{K}$, where $\mathbb{K}$ is the field of scalars, such that $\hat{f} = f \circ T_0$. (See the diagram below.)

$$X \xrightarrow{\;\;T\;\;} Y$$

with maps Q (downward from X to $X/\ker(T)$), f (downward from Y to $\mathbb{K}$), T_0 (diagonal), and $\hat{f}$ (from $X/\ker(T)$ to $\mathbb{K}$).

Therefore,

$$x^*(x) = \hat{f}(x + \ker T) = \hat{f}(Qx) = (f \circ T_0 \circ Q)(x) = (f \circ T)(x),$$

for all $x \in X$, as required. $\square$

We now turn our attention to an alternate formulation of the Open Mapping Theorem. First, we need a definition.

Definition 4.34 Let X and Y be Banach spaces and suppose $T : X \to Y$ is a linear map. The *graph* of T is the subset of $X \times Y$ given by

$$G(T) = \{(x, Tx) : x \in X\}.$$

We call $G(T)$ a *closed graph* if it is closed as a subset of $X \times Y$.

Theorem 4.35 (Closed Graph Theorem) *Let X and Y be Banach spaces and suppose $T : X \to Y$ is a linear map. If T has a closed graph, then T is continuous.*

Proof For $x \in X$ and $y \in Y$, let $\|(x, y)\| = \|x\| + \|y\|$. Under this norm, $X \times Y$ is a Banach space. (See Proposition 3.44.) By assumption, $G(T)$ is closed, and hence also a Banach space. Define a map $S : G(T) \to X$ by

$$S(x, Tx) = x, \quad x \in X.$$

Clearly, S is a bijection and $\|S\| \leq 1$. Thus, by the Bounded Inverse Theorem (Corollary 4.30), we conclude that S^{-1} is bounded.

By definition, $S^{-1}x = (x, Tx)$, and so

$$\|x\| + \|Tx\| = \|(x, Tx)\| = \|S^{-1}x\| \leq \|S^{-1}\| \, \|x\|.$$

Therefore, $\|Tx\| \leq \|S^{-1}\| \, \|x\|$ for all $x \in X$, and consequently $\|T\| \leq \|S^{-1}\|$. It follows that T is bounded, and hence continuous. $\square$

We have now shown that the Open Mapping Theorem implies the Bounded Inverse Theorem (Corollary 4.30), and also that the Bounded Inverse Theorem implies the Closed Graph Theorem (Theorem 4.35). It is also true that the Closed Graph Theorem implies the Open Mapping Theorem, and consequently all three are equivalent. (See Exercise 4.20.)

Example 4.36 In general, the linearity assumption in the Closed Graph Theorem is necessary to prove continuity. Consider the map $f : \mathbb{R} \to \mathbb{R}$ given by

$$f(x) = \begin{cases} 1/x & \text{if } x \neq 0, \\ 0 & \text{if } x = 0. \end{cases}$$

The graph of f is closed, but f is certainly not continuous.

If we restrict our attention to functions $f : [0, 1] \to [0, 1]$, then a closed graph does indeed suffice for continuity. Assume to the contrary that there is a function $f : [0, 1] \to [0, 1]$ with closed graph that is not continuous. Then there exists some $x \in [0, 1]$ and a sequence $(x_n)_{n=1}^{\infty}$ in $[0, 1]$ converging to x such that, for some $\epsilon > 0$,

$$|f(x_n) - f(x)| > \epsilon, \quad n \in \mathbb{N}. \tag{4.10}$$

The interval $[0, 1]$ is compact, and so $(f(x_n))_{n=1}^{\infty}$ has a convergent subsequence, say $\left(f(x_{n_k})\right)_{k=1}^{\infty}$. Suppose y is the limit of this subsequence. By definition, the point $(x_{n_k}, f(x_{n_k}))$ is in $G(f)$ for all $k \in \mathbb{N}$. Since the graph of f is closed, it follows that $(x, y) \in G(f)$, but this implies $y = f(x)$, which contradicts (4.10).

In the preceding paragraph, we considered a function $f : [0, 1] \to [0, 1]$, but this argument works equally well for a function $f : K \to K$, where K is an arbitrary compact set.

4.4 Applications of the Open Mapping Theorem

Consider the torus $\mathbb{T}$. For any $f \in L_1\left(\mathbb{T}, \frac{d\theta}{2\pi}\right)$, we define the Fourier coefficients of f by

$$\hat{f}(n) = \int_0^{2\pi} f(\theta) e^{-in\theta} \frac{d\theta}{2\pi}, \quad n \in \mathbb{Z}.$$

(See Example 4.17.)

For ease of notation, we will abbreviate $L_1\left(\mathbb{T}, \frac{d\theta}{2\pi}\right)$ as $L_1(\mathbb{T})$.

Theorem 4.37 (Riemann-Lebesgue Lemma) *If $f \in L_1(\mathbb{T})$, then $\lim\limits_{|n| \to \infty} \hat{f}(n) = 0$.*

Proof For each $n \in \mathbb{Z}$, define a linear functional on $L_1(\mathbb{T})$ by $\phi_n(f) = \hat{f}(n)$ for all $f \in L_1(\mathbb{T})$. A simple computation shows that $\|\phi_n\| \le 1$ for all $n \in \mathbb{Z}$, and so the sequence of linear functionals is uniformly bounded. If f is a trigonometric polynomial, then there exists some $N \in \mathbb{N}$ such that $\phi_n(f) = 0$ for all $|n| \ge N$. In particular, $\lim_{|n| \to \infty} \phi_n$ exists on a dense subset of $L_1(\mathbb{T})$. By Theorem 4.16, the set $\{f : \lim_{|n| \to \infty} \phi_n(f) \text{ exists}\}$ is a closed linear subspace of $L_1(\mathbb{T})$, and hence the limit exists for all elements of $L_1(\mathbb{T})$.

By Theorem 4.15, the map defined by $\phi(f) = \lim_{|n| \to \infty} \phi_n(f)$ for $f \in L_1(\mathbb{T})$ is a bounded linear functional on $L_1(\mathbb{T})$. We have already established that $\phi(f) = 0$ whenever f is a trigonometric polynomial. Therefore, since ϕ is continuous, and since the trigonometric polynomials are dense in the space of integrable functions, we have that $\phi(f) = 0$ for all $f \in L_1(\mathbb{T})$. This proves the theorem. $\square$

The significance of Theorem 4.37 is that, for any $f \in L_1(\mathbb{T})$, the doubly infinite sequence $\left(\hat{f}(n)\right)_{n \in \mathbb{Z}}$ is always an element of $c_0(\mathbb{Z})$. This may lead one to ask if the converse is true. The next proposition, which makes use of the Open Mapping

Theorem (in the form of the Bounded Inverse Theorem), shows that the converse is not true.

Proposition 4.38 *There exists a sequence* $\xi \in c_0(\mathbb{Z})$ *which is not the Fourier transform of a function in* $L_1(\mathbb{T})$.

Proof Define a map $\mathcal{F} : L_1(\mathbb{T}) \to c_0(\mathbb{Z})$ by

$$\mathcal{F}(f) = \left(\hat{f}(n) \right)_{n \in \mathbb{Z}}, \quad f \in L_1(\mathbb{T}).$$

It suffices to show that $\mathcal{F}$ is not a surjection. The map $\mathcal{F}$ is a bounded linear operator with $\|\mathcal{F}\| = 1$, and $\mathcal{F}$ is injective because Fourier coefficients are unique. Suppose $\mathcal{F}$ does map $L_1(\mathbb{T})$ onto $c_0(\mathbb{Z})$. Then $\mathcal{F}$ is a bounded linear bijection, and hence an isomorphism, by Corollary 4.30 (the Bounded Inverse Theorem). In particular, $\mathcal{F}^{-1}$ is also a bounded linear bijection.

Our assumption that $\mathcal{F}$ is a surjection has led us to conclude that the inverse $\mathcal{F}^{-1} : c_0(\mathbb{Z}) \to L_1(\mathbb{T})$ is a bijection, which implies that $(\mathcal{F}^{-1})^* : L_\infty(\mathbb{T}) \to \ell_1(\mathbb{Z})$ is also a bijection. (See Exercise 4.12.) This, however, is impossible, because $\ell_1(\mathbb{Z})$ is separable and $L_\infty(\mathbb{T})$ is not. (The proofs of separability and nonseparability are similar to those given in Remark 3.14.) We have arrived at a contradiction, and therefore must conclude that $\mathcal{F}$ does not map $L_1(\mathbb{T})$ onto $c_0(\mathbb{Z})$. $\square$

In light of Theorem 4.37 and Proposition 4.38, it is tempting to wonder if, for functions $f \in L^1(\mathbb{T})$, any bounds can be established for the rate at which the sequence $\left(\hat{f}(n) \right)_{n \in \mathbb{Z}}$ decays. It turns out, however, that Fourier coefficients can decay arbitrarily slowly. (See Section I.4 of [20].)

Now consider the sequence space $\ell_p = \ell_p(\mathbb{N})$, where $1 \le p \le \infty$. We suppose $(a_{jk})_{j,k=1}^{\infty}$ is an infinite matrix, where $a_{jk} \in \mathbb{R}$ for each j and k in $\mathbb{N}$.

Proposition 4.39 *Suppose* $(a_{jk})_{j,k=1}^{\infty}$ *is a scalar array such that the series*

$$\eta_j = \sum_{k=1}^{\infty} a_{jk} \xi_k$$

converges for all $\xi = (\xi_k)_{k=1}^{\infty} \in \ell_p$ *and* $j \in \mathbb{N}$. *Furthermore, suppose the sequence* $\eta_\xi = (\eta_j)_{j=1}^{\infty}$ *is an element of* ℓ_p. *If the map* $A : \ell_p \to \ell_p$ *is defined by* $A(\xi) = \eta_\xi$ *for each* $\xi \in \ell_p$, *then* A *is a bounded linear operator on* ℓ_p.

Proof For each $j \in \mathbb{N}$, let

$$\psi_j(\xi) = \lim_{n \to \infty} \sum_{k=1}^{n} a_{jk} \xi_k = \eta_j.$$

By Theorem 4.15, ψ_j is a bounded linear functional on ℓ_p for each $j \in \mathbb{N}$. We use the bounded linear functionals $(\psi_j)_{j=1}^{\infty}$ to define the map $A : \ell_p \to \ell_p$ by

$$A\xi = \left(\psi_j(\xi) \right)_{j=1}^{\infty}, \quad \xi \in \ell_p.$$

By assumption, this map is well defined. Furthermore, A is linear because ψ_j is linear for each $j \in \mathbb{N}$. It remains only to show that A is bounded. In order to do this, we will show that the graph of A is closed and apply the Closed Graph Theorem.

Suppose $(\xi^{(n)})_{n=1}^{\infty}$ is a sequence in ℓ_p such that $\xi^{(n)} \to \xi$ in ℓ_p, and suppose $\zeta = (\zeta_j)_{j=1}^{\infty} \in \ell_p$ is such that $A\xi^{(n)} \to \zeta$ in ℓ_p. For each $j \in \mathbb{N}$, the map ψ_j is a continuous linear functional on ℓ_p, and so $\psi_j(\xi^{(n)}) \to \psi_j(\xi)$. Consequently, it must be that $\zeta_j = \psi_j(\xi)$ for each $j \in \mathbb{N}$, and hence $\zeta = A\xi$. Therefore, the graph of A is closed, and so A is continuous, by the Closed Graph Theorem (Theorem 4.35). □

Proposition 4.39 demonstrates the general principle that if a linear map is properly defined, it will tend to be bounded. The next example provides a further illustration.

Example 4.40 The *Hilbert transform* is the map $H : L_p(\mathbb{R}) \to L_p(\mathbb{R})$ defined by the formula

$$Hf(x) = \int_{-\infty}^{\infty} \frac{1}{x-y} f(y)\,dy, \quad f \in L_p(\mathbb{R}),$$

where $1 < p < \infty$. Provided the map H is well-defined, it will follow directly (by the Closed Graph Theorem) that it is bounded (and hence continuous). As a practical matter, however, the same calculation that shows that H is well-defined also shows that H is bounded. (See Exercise 4.13.)

We now turn our attention back to the topics of Section 3.7: direct sums and quotient spaces. We begin with a definition.

Definition 4.41 Let X be a Banach space with closed subspace V. A map $P : X \to V$ is called a *projection* if $Px = x$ for all $x \in V$. (Note that this is equivalent to the condition $P^2 = P$.)

Theorem 4.42 *Let X be a Banach space with closed subspace V. If $P : X \to V$ is a continuous projection and $W = \ker(P)$, then X and $V \oplus W$ are isomorphic as Banach spaces, and V and X/W are isomorphic as Banach spaces. Conversely, if $X = V \oplus W$, then there exists a continuous projection $P : X \to V$ with $W = \ker(P)$.*

We pause for a moment to recall that a map $T : X \to Y$ is called an *isomorphism* of Banach spaces if it is a continuous linear bijection with a continuous inverse. In such a case, we say X and Y are *isomorphic*. Note that if $(X, \|\cdot\|_\alpha)$ and $(X, \|\cdot\|_\beta)$ are isomorphic as Banach spaces, then the two norms are *equivalent*; that is, there exist positive constants c and C such that

$$c\,\|x\|_\alpha \le \|x\|_\beta \le C\,\|x\|_\alpha, \quad x \in X.$$

Proof of Theorem 4.42 Suppose that $P : X \to V$ is a continuous projection and $W = \ker(P)$. By Proposition 3.49, there exists a linear map $P_0 : X/W \to V$ such that the diagram below commutes, where $Q : X \to X/W$ is the quotient map.

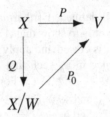

(Recall that W is closed by Proposition 3.49.) The continuity of P_0 follows from the continuity of P and the fact that Q is an open map. Any continuous linear bijection will have a continuous inverse by the Bounded Inverse Theorem (Corollary 4.30), and so P_0 is an isomorphism.

The next step is to show that X is the vector space direct sum of its subspaces V and W. Suppose that $x \in V \cap W$. Since $x \in W$, we have that $Px = 0$. However, P is a projection onto V, and so $x \in V$ implies that $Px = x$. It follows that $x = 0$. Thus, $V \cap W = \{0\}$. Furthermore, every $x \in X$ can be written as the sum of elements from V and W as follows: $x = Px + (x - Px)$. We conclude that $X = V \oplus W$ as vector spaces, by Proposition 3.45.

We now show that X is isomorphic to $V \oplus W$. Specifically, we wish to show $(X, \| \cdot \|)$ is isomorphic to $V \oplus W$ equipped with the norm from Proposition 3.44:

$$\|(v, w)\| = \|v\| + \|w\|, \quad (v, w) \in V \times W.$$

Define $\phi : X \to V \oplus W$ by

$$\phi(x) = (Px, x - Px), \quad x \in X.$$

We know ϕ is well-defined by our earlier remarks. Furthermore, ϕ is linear because P is linear and because of the way the vector space operations are defined in $V \oplus W$. Next, we show that ϕ is a bijection.

Suppose $\phi(x) = (0, 0)$. Then $(Px, x - Px) = (0, 0)$, and so $Px = 0$ and $x - Px = 0$. From this we conclude that $x = Px = 0$, and hence ϕ is injective. To show that ϕ is surjective, let $(v, w) \in V \times W$. Then $P(v) = v$ and $P(w) = 0$. Consequently,

$$\phi(v + w) = (P(v + w), (v + w) - P(v + w)) = (v, v + w - v) = (v, w).$$

Thus, ϕ is a surjection.

We next show that ϕ is continuous by showing that it is bounded:

$$\|\phi(x)\|_{V \oplus W} = \|(Px, x - Px)\|_{V \oplus W} = \|Px\| + \|x - Px\| \le (2\|P\| + 1)\|x\|.$$

Therefore, ϕ is a continuous linear bijection, and consequently an isomorphism, by the Bounded Inverse Theorem (Corollary 4.30).

Now suppose $X = V \oplus W$. Then x has a unique representation of the form $v + w$, where $v \in V$ and $w \in W$. Define $P(x) = v$. Then P is a projection and $W = \ker(P)$. It remains only to show that P is continuous. We will show continuity by means of the Closed Graph Theorem. Suppose $(x_n)_{n=1}^{\infty}$ is a convergent sequence in X such that $(Px_n)_{n=1}^\infty$ converges in V. Then there exists some $x \in X$ and $v \in V$ such that $x_n \to x$ and $Px_n \to v$ as $n \to \infty$. We need to show that $Px = v$.

For each $n \in \mathbb{N}$, we have $x_n - Px_n \in \ker(P) = W$. Since W is closed,

$$\lim_{n \to \infty} x_n - Px_n = x - v \in W.$$

Therefore, $P(x - v) = 0$, and so $Px = Pv$. By assumption, $Pv = v$, and hence $Px = v$, as required. $\qquad \square$

We have seen that whenever a closed subspace V is the image of a continuous projection in X, there is a closed subspace W such that $X = V \oplus W$. For this reason, when V is the image of a projection, we call V a *complemented subspace* of X.

Exercises

Exercise 4.1 Let $\mathbb{R}$ be given the standard topology. Show that the closed set $[0, 1]$ is a G_δ-set. Show that the open set $(0, 1)$ is an F_σ-set.

Exercise 4.2 Show that the space $C[0, 1]$ of continuous functions on the closed interval $[0, 1]$ is not complete in the norm

$$\|f\|_2 = \left(\int_0^1 |f(s)|^2 \, ds \right)^{1/2}, \quad f \in C[0, 1].$$

(The completion of $C[0, 1]$ in the norm $\| \cdot \|_2$ is $L_2(0, 1)$, by Lusin's Theorem.)

Exercise 4.3 Let M and E be complete metric spaces. Suppose $h : M \to E$ is a homeomorphism onto its image (i.e., h is a continuous one-to-one map, and $h^{-1}|_{h(M)}$ is continuous). Show that $h(M)$ is a G_δ-set.

Exercise 4.4 Let X be a Banach space and suppose E is a dense linear subspace which is a G_δ-set. Show that $E = X$.

Exercise 4.5 Show that if Y is a normed space which is homeomorphic to a complete metric space, then Y is a Banach space. (*Hint:* Consider Y as a dense subspace in its completion.)

Exercise 4.6 Let X and Y be Banach spaces and let $T : X \to Y$ be a bounded linear operator. If M is a closed subspace of X, show that either $T(M)$ is first category in Y or $T(M) = Y$.

Exercise 4.7 Let $X = C^{(1)}[0, 1]$ be the space of continuously differentiable functions on $[0, 1]$ and let $Y = C[0, 1]$. Equip both spaces with the supremum norm $\| \cdot \|_\infty$. Define a linear map $T : X \to Y$ by $T(f) = f'$ for all functions $f \in C^{(1)}[0, 1]$. Show that T has closed graph, but T is not continuous. Conclude that $\left(C^{(1)}[0, 1], \| \cdot \|_\infty \right)$ is not a Banach space.

Exercise 4.8 Let $\phi \in C[0, 1]$ be a function which is not identically 0. Show the set $M = \{\phi f : f \in C[0, 1]\}$ is of the first category in $C[0, 1]$ if and only if $\phi(x) = 0$ for some $x \in [0, 1]$.

Exercise 4.9 Let $(a_k)_{k \in \mathbb{Z}}$ be a sequence of complex scalars with only finitely many nonzero terms. Define a trigonometric polynomial $f : \mathbb{T} \to \mathbb{C}$ by $f(\theta) = \sum_{k \in \mathbb{Z}} a_k e^{ik\theta}$.

Show that $\hat{f}(n) = a_n$ for all $n \in \mathbb{Z}$.

Exercise 4.10 Show that there exists a function $f \in L_1(\mathbb{T})$ whose Fourier series fails to converge to f in the L_1-norm. Precisely, show that if

$$S_N f = \sum_{k=-N}^{N} \hat{f}(k) e^{ik\theta},$$

then there exists an $f \in L_1(\mathbb{T})$ such that $\|f - S_N f\|_{L_1(\mathbb{T})}$ does not tend to 0.

Exercise 4.11 Show that the Cesàro means $\frac{1}{N}(S_1 f + \cdots + S_N f)$ converge to f in the L_1-norm for every $f \in L_1(\mathbb{T})$.

Exercise 4.12 Let X and Y be Banach spaces. If $T : X \to Y$ is a bijection, show that the adjoint map $T^* : Y^* \to X^*$ is also a bijection. Conclude that if T is an isomorphism of Banach spaces, then so is T^*.

Exercise 4.13 Show that the Hilbert transform of Example 4.40 is well-defined.

Exercise 4.14 Let $f : [1, \infty) \to \mathbb{R}$ be a continuous function. Suppose $(\xi_n)_{n=1}^{\infty}$ is a strictly increasing sequence of real numbers with $\xi_1 \geq 1$, $\lim_{n \to \infty} \xi_n = \infty$, and $\lim_{n \to \infty} \frac{\xi_{n+1}}{\xi_n} = 1$. If $\lim_{n \to \infty} f(\xi_n x) = 0$ for all $x \geq 1$, then prove that $\lim_{x \to \infty} f(x) = 0$.

Exercise 4.15 Show that $L_2(0, 1)$ is of the first category in $L_1(0, 1)$.

Exercise 4.16 Let (Ω, μ) be a probability space and suppose there exists a sequence of disjoint sets $(E_n)_{n=1}^{\infty}$ such that $\mu(E_n) > 0$ for all $n \in \mathbb{N}$. Show that $L_p(\Omega, \mu) \neq L_q(\Omega, \mu)$ if $1 \leq p < q < \infty$.

Exercise 4.17 Let X be an infinite-dimensional Banach space and suppose V is a closed subspace of X. If both V and X/V are separable, then show X is separable.

Exercise 4.18 Identify the quotient space c/c_0.

Exercise 4.19 Let $1 \leq p \leq \infty$ and let $V = \{(x_j)_{j=1}^{\infty} \in \ell_p : x_{2k} = 0 \text{ for all } k \in \mathbb{N}\}$. Show that ℓ_p/V is isometrically isomorphic to ℓ_p.

Exercise 4.20 Find an example of a map $T : X \to Y$, where X and Y are normed spaces, such that T is a bounded linear bijection, but T^{-1} is not bounded. (*Hint: X and Y cannot both be Banach spaces, or T^{-1} will be bounded by Corollary 4.30.*)

Exercise 4.21 Show that the Closed Graph Theorem (Theorem 4.35) implies the Open Mapping Theorem (Theorem 4.29).

Chapter 5
Consequences of Convexity

In this chapter, we wish to explore the geometric aspects of the Hahn–Banach Theorem. The crucial property, it turns out, is local convexity. We will first recall some notions from general topology and then introduce the concept of a topological vector space. These spaces, which include Banach spaces, are sufficiently complex that we can say something interesting about their structure. Banach spaces are topological vector spaces where the topology is determined by a complete norm, and in this chapter we will get some idea of how they fit into a more general topological framework.

5.1 General Topology

Let E be a set. A *topology* τ on E is a collection of subsets called *open sets* satisfying the following three criteria:

1. The collection τ contains both E and the empty set $\emptyset$.
2. If $\{U_i\}_{i \in I}$ is a (possibly uncountable) family of sets in τ, then $\bigcup_{i \in I} U_i$ is in τ.
3. If U and V are in τ, then $U \cap V$ is in τ.

When E is equipped with a topology τ, we call the pair (E, τ) a *topological space*. When there is no ambiguity, we will suppress the τ and simply write E for the topological space (E, τ), and say U is open in E when $U \in \tau$.

If (E, τ) and (F, τ') are topological spaces, then a function $f : E \to F$ is called *continuous* if $f^{-1}(U)$ is open in E whenever U is open in F. For x a point in E, a *neighborhood* of x is any subset N of E for which there exists an open set U such that $x \in U$ and $U \subset N$.

Example 5.1 Let (E, τ) be a topological space. If every set in E is open, then τ is called the *discrete topology*. If $\tau = \{E, \emptyset\}$, then τ is called the *indiscrete topology*.

Example 5.2 Let (E, τ) and (F, τ') be two topological spaces. The product $E \times F$ can be given the *product topology*, denoted $\tau \times \tau'$, as follows: $W \subseteq E \times F$ is open if

© Springer Science+Business Media, LLC 2014
A. Bowers, N. J. Kalton, *An Introductory Course in Functional Analysis*,
Universitext, DOI 10.1007/978-1-4939-1945-1_5

$$W = \bigcup_{i \in I} (U_i \times V_i),$$

where $U_i \in \tau$ and $V_i \in \tau'$ for each $i \in I$, where I is a (possibly uncountable) index set.

Let (E, τ) be a topological space. The topology τ is said to have a *base* of open sets $\{U_i\}_{i \in I}$ if for each open set $V \in \tau$, there exists an index set $J \subseteq I$ such that $V = \bigcup_{i \in J} U_i$. When τ has a base, we say the base *generates* the topology τ. In Example 5.2, the product topology on $E \times F$ is generated by the base

$$\{U \times V : U \in \tau, \ V \in \tau'\}.$$

Example 5.3 (Product topology). Let I be a (possibly uncountable) index set. For each $i \in I$, let (E_i, τ_i) be a topological space. The *product topology* on the product $\prod_{i \in I} (E_i, \tau_i)$ is a topology with a base consisting of sets of the form

$$U_{i_1} \times \cdots \times U_{i_n} \times \prod_{i \in I \setminus \{i_1, \ldots, i_n\}} E_i,$$

where U_{i_j} is open in E_{i_j} for $i_j \in I$, $n \in \mathbb{N}$, and $j \in \{1, \ldots, n\}$. Observe that all but finitely many elements of the product are the entire space. This example contains Example 5.2 as a special case, because the product in that example is finite.

Example 5.4 Suppose M is a set with a metric d. We will show that the metric d determines a topology on M. For each $x \in M$ and $r > 0$, let

$$B(x, r) = \{z \in M : d(x, z) < r\}.$$

This set is the *open ball about x of radius r*. We declare a subset V of M to be open if for each $x \in V$, there is an $r > 0$ such that $B(x, r) \subseteq V$. The collection of all such open sets forms a topology on M called the *metric topology on M generated by the metric d*, or just the *metric topology* on M, if the metric d is understood.

Suppose V is open in the metric topology. For each $x \in V$, there exists a number $r_x > 0$ such that $B(x, r_x) \subseteq V$. Thus, $V = \bigcup_{x \in V} B(x, r_x)$, and so the collection of open balls forms a base for the metric topology.

Let E be a topological space and let $x \in E$. A *local base at x* is a collection η of open sets, all of which contain x, such that any neighborhood U of x contains an element of η.

In Example 5.4, any point in the metric space M has a local base. For $x \in M$, the collection of open balls $B(x, r)$ for all $r > 0$ forms a local base at x. In fact, if we consider the collection of sets $\eta = \{B(x, 1/n) : n \in \mathbb{N}\}$, then η is a countable local base at x.

A topological space (E, τ) with a countable local base at every point $x \in E$ is called *first countable*. Further, (E, τ) is called *second countable* if τ has a countable base. From Example 5.4 (and the comments following it), we see that any metric space M is first countable; however, M will not be second countable unless M is separable. (See Exercise 5.9.)

Let (E, τ) be a topological space. We say (E, τ) is *metrizable* if there exists a metric d such that d generates the topology τ. That is, if the open balls in (E, d) form a base for the topology τ. We call (E, τ) a *Hausdorff space* if for any distinct points x and y in E there exist open sets U and V in τ such that $x \in U$, $y \in V$, and $U \cap V = \emptyset$.

Example 5.5 Any metrizable space is a Hausdorff space. To see this, suppose (M, d) is a metric space and let x and y be two distinct points in M. Since $x \neq y$, it follows that $d(x, y) > 0$. Let $\delta = d(x, y)$. Furthermore, let $U = B(x, \delta/2)$ and $V = B(y, \delta/2)$. Then U and V are open in the metric topology on M, $x \in U$, $y \in V$, and $U \cap V = \emptyset$.

Example 5.6 Any nonempty set E with the *discrete topology* (see Example 5.1) is metrizable. Define a metric on E by

$$d(x, y) = \begin{cases} 0 & \text{if } x = y, \\ 1 & \text{if } x \neq y, \end{cases} \quad \text{for } (x, y) \in E \times E.$$

It is easy to see that d is, in fact, a metric. This metric is called the *discrete metric* on E and it is not hard to show that d generates the discrete topology on E.

Of particular interest to us is the notion of *compactness*. A topological space is said to be *compact* if any open cover contains a finite open subcover. To be more precise, let X be a topological space. Then X is compact if for any collection $\mathcal{U}$ of open sets such that $X \subseteq \bigcup_{U \in \mathcal{U}} U$ there exists a finite collection $\{U_1, \dots, U_n\}$ of elements in $\mathcal{U}$ such that $X \subseteq U_1 \cup \dots \cup U_n$.

For a (not necessarily compact) topological space, we define a *compact subset* in a similar way: A subset E of a topological space X is *compact* if any cover of E by sets open in X admits a finite subcover of E.

Some well-known properties of compact sets are treated in the exercises at the end of this chapter. (See Exercise 5.2.)

A topological space is said to be *locally compact* if every point has a compact neighborhood. Naturally, all compact spaces are locally compact, but the converse need not be true. For example, the real line $\mathbb{R}$ with its standard topology is locally compact, but not compact.

A notion of fundamental importance in topology is that of a *convergent sequence*. If X is a topological space, and $(x_n)_{n=1}^{\infty}$ is a sequence of elements from X, then $(x_n)_{n=1}^{\infty}$ is said to *converge* to a point $x \in X$ if for every open neighborhood U of x there exists an $N \in \mathbb{N}$ such that $x_n \in U$ for all $n \geq N$. In such a case, we say x is the limit of the sequence $(x_n)_{n=1}^{\infty}$ and we write $x = \lim_{n \to \infty} x_n$. (Note that this notion of a limit agrees with the standard definition of a limit in a metric space.)

In general, the limit of a sequence need not be unique. The spaces we consider, however, are Hausdorff spaces, and limits are necessarily unique in a Hausdorff space. (See Exercise 5.8.)

A subset U of a topological space X is called *sequentially open* if every sequence $(x_n)_{n=1}^\infty$ that converges to a point in U is eventually in U. That is, if there exists some $N \in \mathbb{N}$ such that $x_n \in U$ for all $n \geq N$. We call X a *sequential space* if every sequentially open set is open. Any first countable topological space is a sequential space. In particular, any metric space is a sequential space.

5.2 Topological Vector Spaces

We now consider topological spaces with additional structure, namely an underlying linear structure.

Let X be a vector space over the field $\mathbb{K}$ (which is either $\mathbb{R}$ or $\mathbb{C}$). A topology τ on X is called a *vector topology* if the maps

$$(\lambda, x) \mapsto \lambda x, \quad \lambda \in \mathbb{K}, \ x \in X,$$

and

$$(x_1, x_2) \mapsto x_1 + x_2, \quad (x_1, x_2) \in X \times X,$$

are both continuous. That is, if both scalar multiplication and addition are continuous in the topology on X. In this case, (X, τ) is called a *topological vector space*.

Example 5.7 Any normed vector space X is a topological vector space, where the topology is given by the base of open balls:

$$x + \lambda(\operatorname{int} B_X), \quad \lambda > 0, \ x \in X.$$

Equivalently, the topology on X is generated by the metric d given by the formula $d(x, y) = \|x - y\|$ for $(x, y) \in X \times X$.

A vector topology is determined by a base of neighborhoods at the origin, since sets can be translated and scaled continuously. We will denote the origin by 0. Let η be a base of neighborhoods of the origin in a topological vector space (X, τ). A set $V \in \eta$ is called *absorbent* if $X = \bigcup_{n=1}^\infty n V$. A set $V \in \eta$ is called *balanced* if $\lambda V \subseteq V$ for all scalars λ such that $|\lambda| \leq 1$.

Lemma 5.8 *In a topological vector space, any open neighborhood of the origin is absorbent.*

Proof Let X be a topological vector space. Suppose V is an open neighborhood of 0 and let $x \in X$. Scalar multiplication is continuous, and so the map $\lambda \mapsto \lambda x$ is continuous. Consequently, the set $\{\lambda : \lambda x \in V\}$ is open in $\mathbb{K}$. By assumption, V is a neighborhood of 0, and so $0 \in \{\lambda : \lambda x \in V\}$. We have established that the set $\{\lambda : \lambda x \in V\}$ is open in $\mathbb{K}$ and contains 0. Thus, it must contain $\frac{1}{n}$ for a sufficiently large $n \in \mathbb{N}$. We conclude that $\frac{x}{n} \in V$, and consequently $x \in nV$. Therefore, V is absorbent. $\qquad\square$

Proposition 5.9 *Any topological vector space has a base of neighborhoods η of the origin such that for all $V \in \eta$: (i) V is balanced, (ii) V is absorbent, and (iii) there exists $W \in \eta$ such that $W + W \subseteq V$.*

Proof Let (X, τ) be a topological vector space and let U be a neighborhood of the origin. Let $s : \mathbb{K} \times X \to X$ be scalar multiplication, so that $s(\lambda, x) = \lambda x$ for all $\lambda \in \mathbb{K}$ and $x \in X$. By assumption, s is continuous. Thus, since U is open in X, the preimage $s^{-1}(U)$ is open in $\mathbb{K} \times X$. Certainly, $(0, 0) \in s^{-1}(U)$, and so there exists some $\delta > 0$ and an open neighborhood W of 0 in X such that $\delta B_{\mathbb{K}} \times W \subseteq s^{-1}(U)$. Therefore, $s(\delta B_{\mathbb{K}} \times W) \subseteq U$, and hence $\alpha W \subseteq U$ for all $|\alpha| \leq \delta$. Let

$$V = \bigcup_{\alpha \in \delta B_{\mathbb{K}}} \alpha W.$$

Then V is open, balanced, and contained in U. For each open neighborhood of 0, such a V can be constructed. Let η be the collection of all such balanced sets. Then *(i)* follows from the construction and *(ii)* follows from Lemma 5.8.

It remains to verify *(iii)*. Let $V \in \eta$. By the continuity of addition, there exist two open neighborhoods U_1 and U_2 of $0 \in X$ such that $U_1 + U_2 \subseteq V$. Let $U = U_1 \cap U_2$. Then U is an open neighborhood of 0 such that $U + U \subseteq V$. As demonstrated earlier in this proof, U contains a subset $W \in \eta$, and this W is the required set. $\square$

Proposition 5.10 *Let X be a topological vector space with η a base of open sets about the origin. Then X is a Hausdorff space if and only if $\bigcap_{V \in \eta} V = \{0\}$.*

Proof Without loss of generality, we may assume that η satisfies the conclusions of Proposition 5.9.

Assume X is a Hausdorff space. Certainly $0 \in \bigcap_{V \in \eta} V$. Suppose $x \neq 0$. We will show that $x \notin \bigcap_{V \in \eta} V$. Since X is a Hausdorff space, there are open sets U and W such that $0 \in U$ and $x \in W$ and $U \cap W = \emptyset$. By assumption, η is a base of open sets about the origin, and consequently there exists a set $V_0 \in \eta$ such that $V_0 \subseteq U$. It follows that $x \notin V_0$, and so $x \notin \bigcap_{V \in \eta} V$. Therefore, $\bigcap_{V \in \eta} V = \{0\}$.

Now assume $\bigcap_{V \in \eta} V = \{0\}$. We will show that X is a Hausdorff space. Let x and y be elements of X that cannot be separated by disjoint open sets. Let $V \in \eta$. By Proposition 5.9, there exists a set $W \in \eta$ such that $W + W \subseteq V$. By assumption, $x + W$ and $y + W$ are not disjoint. Then there exist elements w_1 and w_2 in W such that

$$x + w_1 = y + w_2.$$

Therefore, $x - y = w_2 - w_1 \in W - W$. The set W is balanced, and so we conclude $x - y \in W + W \subseteq V$. This is true for every $V \in \eta$, and so $x - y \in \bigcap_{V \in \eta} V = \{0\}$. It follows that $x = y$, and consequently X is a Hausdorff space. $\square$

In our discussions of normed spaces, a key notion was that of the dual space. In the more general context of topological vector spaces, this will remain true.

Definition 5.11 Let X be a topological vector space. The *dual space* X^* consists of all continuous linear scalar-valued functionals on X.

5.3 Some Metrizable Examples

In this section, we consider some examples of real topological vector spaces which are metrizable, but do not have a norm structure.

Example A: $L_p(0, 1)$, $0 < p < 1$

If $0 < p < \infty$, the symbol $L_p(0, 1)$ denotes the collection of all (equivalence classes of) Lebesgue measurable real-valued functions f on $[0, 1]$ such that

$$\|f\|_p = \left(\int_0^1 |f(t)|^p \, dt \right)^{1/p} < \infty.$$

If $p \geq 1$, then $L_p(0, 1)$ is a Banach space. If $0 < p < 1$, however, then $\| \cdot \|_p$ does not determine a norm, because it is no longer subadditive. On the other hand, if $0 < p < 1$, then it is true that

$$\|f + g\|_p^p \leq \|f\|_p^p + \|g\|_p^p.$$

This fact follows from the proposition below.

Proposition 5.12 *If* $\{a, b\} \subseteq \mathbb{R}$ *and* $0 < p < 1$, *then* $|a + b|^p \leq |a|^p + |b|^p$.

Proof Without loss of generality, assume that a and b are nonnegative real numbers such that $a + b = 1$. Let $a = t$ and $b = 1 - t$, and let $f(t) = t^p + (1 - t)^p$. We will show that $f(t) \geq 1$ for all $t \in [0, 1]$. Since $f(0) = f(1) = 1$, it will suffice to show that f is a concave function.

We require only techniques of elementary differential calculus. Calculating the first derivative of f, we have $f'(t) = pt^{p-1} - p(1-t)^{p-1}$, and so f has one critical point, which is at $t = 1/2$. Differentiating a second time, we have

$$f''(t) = p(p - 1)t^{p-2} + p(p - 1)(1 - t)^{p-2}.$$

Therefore, $f''(1/2) = 2^{3-p} p(p-1) < 0$, and so f has a local maximum at $t = 1/2$. The result follows.   □

From the above proposition, we conclude that $\|f + g\|_p^p \leq \|f\|_p^p + \|g\|_p^p$ whenever $0 < p < 1$. Consequently,

$$d(f, g) = \|f - g\|_p^p$$

determines a metric on $L_p(0, 1)$ when $0 < p < 1$. It follows that if $0 < p < 1$, then $L_p(0, 1)$ is a metrizable space, if not a normed space.

The metric d is even complete. The proof of this fact is similar to the case when $1 \leq p < \infty$. Observe that the proof of the Cauchy Summability Criterion (Lemma 2.24) requires only subadditivity of the norm, a property which is shared with $\| \cdot \|_p^p$ when $0 < p < 1$. Consequently, we can use Lemma 2.24 to prove that $L_p(0, 1)$ is complete when $0 < p < 1$. The details of the proof are left to the reader. (See Exercise 5.12.)

Let $B = \{f : \|f\|_p < 1\}$. Then the collection of open sets $(2^{-n}B)_{n=1}^\infty$ determines a countable base at 0 which satisfies the conclusions of Proposition 5.9. The first two properties are clear. To see property *(iii)*, simply observe that $2^{-N}B + 2^{-N}B \subseteq B$ whenever $N > 1/p$.

We will now compute $L_p(0,1)^*$ for $0 < p < 1$. Suppose ϕ is a continuous linear functional on $L_p(0,1)$. Then ϕ is bounded on ∂B, so that

$$\|\phi\| = \sup_{\|f\|_p = 1} |\phi(f)| < \infty.$$

This should be taken as the definition of $\|\phi\|$ in this context. The function ϕ is a linear functional, but not on a normed space, and consequently the notation $\|\phi\|$ has not yet been given a meaning.

Let $f \in L_p(0,1)$ be such that $\|f\|_p = 1$. The map

$$t \mapsto \int_0^t |f(s)|^p\, ds, \quad t \in [0,1],$$

is continuous with range $[0,1]$. Therefore, by the Intermediate Value Theorem, there exists some $a \in [0,1]$ such that $\int_0^a |f(s)|^p\, ds = 1/2$.

Define two functions g and h in $L_p(0,1)$ by

$$g = f\,\chi_{(0,a)} \quad \text{and} \quad h = f\,\chi_{(a,1)}.$$

By the choice of a,

$$\|g\|_p = \left(\int_0^a |f(s)|^p\, ds \right)^{1/p} = \left(\frac{1}{2} \right)^{1/p},$$

and similarly, $\|h\|_p = (1/2)^{1/p}$.

By the linearity of ϕ, together with the definition of $\|\phi\|$, we have the two bounds $|\phi(g)| \le \|\phi\|(1/2)^{1/p}$ and $|\phi(h)| \le \|\phi\|(1/2)^{1/p}$. Thus, again using the linearity of ϕ,

$$|\phi(f)| \le 2 \cdot \|\phi\|(1/2)^{1/p} = \|\phi\|\, 2^{1-\frac{1}{p}}.$$

Taking the supremum over all functions $f \in L_p(0,1)$ with $\|f\|_p = 1$, we have

$$\|\phi\| \le \|\phi\|\, 2^{1-\frac{1}{p}}.$$

However, this can happen only if $\phi = 0$. This implies that $L_p(0,1)^* = \{0\}$.

The preceding remark guarantees that $L_p(0,1)$ does not satisfy a Hahn–Banach Theorem if $0 < p < 1$. On the other hand, since $L_p(0,1)$ is a complete metric space, even when $0 < p < 1$, we can apply the Baire Category Theorem (Theorem 4.1). It is also possible to prove a version of the Open Mapping Theorem (Theorem 4.29) and the Closed Graph Theorem (Theorem 4.35) for these spaces.

Example B: $L_0(0, 1)$

We denote by $L_0(0, 1)$ the set of all (equivalence classes of) scalar-valued Lebesgue measurable functions on $[0, 1]$. (As usual, we identify functions if they agree almost everywhere.) The topology on $L_0(0, 1)$ is determined by convergence in Lebesgue measure. More precisely, we define a set to be open when it is sequentially open, and a sequence converges when it converges in Lebesgue measure. Recall that a sequence $(f_n)_{n=1}^\infty$ of measurable functions *converges in Lebesgue measure* to a measurable function f if for every $\epsilon > 0$,

$$\lim_{n \to \infty} m\{t : |f(t) - f_n(t)| \geq \epsilon\} = 0,$$

where m is Lebesgue measure on $[0, 1]$.

We claim the topology on $L_0(0, 1)$ is metrizable and is induced by the metric

$$d(f, g) = \int_0^1 \frac{|f(t) - g(t)|}{1 + |f(t) - g(t)|} \, dt,$$

where f and g are measurable functions. The only property of a metric that is not immediate is the triangle inequality. In order to verify this, it suffices to show that the function $\phi(x) = x/(1 + x)$ is a nondecreasing subadditive function on $[0, \infty)$.

A simple application of the quotient rule reveals that $\phi'(x) = 1/(1 + x)^2$, and so ϕ is strictly increasing for all $x \geq 0$. To see that ϕ is subadditive on $[0, \infty)$, observe that

$$\phi(x+y) = \frac{x+y}{1+x+y} = \frac{x}{1+x+y} + \frac{y}{1+x+y} \leq \frac{x}{1+x} + \frac{y}{1+y} = \phi(x) + \phi(y),$$

because $x \geq 0$ and $y \geq 0$. Given these properties, the triangle inequality follows readily from the fact that $d(f, g) = \int_0^1 \phi(|f(t) - g(t)|) \, dt$.

To see that the topology on $L_0(0, 1)$ coincides with that induced by the metric d, it suffices to show that the same sequences converge in each topology (since both spaces are sequential spaces). Suppose the sequence $(f_n)_{n=1}^\infty$ of measurable functions converges in the metric d to a measurable function f. Then $d(f, f_n) \to 0$ as $n \to \infty$. Therefore, $\frac{|f - f_n|}{1 + |f - f_n|} \to 0$ in the L_1-norm, and hence in measure. It follows that $f_n \to f$ in measure, as required.

The reverse implication, that convergence in measure implies convergence in d, is true by the Lebesgue Dominated Convergence Theorem. (We state the Lebesgue Dominated Convergence Theorem in Theorem A.17 for almost everywhere convergence, but it remains valid for sequences that converge in measure on a σ-finite measure space.)

It remains to show that the metric d is complete. Let

$$\|f\|_0 = \int_0^1 \frac{|f(t)|}{1 + |f(t)|} \, dt, \quad f \in L_0(0, 1).$$

Observe that $d(f, g) = \|f - g\|_0$ for all measurable functions f and g on $[0, 1]$. Certainly, $\|\cdot\|_0$ is not a norm (it is not homogeneous), but it does satisfy the triangle inequality, because $\|f\|_0 = \int_0^1 \phi(|f(t)|)\, dt$ and ϕ is subadditive on $[0, \infty)$. Consequently, we may use Lemma 2.24 (the Cauchy Summability Criterion) to prove that d is a complete metric space (because the proof of Lemma 2.24 does not require homogeneity of the norm).

Suppose $(f_n)_{n=1}^\infty$ is a sequence of measurable functions such that $\sum_{n=1}^\infty \|f_n\|_0 < \infty$. Then, by Fubini's Theorem,

$$\sum_{n=1}^\infty \|f_n\|_0 = \sum_{n=1}^\infty \int_0^1 \frac{|f_n(t)|}{1 + |f_n(t)|}\, dt = \int_0^1 \left(\sum_{n=1}^\infty \frac{|f_n(t)|}{1 + |f_n(t)|} \right) dt < \infty.$$

It follows that, for almost every $t \in [0, 1]$, there exists some $M_t > 0$ such that $\sum_{n=1}^\infty \frac{|f_n(t)|}{1+|f_n(t)|} \le M_t$. Consequently, by the subadditivity of ϕ, for every $N \in \mathbb{N}$,

$$\phi\left(\sum_{n=1}^N |f_n(t)| \right) \le \sum_{n=1}^N \phi(|f(t)|) \le M_t < \infty \quad \text{a.e.}(t).$$

Since ϕ is a strictly increasing function on the interval $[0, \infty)$, we conclude that, for almost every t, the sequence $(\sum_{n=1}^N |f_n(t)|)_{N=1}^\infty$ converges. Therefore, $(\sum_{n=1}^N f_n)_{N=1}^\infty$ converges almost everywhere, and hence in measure. Therefore, $L_0(0, 1)$ is a complete metric space.

As was the case in Example A (where $0 < p < 1$), the dual space of $L_0(0, 1)$ is trivial; that is, $L_0(0, 1)^* = \{0\}$. We leave the verification of this fact as an exercise. (See Exercise 5.14.)

Example C: $\omega = \mathbb{R}^{\mathbb{N}}$

Let J be a (possibly uncountable) index set. Let $\mathbb{R}^J$ denote the product space $\prod_{j \in J} \mathbb{R}_j$, where $\mathbb{R}_j = \mathbb{R}$ for each $j \in J$. An element x in $\mathbb{R}^J$ is a function $x : J \to \mathbb{R}$, where $x(j) \in \mathbb{R}(= \mathbb{R}_j)$ for each $j \in J$.

When the space $\mathbb{R}^J$ is equipped with the product topology, it becomes a topological vector space. The vector space operations are done pointwise; that is, if x and y are elements in $\mathbb{R}^J$, then $(x + y)(j) = x(j) + y(j)$ for each $j \in J$. Convergence, too, is pointwise: $x_n \to x$ in $\mathbb{R}^J$ as $n \to \infty$ if $x_n(j) \to x(j)$ in $\mathbb{R}_j$ as $n \to \infty$ for each $j \in J$.

If J is an *uncountable* index set, then $\mathbb{R}^J$ is not metrizable, since $\mathbb{R}^J$ with the product topology is not first countable; i.e., it does not have a countable local base at 0. (See Example 5.4.)

In this example, we are interested in *countable* index sets, and so we let $J = \mathbb{N}$. We denote $\mathbb{R}^{\mathbb{N}}$ by the Greek letter ω. Generally, we think of ω as the collection of all sequences in $\mathbb{R}$. If $\xi \in \omega$, we let $\xi_k = \xi(k)$ for each $k \in \mathbb{N}$, and we write $\xi = (\xi_k)_{k=1}^\infty$. In this context, the vector space operations are done coordinate-wise. Convergence is also now viewed coordinate-wise, so that $\xi^{(n)} \to \xi$ in ω as $n \to \infty$ if $\xi_k^{(n)} \to \xi_k$ in $\mathbb{R}$ as $n \to \infty$ for each $k \in \mathbb{N}$.

Unlike $\mathbb{R}^J$ when J is uncountable, the space ω is first countable. A base of neighborhoods at the origin is formed by sets of the type

$$(-\epsilon_1, \epsilon_1) \times \cdots \times (-\epsilon_n, \epsilon_n) \times \mathbb{R} \times \mathbb{R} \times \cdots, \qquad (5.1)$$

where $n \in \mathbb{N}$ and $\epsilon_i > 0$ for each $i \in \{1, \dots, n\}$. If we denote elements of ω by $\xi = (\xi_k)_{k=1}^\infty$, then the set in (5.3.1) can be written

$$\{\xi : |\xi_1| < \epsilon_1, \cdots, |\xi_n| < \epsilon_n\}.$$

To identify a countable base, consider the sets with $\epsilon_i = \frac{1}{k}$, where $k \in \mathbb{N}$, for all $i \in \{1, \dots, n\}$ and $n \in \mathbb{N}$.

Not only is ω first countable, but it is also metrizable. Recall that ω was defined to be $\prod_{k=1}^\infty \mathbb{R}_k$. Denote the metric on $\mathbb{R}_k$ by d_k. We define a metric d on ω by

$$d(\xi, \eta) = \sum_{k=1}^\infty \frac{1}{2^k} \frac{d_k(\xi_k, \eta_k)}{1 + d_k(\xi_k, \eta_k)},$$

where $\xi = (\xi_k)_{k=1}^\infty$ and $\eta = (\eta_k)_{k=1}^\infty$.

We now wish to identify the space dual to ω. To that end, we prove the following proposition.

Proposition 5.13 *Let X be a topological vector space and let $\mathbb{K}$ be the field of scalars. A linear functional $f : X \to \mathbb{K}$ is continuous if and only if there exists a neighborhood V of 0 such that the set $f(V)$ is bounded in $\mathbb{K}$.*

Proof Let $U_\mathbb{K}$ be the open unit ball in $\mathbb{K}$. If f is continuous, then $f^{-1}(U_\mathbb{K})$ is an open neighborhood of 0, and $f(f^{-1}(U_\mathbb{K})) \subseteq B_\mathbb{K}$ is bounded in $\mathbb{K}$.

Now suppose V is a neighborhood of 0 such that $f(V)$ is bounded in $\mathbb{K}$. By definition, there is some $M > 0$ such that $f(V) \subseteq M U_\mathbb{K}$. Let $\epsilon > 0$ be given. Then $f(\frac{\epsilon}{M} V) \subseteq \epsilon U_\mathbb{K}$. Therefore, $|f(x)| < \epsilon$ whenever $x \in \frac{\epsilon}{M} V$. In other words, f is continuous at zero. Continuity then follows from the linearity of f. $\qquad\square$

We can now use the preceding proposition to identify the continuous linear functionals on ω. Let $f \in \omega^*$. By Proposition 5.13, there must be some neighborhood V of 0 such that $f(V)$ is bounded in $\mathbb{R}$. Without loss of generality, we may assume V is a basic set, say $V = \{\xi : |\xi_1| < \epsilon_1, \cdots, |\xi_n| < \epsilon_n\}$ for some $n \in \mathbb{N}$.

The set $f(V)$ is bounded, and so there exists some $M > 0$ such that $|f(\xi)| \leq M$ for any $\xi \in V$. Let $\xi = (0, \dots, 0, \xi_{n+1}, \dots)$. Then $\xi \in V$, and so too is any constant multiple of ξ. Therefore, for any $K > 0$, we have that $|f(K\xi)| \leq M$, and hence $|f(\xi)| \leq M/K$, by the linearity of f. Since this inequality holds for all $K > 0$, it must be that $f(\xi) = 0$. This is true for any $\xi \in V$ having $\xi_i = 0$ for all $i \in \{1, \dots, n\}$. Thus, because f is linear, it follows that $f(\xi) = f(\xi')$ for any ξ and ξ' that agree on the first n coordinates.

Define a function $g : \mathbb{R}^n \to \mathbb{R}$ by $g(\xi_1, \dots, \xi_n) = f(\xi_1, \dots, \xi_n, 0, \dots)$. Since f is linear and continuous, it follows that $g \in (\mathbb{R}^n)^*$. Consequently, there exists $\alpha_k \in \mathbb{R}$

for each $k \in \{1, \dots, n\}$ such that

$$g(\xi_1, \dots, \xi_n) = \sum_{k=1}^{n} \alpha_k \xi_k, \quad (\xi_k)_{k=1}^{n} \in \mathbb{R}^n.$$

Since the value of $f(\xi)$ depends only on the first n coordinates of ξ, we conclude that

$$f(\xi) = \sum_{k=1}^{n} \alpha_k \xi_k, \quad \xi = (\xi_k)_{k=1}^{\infty} \in \omega.$$

5.4   The Geometric Hahn–Banach Theorem

In this section, we will meet the Hahn–Banach Theorem without the advantages of a norm structure. The key property a space must have, we shall see, is *local convexity*.

Definition 5.14   Let X be a real or complex vector space. A subset V of X is called *convex* if given any x and y in V, we have $(1 - t)x + ty \in V$ for all $t \in [0, 1]$. That is, if two points are in V, then the line segment joining them is also in V. A balanced convex set is called *absolutely convex*.

Lemma 5.15   *Let X be a real or complex vector space. A subset V of X is absolutely convex if and only if $\alpha x + \beta y \in V$ whenever x and y are in V and α and β are scalars such that $|\alpha| + |\beta| \le 1$.*

Proof   We first observe that V is absolutely convex if the latter condition holds: to show balance, take $\beta = 0$; to show convexity, let $\alpha = t - 1$ and $\beta = t$.

Now suppose V is absolutely convex. Let x and y be in V and suppose α and β are scalars such that $|\alpha| + |\beta| \le 1$. We wish to show $\alpha x + \beta y \in V$. Observe that

$$\alpha x + \beta y = \frac{\alpha}{\alpha + \beta}(\alpha + \beta)x + \frac{\beta}{\alpha + \beta}(\alpha + \beta)y.$$

Since V is balanced, $x' = (\alpha + \beta)x$ and $y' = (\alpha + \beta)y$ are both elements of V. Thus, by convexity,

$$\alpha x + \beta y = \frac{\alpha}{\alpha + \beta}x' + \frac{\beta}{\alpha + \beta}y' \in V.$$

This completes the proof.   □

Definition 5.16   A topological vector space is *locally convex* if there is a base of neighborhoods of 0 consisting of convex sets.

By Proposition 5.9, we can always take the elements of a base in a locally convex topological vector space to be balanced, and hence absolutely convex.

Example 5.17   Any normed space is locally convex. It is easy to see that balls with center at the origin are convex.

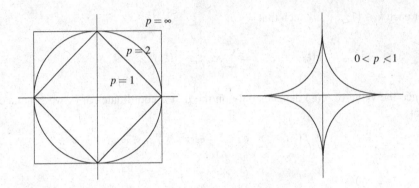

Fig. 5.1 Closed unit balls in ℓ_p^2 for various values of p

Example 5.18 Consider the space ℓ_p^2 of ordered pairs in the $\|\cdot\|_p$ norm for $p > 0$. (We use the term "norm" here even though it is not a norm when $0 < p < 1$.) If $p \geq 1$, then the unit ball is convex and balanced; however, if $0 < p < 1$, then the unit ball is balanced, but not convex. (See Fig. 5.1.)

Example 5.19 Let $X = L_p(0, 1)$ for $0 < p < 1$. (See Example A in Sect. 5.3.) We claim that the only nonempty open convex subset of X is X. To show this, let V be a nonempty open convex subset of X. Without loss of generality, assume $0 \in V$. Then there exists some $\delta > 0$ such that $\delta B \subseteq V$, where $B = \{f : \|f\|_p < 1\}$. (We remind the reader that $\|\cdot\|_p$ is not a norm in this case.)

Choose any $f \in X$. Because $p < 1$, there is some $n \in \mathbb{N}$ such that $n^{p-1}\|f\|_p^p < \delta$. Pick real numbers $\{t_0, t_1, \ldots, t_n\}$ so that $0 = t_0 < t_1 < \cdots < t_n = 1$ and such that

$$\int_{t_{k-1}}^{t_k} |f(s)|^p \, ds = \frac{1}{n}\|f\|_p^p, \quad k \in \{1, \ldots, n\}.$$

For each $k \in \{1, \ldots, n\}$, let $g_k = n f \chi_{(t_{k-1}, t_k]}$. Then $\|g_k\|_p^p = n^{p-1}\|f\|_p^p < \delta$, and therefore $g_k \in \delta B \subseteq V$. This is true for each $k \in \{1, \ldots, n\}$, and so $\{g_1, \ldots, g_n\} \subseteq V$. Observe that $f = \frac{1}{n}(g_1 + \cdots + g_n)$. Since V is convex, it follows that $f \in V$. The choice of $f \in X$ was arbitrary, and so $V = X$, as required.

Note that this argument implies that $L_p(0, 1)^* = \{0\}$, a fact we first observed in Example A in Sect. 5.3.

The next theorem is a geometric version of the Hahn–Banach Theorem. This version of the theorem is not set in the context of a complete normed space, but in that of a locally convex topological vector space.

Theorem 5.20 (Hahn–Banach Separation Theorem) *Let E be a real locally convex topological vector space. Let K be a closed nonempty convex subset of E. If $x_0 \notin K$, then there exists a continuous linear functional f on E such that*

$$f(x_0) > \sup_{y \in K} f(y).$$

Proof Without loss of generality, we may assume $0 \in K$. (If not, use a translation.) Since K is closed and $x_0 \notin K$, there exists some open neighborhood N of x_0 such that $N \cap K = \emptyset$. It follows that there exists an absolutely convex open neighborhood W of 0 such that $(x_0 + W) \cap K = \emptyset$. This implies that $x_0 \notin K + W$, for otherwise there would exist some $k \in K$ and $w \in W$ such that $x_0 - w = k$, contradicting the fact that the intersection of $x_0 + W$ with K is empty. (Here we use the fact that $W = -W$, because W is balanced.)

Let $V = K + \frac{1}{2}W$. Then V is a convex neighborhood of 0. Define a function $p : E \to \mathbb{R}$ by

$$p(x) = \inf\{\lambda > 0 : x \in \lambda V\}, \quad x \in E.$$

Recall that every neighborhood of 0 is absorbent. In particular V is absorbent, and so $p(x) < \infty$ for all $x \in E$. We claim p is sublinear. For any $x \in E$ and $\alpha \geq 0$,

$$p(\alpha x) = \inf\{\lambda > 0 : \alpha x \in \lambda V\} = \alpha \inf\left\{\frac{\lambda}{\alpha} > 0 : x \in \frac{\lambda}{\alpha}V\right\} = \alpha p(x).$$

This proves positive homogeneity. It remains to show that p is subadditive.

Let x and y be in E and let $\epsilon > 0$. Because $p(x)$ and $p(y)$ are infima, there exist real numbers $\lambda > 0$ and $\mu > 0$ such that $p(x) < \lambda < p(x) + \frac{\epsilon}{2}$ and $p(y) < \mu < p(y) + \frac{\epsilon}{2}$. By the definition of p, we have that $\frac{x}{\lambda} \in V$ and $\frac{y}{\mu} \in V$. By the convexity of V,

$$\frac{x+y}{\lambda+\mu} = \frac{\lambda}{\lambda+\mu}\left(\frac{x}{\lambda}\right) + \frac{\mu}{\lambda+\mu}\left(\frac{y}{\mu}\right) \in V.$$

Therefore,

$$p(x + y) \leq \lambda + \mu < p(x) + p(y) + \epsilon.$$

The choice of ϵ was arbitrary, and so $p(x+y) \leq p(x) + p(y)$, as required. Therefore p is sublinear.

By definition, $p(x) \leq 1$ for all $x \in V$. We now show $p(x_0) > 1$. Suppose to the contrary that $p(x_0) \leq 1$. It follows that $\frac{x_0}{\lambda} \in V$ for all $\lambda \geq 1$. Since $\frac{x_0}{\lambda} \to x_0$ as $\lambda \to 1$, we conclude that $x_0 \in \overline{V}$, and consequently $(x_0 + \frac{1}{2}W) \cap V \neq \emptyset$.

Recall that $V = K + \frac{1}{2}W$. Thus, $(x_0 + \frac{1}{2}W) \cap (K + \frac{1}{2}W) \neq \emptyset$, and so there exists an element $k \in K$ and elements w_1 and w_2 in W such that $x_0 + \frac{1}{2}w_1 = k + \frac{1}{2}w_2$. Hence,

$$x_0 = k + \frac{1}{2}w_2 - \frac{1}{2}w_1 \in K + \frac{1}{2}W - \frac{1}{2}W.$$

Because W is absolutely convex, we have that $\frac{1}{2}W - \frac{1}{2}W \subseteq W$. From this we conclude that $x_0 \in K + W$. This is a contradiction, and so it must be that $p(x_0) > 1$.

We now make use of Exercise 3.9. There exists a linear functional f on E such that $f \leq p$ and $f(x_0) > 1$. Because $K \subseteq V$, and because $p(x) \leq 1$ for all $x \in V$, we have

$$\sup_{y \in K} f(y) \leq 1 < f(x_0).$$

It remains to show that f is continuous. We will demonstrate this by showing that f is bounded on some neighborhood of zero and applying Proposition 5.13. Since

$0 \in K$, we have that $\frac{1}{2}W \subseteq K + \frac{1}{2}W = V$. By construction, $f(x) \leq p(x) \leq 1$ for all $x \in V$, and hence $f(x) \leq 1$ for all $x \in \frac{1}{2}W$. The set W is balanced, and thus $|f(x)| \leq 1$ for all $x \in \frac{1}{2}W$. Therefore, we have demonstrated that $f(\frac{1}{2}W) \subseteq [-1, 1]$. Consequently, the linear functional f is continuous, by Proposition 5.13. $\qquad\square$

Example 5.21 Suppose E is a real locally convex topological vector space and K is a closed linear subspace of E. If $x_0 \notin K$, then, by Theorem 5.20, there exists a continuous linear functional f on E such that $f(K) = 0$ and $f(x_0) > 0$. (See Exercise 5.20.)

There is also a version of Theorem 5.20 for complex topological vector spaces.

Theorem 5.22 *Let E be a complex locally convex topological vector space. Let K be a closed nonempty convex subset of E. If $x_0 \notin K$, then there exists a continuous linear functional f on E such that*

$$\Re(f(x_0)) > \sup_{x \in K} \Re(f(x)).$$

Proof Ignoring multiplication by complex scalars, we may treat E as a vector space over $\mathbb{R}$. Therefore, by Theorem 5.20, there exists a real linear functional g on E such that $g(x_0) > \sup_{x \in K} g(x)$. Now, define a complex linear functional on E by $f(x) = g(x) - ig(ix)$ for all $x \in E$. The functional f is the desired continuous linear functional on E. $\qquad\square$

Definition 5.23 Let X be a vector space and let $\mathbb{K}$ denote the scalar field. A function $p : X \to \mathbb{R}$ is called a *semi-norm* if the following three conditions are satisfied:

  (i) $p(x) \geq 0$ for all $x \in X$,
 (ii) $p(x + y) \leq p(x) + p(y)$ for all $\{x, y\} \subseteq X$, and
(iii) $p(\alpha x) = |\alpha|\, p(x)$ for all $\alpha \in \mathbb{K}$ and $x \in X$.

What distinguishes a semi-norm from a norm is that a semi-norm p may satisfy $p(x) = 0$ even when $x \neq 0$. As in the case of a norm, we call the property in *(ii)* *subadditivity* (or the *triangle inequality*) and we call the property in *(iii)* *homogeneity*.

Theorem 5.24 *Suppose $\{p_\alpha\}_{\alpha \in A}$ is a family of semi-norms on a vector space X.*
    Let
$$V(\alpha, n) = \{x : p_\alpha(x) < 1/n\}, \quad \alpha \in A, \, n \in \mathbb{N}.$$

If η is the collection of all finite intersections of the sets $V(\alpha, n)$, where $\alpha \in A$ and $n \in \mathbb{N}$, then η determines a locally convex vector topology on X in which the elements of η form an absolutely convex base of neighborhoods at 0.

Proof We define a topology on X by declaring a set $E \subseteq X$ to be open if and only if E is a (possibly empty) union of translates of elements in η. This defines a topology for which all members of η are absolutely convex (that is, convex and balanced).

It remains to show that addition and scalar multiplication are continuous. Let U be an open neighborhood of 0 in X. Without loss of generality, we may assume U is an element of η. Thus,

$$U = V(\alpha_1, n_1) \cap \cdots \cap V(\alpha_k, n_k) \tag{5.2}$$

for $\{\alpha_1, \ldots, \alpha_k\} \subseteq A$ and $\{n_1, \ldots, n_k\} \subseteq \mathbb{N}$. If $V = V(\alpha_1, 2n_1) \cap \cdots \cap V(\alpha_k, 2n_k)$, then $V + V \subseteq U$ (because p_α is subadditive for every $\alpha \in A$). Therefore, addition is continuous.

Now, let $x \in X$ and $\kappa \in \mathbb{K}$, where $\mathbb{K}$ is the scalar field. A basic open neighborhood of κx can be written as $\kappa x + U$, where U is written as in (5.2). We will show there exists an open neighborhood W of x and a $\delta > 0$ such that $\lambda W \subseteq \kappa x + U$ for all $|\kappa - \lambda| < \delta$.

Let $V = V(\alpha_1, 2n_1) \cap \cdots \cap V(\alpha_k, 2n_k)$, as above. Since V is an open neighborhood of 0, it is absorbent. Thus, there exists some $n \in \mathbb{N}$ such that $x \in nV$. Let

$$\delta = \frac{1}{n} \quad \text{and} \quad W = x + \frac{n}{1 + |\kappa|n} V.$$

Suppose $w \in W$ and $\lambda \in B(\kappa, \delta)$. Then

$$\kappa x - \lambda w = (\kappa - \lambda)x + \lambda(x - w).$$

Observe that $x = nv_1$ and $w - x = \frac{n}{1+|\kappa|n}v_2$ for some choice of v_1 and v_2 in V. Hence,

$$\kappa x - \lambda w = (\kappa - \lambda)nv_1 - \frac{\lambda n}{1 + |\kappa|n}v_2.$$

Therefore, because V is balanced,

$$\kappa x - \lambda w \in |\kappa - \lambda| nV + \frac{|\lambda|n}{1 + |\kappa|n} V \subseteq V + V \subseteq U.$$

It follows that scalar multiplication is continuous, and so the proof is complete. $\square$

Definition 5.25 Suppose X is a topological vector space and let U be an absorbent subset of X. The *Minkowski functional* of U on X is the function $p_U : X \to \mathbb{R}$ defined by

$$p_U(x) = \inf\{\lambda > 0 : x \in \lambda U\}, \quad x \in X.$$

Note that $p_U(x) < \infty$ for all $x \in X$, because U is absorbent.

Suppose that X is a locally convex topological vector space. Then X has a base of neighborhoods of 0 that are absolutely convex. Such sets are absorbent, and so each such set will give rise to a well-defined Minkowski functional.

Proposition 5.26 *Let X be a topological vector space and let U be an absorbent absolutely convex subset of X. The Minkowski functional p_U is a semi-norm on X.*

Proof Certainly $p_U(x) \geq 0$ for each $x \in X$, by the definition of p_U.

To show the subadditivity of p_U, we will use the convexity of U. Let x and y be elements in X and let $\epsilon > 0$. By the definition of p_U, there exist numbers $\lambda_1 > 0$

and $\lambda_2 > 0$ such that $p_U(x) < \lambda_1 < p_U(x) + \frac{\epsilon}{2}$ and $p_U(y) < \lambda_2 < p_U(y) + \frac{\epsilon}{2}$. It is necessarily the case that x/λ_1 and y/λ_2 are both in U. By the convexity of U,

$$\frac{x+y}{\lambda_1 + \lambda_2} = \frac{\lambda_1}{\lambda_1 + \lambda_2}\left(\frac{x}{\lambda_1}\right) + \frac{\lambda_2}{\lambda_1 + \lambda_2}\left(\frac{y}{\lambda_2}\right) \in U.$$

Therefore,

$$p_U(x+y) \le \lambda_1 + \lambda_2 < p_U(x) + p_U(y) + \epsilon.$$

The choice of ϵ was arbitrary, and so $p_U(x+y) \le p_U(x) + p_U(y)$. (Compare to the proof of Theorem 5.20.)

Finally, we show homogeneity. Let $\alpha \in \mathbb{K}$, where $\mathbb{K}$ is the field of scalars. Computing directly, we have

$$p_U(\alpha x) = \inf\{\lambda > 0 : \alpha x \in \lambda U\} = |\alpha| \inf\left\{\frac{\lambda}{|\alpha|} > 0 \ : \ x \in \frac{\lambda}{|\alpha|} \cdot \mathrm{sign}(\alpha)U\right\}.$$

Since U is balanced, $\mathrm{sign}(\alpha)U = U$. Letting $\lambda' = \lambda/|\alpha|$,

$$p_U(\alpha x) = |\alpha| \inf\left\{\lambda' > 0 : x \in \lambda' U\right\} = |\alpha|\, p_U(x).$$

Therefore, p_U is a semi-norm on X, as claimed.   □

If X is a locally convex topological vector space, then there exists a base of absolutely convex neigborhoods of 0, say η. By Proposition 5.26, the Minkowski functional p_U is a semi-norm on X for each $U \in \eta$. By Theorem 5.24, the family of semi-norms $\{p_U\}_{U\in\eta}$ generates a locally convex vector topology on X. We leave it as an exercise to show that the topology generated by $\{p_U\}_{U\in\eta}$ is, in fact, the original topology. (See Exercise 5.17.)

So far, we have considered general topological vector spaces. We now focus our attention on topological vector spaces that have a complete norm structure—that is, Banach spaces. We have already said much about the norm topology of a Banach space X. We now consider a new topology on X, the so-called *weak topology*.

Definition 5.27 Let X be a topological vector space. The *weak topology* on X (or the *w-topology*) is defined by a base of neighborhoods at 0 of the form

$$W(x_1^*, \dots, x_n^*; \epsilon) = \{x : |x_i^*(x)| < \epsilon, \ 1 \le i \le n\},$$

where $\epsilon > 0$ and $\{x_1^*, \dots, x_n^*\} \subseteq X^*$ for $n \in \mathbb{N}$.

The weak topology on X is the topology it inherits as a subspace of the space $\mathbb{K}^{X^*}$ with the product topology. The space $\mathbb{K}^{X^*}$ is the collection of all functions from X^* into the scalar field $\mathbb{K}$, and we identify X with a subspace of $\mathbb{K}^{X^*}$ by identifying $x \in X$ with $\hat{x} \in \mathbb{K}^{X^*}$ via the relationship $\hat{x}(x^*) = x^*(x)$ for all $x^* \in X^*$.

To distinguish between the norm and weak topologies on X, we will frequently denote X with the norm topology by $(X, \|\cdot\|)$ and X with the weak topology by (X, w). The weak and norm topologies are generally quite different. Any weakly open set is necessarily open in the norm topology (the basic sets are intersections of preimages of open sets under continuous maps), but not every set open in the norm topology will be weakly open. (We will demonstrate this shortly.)

The weak topology on X generally has fewer open sets, and so it is "harder" for a function on (X, w) to be continuous than a function on $(X, \|\cdot\|)$. For example, consider the identity map Id_X on X. The map $\mathrm{Id}_X : (X, \|\cdot\|) \to (X, w)$ is always continuous, but $\mathrm{Id}_X : (X, w) \to (X, \|\cdot\|)$ need not be. Indeed, if both maps are continuous, then the topologies must coincide, and then X must be finite-dimensional. (See Proposition 5.30.)

Let us consider which sequences converge in X with the weak topology. Without loss of generality, we may consider only those sequences converging to 0. If a sequence $(x_n)_{n=1}^{\infty}$ converges to 0 in the weak topology on X, we say that $(x_n)_{n=1}^{\infty}$ *converges weakly* to 0 (or $x_n \to 0$ *weakly*). The sequence $(x_n)_{n=1}^{\infty}$ converges weakly to 0 precisely when it converges coordinate-wise to 0 in $\mathbb{K}^{X^*}$. That is to say, $x_n \to 0$ weakly if and only if

$$\lim_{n \to \infty} x^*(x_n) = 0, \quad \text{for all } x^* \in X^*.$$

In other words, x_n converges to 0 in the weak topology if and only if every weak neighborhood of the origin eventually contains the sequence $(x_n)_{n=1}^{\infty}$.

A sequence converges to 0 in the norm topology if and only if every "strong" neighborhood of the origin eventually contains the sequence. However, the norm topology has more open neighborhoods about 0 than the weak topology. Consequently, it is more "difficult" for a sequence to converge in the norm topology than to converge in the weak topology.

Example 5.28 Consider ℓ_p for $1 \leq p < \infty$. For each $n \in \mathbb{N}$, let e_n be the sequence with 1 in the nth coordinate, and 0 elsewhere. If m and n are elements of $\mathbb{N}$ such that $m \neq n$, then $\|e_m - e_n\|_{\ell_p} = 2^{1/p}$. Consequently, the sequence $(e_n)_{n=1}^{\infty}$ does not converge in the norm topology. On the other hand, if $x^* = (x_n^*)_{n=1}^{\infty}$ is a sequence in $(\ell_p)^* = \ell_q$, where $p > 1$ and q is the exponent conjugate to p, then

$$\lim_{n \to \infty} x^*(e_n) = x_n^* = 0.$$

Since this is true for all $x^* \in \ell_q$, we conclude that $e_n \to 0$ weakly.

The above conclusion does not remain true when $p = 1$. In this case, $q = \infty$. Let $e = (1, 1, 1, \dots)$ be the constant sequence with all terms equal to 1. This sequence is bounded, and so $e \in \ell_\infty = (\ell_1)^*$. For each $n \in \mathbb{N}$, we have that $e(e_n) = 1$, and so $e_n \not\to 0$ in the weak topology in this case.

Example 5.29 Consider the Banach space $L_p(\mathbb{T})$ of p-integrable complex-valued functions on the torus $\mathbb{T} = [0, 2\pi)$, where $1 \leq p < \infty$. For each $n \in \mathbb{N}$, define a function $f_n : \mathbb{T} \to \mathbb{C}$ by $f_n(\theta) = e^{in\theta}$, where $\theta \in \mathbb{T}$. Let $\Lambda \in L_p(\mathbb{T})^*$. By duality, there exists some $g \in L_q(\mathbb{T})$, where $1/p + 1/q = 1$, such that

$$\Lambda(f) = \int_{\mathbb{T}} f(\theta) g(\theta) \frac{d\theta}{2\pi}, \quad f \in L_p(\mathbb{T}).$$

Therefore,

$$\lim_{n \to \infty} \Lambda(f_n) = \lim_{n \to \infty} \left(\int_{\mathbb{T}} e^{in\theta} g(\theta) \frac{d\theta}{2\pi} \right) = \lim_{n \to \infty} \hat{g}(-n) = 0.$$

This last equality follows from the Riemann–Lebesgue Lemma (Theorem 4.37). Therefore, $\lim_{n \to \infty} \Lambda(f_n) = 0$ for all $\Lambda \in L_q(\mathbb{T})$, and so $f_n \to 0$ weakly. However, $\|f_n\|_{L_p(\mathbb{T})} = 1$ for all $n \in \mathbb{N}$, and so $f_n \not\to 0$ in the norm topology.

If X is a finite-dimensional Banach space, then all linear functionals are continuous.

Proposition 5.30 *Let X be a Banach space. The following are equivalent:*

(i) *$\dim(X) < \infty$,*
(ii) *the weak topology on X coincides with the norm topology on X, and*
(iii) *the weak topology on X is metrizable.*

Proof The implications *(i)* $\Rightarrow$ *(ii)* $\Rightarrow$ *(iii)* are clear. It remains to show *(iii)* $\Rightarrow$ *(i)*. Assume the weak topology on X is metrizable.

Then (X, w) is first countable, and so there exists a weak base of neighborhoods $(W_n)_{n=1}^{\infty}$ at the origin of the form

$$W_n = \{x : |x_{n,j}^*(x)| \leq \epsilon_n, \ 1 \leq j \leq N_n\},$$

where $x_{n,j}^* \in X^*$, $\epsilon_n > 0$, and $N_n \in \mathbb{N}$, for all $n \in \mathbb{N}$ and all $j \in \{1, \ldots, N_n\}$.

For each $n \in \mathbb{N}$, define

$$E_n = \text{span}\{x_{n,j}^* : 1 \leq j \leq N_n\}.$$

Fix some $x^* \in X^*$. The set $\{x : |x^*(x)| \leq 1\}$ is a weak neighborhood of 0 in X, and consequently must contain W_n for some $n \in \mathbb{N}$. For this fixed n, define a linear map $T : X \to \mathbb{K}^{N_n}$, where $\mathbb{K}$ is the scalar field, by

$$T(x) = (x_{n,1}^*(x), \ldots, x_{n,N_n}^*(x)), \quad x \in X.$$

We claim $x^* \in (\ker T)^{\perp}$. (Recall Definition 3.50.) To verify this, suppose $y \in \ker(T)$. By the definition of T, we have that $x_{n,j}^*(y) = 0$ for all $j \in \{1, \ldots, N_n\}$. Naturally, if $\lambda \in \mathbb{K}$, then it follows that $x_{n,j}^*(\lambda y) = 0$ for all $j \in \{1, \ldots, N_n\}$. Consequently, $\lambda y \in W_n$ for all $\lambda \in \mathbb{K}$. By design, $W_n \subseteq \{x : |x^*(x)| \leq 1\}$, and so $|x^*(\lambda y)| \leq 1$ for all $\lambda \in \mathbb{K}$. This can occur only if $|x^*(y)| \leq 1/\lambda$ for all $\lambda \in \mathbb{K}$, and thus $x^*(y) = 0$. This remains true for any $y \in \ker(T)$, and so we have that $x^* \in (\ker T)^{\perp}$.

By Lemma 4.33, there then exists some $f \in (\mathbb{K}^{N_n})^*$ such that $x^*(x) = (f \circ T)(x)$ for all $x \in X$. Since $\mathbb{K}^{N_n}$ is finite-dimensional, there exists a finite sequence $(a_j)_{j=1}^{N_n}$

such that

$$f(\xi_1, \ldots, \xi_{N_n}) = \sum_{j=1}^{N_n} a_j \xi_j, \quad (\xi_j)_{j=1}^{N_n} \in \mathbb{K}^{N_n}.$$

Therefore,

$$x^*(x) = f(x_{n,1}^*(x), \ldots, x_{n,N_n}^*(x)) = \sum_{j=1}^{N_n} a_j x_{n,j}^*(x), \quad x \in X,$$

and so $x^* \in E_n$.

We have shown that each $x^* \in X^*$ is in E_n for some $n \in \mathbb{N}$. We therefore conclude that $X^* = \bigcup_{n=1}^{\infty} E_n$. For each $n \in \mathbb{N}$, the space E_n is finite-dimensional, and so is closed. Therefore, by Theorem 4.7 (the complementary version of the Baire Category Theorem), there exists some $n \in \mathbb{N}$ such that $\text{int}(E_n) \neq \emptyset$. We conclude that E_n is an open neighborhood of the origin in X^*, and consequently is absorbent. Therefore, $X^* = \bigcup_{k=1}^{\infty} k E_n = E_n$. Thus, the space X^* is finite-dimensional, and so X is finite-dimensional, as well.   □

Proposition 5.31 *Let X be a Banach space. Then:*

(i) *The weak topology on X is a Hausdorff topology.*

(ii) *A linear functional is continuous in the weak topology if and only if it is continuous in the norm topology.*

Proof (i) Assume x_1 and x_2 are elements in X such that $x_1 \neq x_2$. By the Hahn–Banach Separation Theorem (Theorem 5.20), there exists an $x^* \in X^*$ such that $\epsilon = x^*(x_2 - x_1) > 0$. Therefore, the set $\{x : |x^*(x) - x^*(x_1)| < \epsilon/2\}$ is a weak neighborhood of x_1, the set $\{x : |x^*(x) - x^*(x_2)| < \epsilon/2\}$ is a weak neighborhood of x_2, and these two neighborhoods are disjoint. Hence, (X, w) is a Hausdorff topological space.

(ii) If a linear functional f is continuous in the weak topology on X, then $f^{-1}(V)$ is a weakly open set whenever V is an open set in the scalar field. But the norm topology contains all of the weakly open sets, so $f^{-1}(V)$ is open in the norm topology. Therefore, f is continuous in the norm topology on X. (The idea is that it is "easier" to be continuous in the norm topology, because there are more open sets.)

Now, suppose f is a norm continuous linear functional. Then $f \in X^*$, and so the set $\{x : |f(x)| < \epsilon\}$ is a weak neighborhood of 0 (by the definition of the weak topology). Thus, f is continuous in the weak topology on X.   □

The weak topology on X is the *weakest* topology on X such that all norm continuous linear functionals remain continuous. When we say a topology is *weaker*, we mean that it contains fewer open sets. The norm topology on X is *stronger* than the weak topology on X, because it contains more open sets. Weakly open sets are open in the norm topology, but the converse need not be true. A function is continuous if the preimage of any open set is open. The stronger the topology on the domain, the easier it is for a function to be continuous, because with more open sets, it is more likely that a given preimage is open.

Definition 5.32 Let X be a topological vector space. The *weak* * *topology* on X^* (or the *w***-topology*) is defined by a base of neighborhoods at 0 of the form

$$W^*(x_1,\dots,x_n;\epsilon) = \{x^* : |x^*(x_i)| < \epsilon, \ 1 \le i \le n\},$$

where $\epsilon > 0$ and $\{x_1,\dots,x_n\} \subseteq X$ for $n \in \mathbb{N}$.

The weak* topology on X^* is the topology inherited from viewing X^* as a subspace of $\mathbb{K}^X$, the space of all scalar-valued functions on X. As before, we endow $\mathbb{K}^X$ with the product topology. We use (X^*, w^*) to denote X^* with the weak* topology.

Observe that any $x \in X$ can be thought of as a linear functional on X^* via the mapping $x \mapsto \phi_x$, where $\phi_x(x^*) = x^*(x)$ for all $x^* \in X^*$. The weak* topology on X^* is the weakest topology on X^* for which the linear functionals ϕ_x are continuous for all $x \in X$.

The Banach space X^* has also a weak topology that is induced by it's dual space $(X^*)^* = X^{**}$ (the *bidual* of X). The weak* topology on X^* is weaker than the weak topology on X^*, because it requires fewer members in X^{**} to be continuous. (Only those coming from X.)

Example 5.33 Consider the sequence space ℓ_1. In Example 5.28, we saw that ℓ_1 had a weak topology induced upon it by $\ell_1^* = \ell_\infty$. In this weak topology, we saw that the sequence $(e_n)_{n=1}^\infty$ did not converge to 0 (because $e(e_n) = 1$ for all $n \in \mathbb{N}$, where $e = (1, 1, \dots)$ is the constant sequence with all terms equal to 1). The space ℓ_1 can also be given a weak* topology as the dual space of c_0.

Suppose $\xi = (\xi_k)_{k=1}^\infty$ is an element of c_0. Since c_0 consists of sequences that converge to 0, it follows that $e_n(\xi) = \xi_n \to 0$ as $n \to \infty$. This is true for every $\xi \in c_0$, and so the sequence $(e_n)_{n=1}^\infty$ converges to 0 in the weak* topology on ℓ_1.

In this example we have found a sequence which converges in the weak* topology on ℓ_1, but not in the weak topology on ℓ_1. This happens because the weak* topology has fewer open sets than the weak topology. (That is to say, the weak* topology is weaker than the weak topology).

Proposition 5.34 *Let X be a Banach space. Then:*

(i) *The weak* topology on X^* is a Hausdorff topology.*

(ii) *A linear functional f on X^* is weak*-continuous if and only if there exists some $x \in X$ such that $f(x^*) = x^*(x)$ for all $x^* \in X^*$. (In other words, $(X^*, w^*)^* = X$.)*

Proof (i) Let x_1^* and x_2^* be elements in X^* such that $x_1^* \ne x_2^*$. Then there exists some $x \in X$ such that $x^*(x) \ne x_2^*(x)$. (Otherwise they would be the same as linear functionals on X.) If $\epsilon = |(x_1^* - x_2^*)(x)|$, then the sets $\{x^* : |x^*(x) - x_1^*(x)| < \epsilon/2\}$ and $\{x^* : |x^*(x) - x_2^*(x)| < \epsilon/2\}$ are disjoint weak*-open sets containing x_1^* and x_2^*, respectively.

(ii) Certainly, if $f(x^*) = x^*(x)$ for all $x^* \in X^*$, then f is continuous in the weak* topology. It remains only to show that any weak*-continuous linear functional on X^* can be achieved in this way.

Assume f is a continuous linear functional on (X^*, w^*). By Proposition 5.13, there exists a basic neighborhood of (X^*, w^*) on which f is bounded. Thus, there exists a real number $\epsilon > 0$ and a finite set $\{x_1, \ldots, x_n\} \subseteq X$ such that $|f(x^*)| \leq 1$ for all $x^* \in W^*(x_1, \ldots, x_n; \epsilon)$.

Define a map $T : X^* \to \mathbb{K}^n$, where $\mathbb{K}$ is the scalar field, by

$$T(x^*) = (x^*(x_1), \ldots, x^*(x_n)), \quad x^* \in X^*.$$

Suppose $x^* \in \ker(T)$. Then $x^*(x_j) = 0$ for each $j \in \{1, \ldots, n\}$. Thus, for any $\lambda \in \mathbb{K}$, we have that $x^*(\lambda x_j) = 0$ for $j \in \{1, \ldots, n\}$, and so $(\lambda x^*) \in W^*(x_1, \ldots, x_n; \epsilon)$. It follows that $|f(\lambda x^*)| \leq 1$, and consequently $|f(x^*)| \leq 1/|\lambda|$ for all $\lambda \neq 0$. From this we conclude that $f(x^*) = 0$, and hence $f \in (\ker T)^\perp$. By Lemma 4.33, then, there exists a bounded linear functional $\phi : \mathbb{K}^n \to \mathbb{K}$ such that $f = \phi \circ T$. Therefore, there exists a finite collection of scalars $\{a_1, \ldots, a_n\} \subseteq \mathbb{K}$ such that

$$f(x^*) = \phi(Tx) = \phi(x^*(x_1), \ldots, x^*(x_n)) = \sum_{j=1}^{n} a_j x^*(x_j), \quad x^* \in X^*.$$

The desired element of X is $x = \sum_{j=1}^{n} a_j x_j$. $\qquad\square$

Remark 5.3 In Example 5.33 we saw that the weak* topology may be strictly weaker than the weak topology. If X is a *reflexive* space (recall Definition 3.33), however, then the weak and weak* topologies coincide.

Shortly, we will prove Proposition 5.37 which (in some sense) demonstrates that it is "hard" to be compact in a normed space. Before we state and prove this proposition, however, we need a lemma, which is of independent interest.

Lemma 5.36 *All norms on a finite-dimensional vector space are equivalent.*

Proof Let X be a finite-dimensional vector space over the scalar field $\mathbb{K}$. Choose $x_1, \ldots, x_n$ in X so that $X = \mathrm{span}\{x_1, \ldots, x_n\}$. We recall that each element of X has a unique representation of the form $\sum_{i=1}^{n} \alpha_i x_i$, where $\alpha_i \in \mathbb{K}$ for each $i \in \{1, \ldots, n\}$. Define a norm $|||\cdot|||$ on X as follows:

$$\left\lVert\left\lVert\left\lVert \sum_{i=1}^{n} \alpha_i x_i \right\rVert\right\rVert\right\rVert = \sum_{i=1}^{n} |\alpha_i|.$$

It is straightforward to show that this does indeed define a norm on X.

Now, let $\| \cdot \|$ be another norm on X. We will find positive constants c and C such that $c|||x||| \leq \|x\| \leq C|||x|||$ for all $x \in X$. By the triangle inequality,

$$\left\lVert \sum_{i=1}^{n} \alpha_i x_i \right\rVert \leq \sum_{i=1}^{n} |\alpha_i| \cdot \|x_i\| \leq (\max_i \|x_i\|)\left(\sum_{i=1}^{n} |\alpha_i| \right) = (\max_i \|x_i\|)\left\lVert\left\lVert\left\lVert \sum_{i=1}^{n} \alpha_i x_i \right\rVert\right\rVert\right\rVert.$$

$$(5.3)$$

Thus, we may choose $C = \max_i \|x_i\|$.

Next, define a set

$$S = \left\{ (\alpha_1, \ldots, \alpha_n) : \sum_{i=1}^{n} |\alpha_i| = 1 \right\}.$$

Observe that S is a closed and bounded subset of $\mathbb{K}^n$. Therefore, S is compact by the Heine–Borel Theorem. Define a function $f : S \to \mathbb{R}^+$ by

$$f(\alpha_1, \ldots, \alpha_n) = \left\| \sum_{i=1}^{n} \alpha_i x_i \right\|.$$

We claim that the function f is continuous. To see this, observe that

$$\left| f(\alpha_1, \ldots, \alpha_n) - f(\beta_1, \ldots, \beta_n) \right| = \left| \left\| \sum_{i=1}^{n} \alpha_i x_i \right\| - \left\| \sum_{i=1}^{n} \beta_i x_i \right\| \right| \leq \left\| \sum_{i=1}^{n} \alpha_i x_i - \sum_{i=1}^{n} \beta_i x_i \right\|$$

$$= \left\| \sum_{i=1}^{n} (\alpha_i - \beta_i) x_i \right\| \leq \sum_{i=1}^{n} |\alpha_i - \beta_i| \|x_i\| \leq \left(\sum_{i=1}^{n} |\alpha_i - \beta_i|^2 \right)^{1/2} \left(\sum_{i=1}^{n} \|x_i\|^2 \right)^{1/2}.$$

The last inequality follows from the Cauchy–Schwarz Inequality. From this, it follows that f is continuous.

By the Extreme Value Theorem, since f is continuous on a compact set, the function f attains a minimum value on the set S. Let c be that minimum value. Then $f(\alpha_1, \ldots, \alpha_n) \geq c$ for all $(\alpha_1, \ldots, \alpha_n)$ in S. This means that $\left\| \sum_{i=1}^{n} \alpha_i x_i \right\| \geq c$ for all $(\alpha_1, \ldots, \alpha_n)$ in $\mathbb{K}^n$ such that $\sum_{i=1}^{n} |\alpha_i| = 1$. Alternately, for any $(\alpha_1, \ldots, \alpha_n)$ in $\mathbb{K}^n$,

$$\left\| \sum_{i=1}^{n} \alpha_i x_i \right\| \geq c \sum_{i=1}^{n} |\alpha_i| = c \left\| \left\| \sum_{i=1}^{n} \alpha_i x_i \right\| \right\|. \tag{5.4}$$

Combining (5.3) and (5.4), we conclude that the two norms are equivalent. Since the norm $\| \cdot \|$ was arbitrary, it follows that all norms on X are equivalent. $\qquad \square$

Proposition 5.37 *Suppose X is a Banach space (or just a normed linear space). Then B_X is compact in the norm topology on X if and only if $\dim(X) < \infty$.*

Proof Suppse X is a finite-dimensional normed vector space. By Lemma 5.36, X is homeomorphic to $\mathbb{K}^n$ with the Euclidean norm. Therefore, B_X is compact by the Heine–Borel Theorem.

Next, suppose B_X is compact in the norm topology on X. Denote by $B(x, r)$ the open ball of radius r centered at $x \in X$. Since B_X is compact, there exists a finite sequence $\{x_1, \ldots, x_k\}$ of elements in X, such that

$$B_X \subseteq \bigcup_{j=1}^{k} B\left(x_j, \frac{1}{2}\right) = \bigcup_{j=1}^{k} \left(x_j + \frac{1}{2} B_X\right). \tag{5.5}$$

Let $F = \mathrm{span}\{x_1, \ldots, x_k\}$. Then (5.5) implies that $B_X \subseteq F + \frac{1}{2} B_X$. This is a recursive statement, and so we apply it to itself to get

$$B_X \subseteq F + \frac{1}{2}\left(F + \frac{1}{2} B_X\right) = F + \frac{1}{2} F + \frac{1}{4} B_X = F + \frac{1}{4} B_X.$$

Continuing recursively, we have $B_X \subseteq F + \frac{1}{2^n} B_X$ for all $n \in \mathbb{N}$. Therefore,

$$B_X \subseteq \bigcap_{n=1}^{\infty} \left(F + \frac{1}{2^n} B_X\right).$$

However, F is closed, as a consequence of Lemma 5.36 (because F is finite-dimensional). Thus,

$$\bigcap_{n=1}^{\infty} \left(F + \frac{1}{2^n} B_X\right) = F,$$

and so $B_X \subseteq F$. Since B_X is absorbent, we have $X = \bigcup_{n=1}^{\infty} n B_X \subseteq \bigcup_{n=1}^{\infty} n F$. But F is a vector space, and so $X \subseteq F$. Therefore, $X = F$, as required. $\square$

While the unit ball in a Banach space can be compact in the norm topology only if the space is finite-dimensional, the unit ball in the weak* topology will always be compact. Before proving this statement, known as the Banach-Alaoglu Theorem, let us recall a theorem from general topology.

Theorem 5.38 (Tychonoff's Theorem) *Let I be an arbitrary index set. If $\{K_i\}_{i \in I}$ is a collection of compact topological spaces, then $\prod_{i \in I} K_i$ is compact in the product topology.*

We will not prove this theorem, but we do wish to point out it relies on the Axiom of Choice. We are now ready to state and prove the Banach-Alaoglu Theorem.

Theorem 5.39   (Banach-Alaoglu Theorem) *If X is a Banach space, then B_{X^*} is compact in the weak* topology on X^*.*

Proof Let X be a Banach space over the scalar field $\mathbb{K}$. Recall that X^* in the weak* topology is achieved by viewing X^* as a subspace of $\mathbb{K}^X = \prod_{x \in X} \mathbb{K}$ in the product topology. We make this explicit by defining the map $\phi : X^* \to \mathbb{K}^X$ by

$$\phi(x^*) = (x^*(x))_{x \in X}, \quad x^* \in X^*.$$

If $x^* \in B_{X^*}$, then for each $x \in X$, we have $|x^*(x)| \leq \|x\|$. Consequently,

$$\phi(B_{X^*}) \subseteq \prod_{x \in X} \|x\| B_{\mathbb{K}},$$

where $B_{\mathbb{K}}$ is the closed unit ball in the scalar field $\mathbb{K}$ and $\|x\| B_{\mathbb{K}}$ is the closed ball of radius $\|x\|$ centered at the origin. The product $A = \prod_{x \in X} \|x\| B_{\mathbb{K}}$ is compact, by Tychonoff's Theorem. There is no reason the image of B_{X^*} would be all of A, but it is a closed subspace. Indeed, the image is precisely the collection of elements in the following set:

$$\bigcap_{\substack{\{\alpha_1, \alpha_2\} \subseteq \mathbb{R} \\ \{x_1, x_2\} \subseteq X}} \left\{ f : f(\alpha_1 x_1 + \alpha_2 x_2) = \alpha_1 f(x_1) + \alpha_2 f(x_2) \right\} \bigcap \prod_{x \in X} \|x\| B_{\mathbb{K}}.$$

(The first set of relations ensures $f \in \mathbb{K}^X$ is linear, while the second ensures it is bounded.) Therefore, $\phi(B_{X^*})$ is a closed subset of the compact set A, and hence $\phi(B_{X^*})$ is compact in the product topology on $\mathbb{K}^X$. It follows that B_{X^*} is compact in the weak* topology on X^*, as required. $\square$

The Banach–Alaoglu Theorem as given here is due to Leonidas Alaoglu [1], although the result was known to Banach. Banach did not have the notions of general topology available to him, and so he could not formulate it in this way.

5.5 Goldstine's Theorem

Let X be a Banach space. Recall that X can be thought of as a subspace of it's bidual X^{**}. The space X^{**} is the dual space for X^*, and as such can be given a weak* topology. The weak* topology on X^{**} is the weakest topology under which elements of X^* define continuous functions on X^{**}. If we restrict to the subspace X, then the weakest topology under which elements of X^* are continuous is the weak topology on X. Therefore

$$(X^{**}, w^*)|_X = (X, w).$$

In other words, the restriction of the weak* topology on X^{**} to X is the weak topology on X.

Theorem 5.40 (Goldstine's Theorem) *If X is a Banach space, then B_X is weak*-dense in $B_{X^{**}}$.*

Proof Let X be a Banach space. For simplicity, we will assume X is real. (If X is complex, the argument is similar.) Denote the closure of B_X in the weak* topology on X^{**} by $\overline{B_X}^{(w^*)}$. Our goal is to show the equality $\overline{B_X}^{(w^*)} = B_{X^{**}}$.

By the Banach–Alaoglu Theorem (Theorem 5.39), the set $B_{X^{**}}$ is a compact (and hence closed) set in the weak* topology on X^{**}. Therefore, since $X \subseteq X^{**}$, we see that $\overline{B_X}^{(w^*)} \subseteq B_{X^{**}}$.

Suppose $x_0^{**} \in B_{X^{**}} \setminus \overline{B_X}^{(w^*)}$. By the Hahn–Banach Separation Theorem (Theorem 5.20), there exists a weak*-continuous linear functional f on X^{**} such that

$$f(x_0^{**}) > \sup \{ f(u^{**}) : u^{**} \in \overline{B_X}^{(w^*)} \}. \tag{5.6}$$

By Proposition 5.34, since f is continuous in the weak* topology on X^{**}, there exists an $x^* \in X^*$ such that $f(x^{**}) = x^{**}(x^*)$ for all $x^{**} \in X^{**}$. Therefore, (5.5.1) becomes

$$x_0^{**}(x^*) > \sup\{u^{**}(x^*) : u^{**} \in \overline{B_X}^{(w^*)}\} \geq \sup\{x^*(x) : x \in B_X\} = \|x^*\|.$$

This implies $\|x_0^{**}\| > 1$, contradicting the assumption that $x_0^{**} \in B_{X^{**}}$. The result follows. □

In the proof of Goldstine's Theorem, we assumed that X was a real Banach space for the sake of simplicity. The argument is similar when the Banach space is complex, but instead of Theorem 5.20, which is the Hahn–Banach Separation Theorem for real spaces, we use Theorem 5.22, which is the Hahn–Banach Separation Theorem for complex spaces, and we replace f with $\Re(f)$.

Theorem 5.41 *A Banach space X is reflexive if and only if the closed unit ball B_X is weakly compact.*

Proof Assume first that X is reflexive. Then $B_X = B_{X^{**}}$. By the Banach–Alaoglu Theorem (Theorem 5.39), the set $B_{X^{**}}$ is compact in the weak* topology on X^{**}. Since X is reflexive, the weak* topology on X^{**} coincides with the weak topology on X. Therefore, B_X is compact in the weak topology on X.

Now, assume instead that B_X is weakly compact. The weak topology on X is the restriction of the weak* topology on X^{**}, and so B_X is compact (and hence closed) in the weak* topology on X^{**}. By Goldstine's Theorem (Theorem 5.40), we conclude that $B_X = B_{X^{**}}$, since B_X is closed and dense in $B_{X^{**}}$. Therefore, X is reflexive. □

Proposition 5.42 *Suppose X and Y are Banach spaces (or simply normed linear spaces). If $T : X \to Y$ is a linear map, then the following are equivalent:*

(i) *T is bounded (i.e., norm-to-norm continuous).*
(ii) *T is $(X, \| \cdot \|)$ to (Y, w) continuous.*
(iii) *T is (X, w) to (Y, w) continuous.*

Proof Certainly *(iii)* implies *(ii)*. We will show that *(ii)* implies *(i)*, and then *(i)* implies *(iii)*.

Assume *(ii)*. We wish to show that $T(B_X)$ is bounded in the norm topology on Y. Let $y^* \in Y^*$. Then y^* is continuous in the weak topology on Y. Consequently, since T is norm-to-weak continuous, the functional $y^* \circ T$ is continuous in the norm topology on X. Thus, $y^* \circ T \in X^*$, and so

$$\sup_{\|x\| \leq 1} |y^*(Tx)| = \sup_{\|x\| \leq 1} |(y^* \circ T)(x)| < \infty. \tag{5.7}$$

Since (5.5.2) holds for each $y^* \in Y^*$, we conclude that the set $T(B_X)$ is weakly bounded in Y. Therefore, $T(B_X)$ is bounded in the norm topology, by Theorem 4.12.

Now assume *(i)*. Consider a weak neighborhood in Y, say

$$W_Y = W_Y(y_1^*, \ldots, y_n^*; \epsilon) = \{y : |y_j^*(y)| < \epsilon, \ 1 \leq j \leq n\},$$

for $\{y_1^*, \ldots, y_n^*\} \subseteq Y^*$ and $\epsilon > 0$. Suppose $x \in X$ is such that $Tx \in W_Y$. Then for each $j \in \{1, \ldots, n\}$, we have $|y_j^*(Tx)| < \epsilon$. Recall that the adjoint operator T^* was defined so that $T^* \circ y^* = y^* \circ T$. Therefore, $|T^* y_j^*(x)| < \epsilon$ for all $j \in \{1, \ldots, n\}$, and so it follows that $x \in W_X = W_X(T^* y_1^*, \ldots, T^* y_n^*; \epsilon)$, a weak neighborhood of X. We conclude that $T^{-1}(W_Y) \subseteq W_X$. Equality is obtained by running through the same argument in reverse, and so T is weak-to-weak continuous, as required.   □

Suppose that $T : X \to Y$ is a bounded linear mapping between real Banach spaces. If X is reflexive, then $T(B_X)$ is weakly compact, and hence norm-closed in Y. This is not true in general (i.e., for non-reflexive spaces X). Consider any $x^* \in X^*$ with $\|x^*\| = 1$. Then $x^*(B_X)$ could be either $(-1, 1)$ or $[-1, 1]$. If X is reflexive, then the second interval (the closed one) is the only option.

Example 5.43 Consider the real Banach space $X = \ell_1$. Recall that $\ell_1^* = \ell_\infty$. Let ξ in ℓ_∞ be the bounded sequence $\xi = (1 - 1/n)_{n=1}^\infty$. Now suppose $x = (x_n)_{n=1}^\infty$ is any element in B_{ℓ_1}, so that $\sum_{n=1}^\infty |x_n| \leq 1$. Then

$$|\xi(x)| = \Big| \sum_{n=1}^\infty \xi_n x_n \Big| \leq \sum_{n=1}^\infty \xi_n |x_n| < 1.$$

Since $\|\xi\|_{\ell_\infty} = 1$, we have a norm-one element $\xi \in \ell_1^*$ such that $\xi(B_{\ell_1}) = (-1, 1)$.

We see that a linear functional on a reflexive Banach space attains its maximum value on the closed unit ball. It was a long standing question whether or not this property characterized reflexive spaces. In 1964, R.C. James showed that it did when he proved the statement: *If every bounded linear functional on X attains its maximum value on the closed unit ball, then X is reflexive* [19].

We conclude this section with a result about the adjoint operator.

Proposition 5.44 *If $T : X \to Y$ is a bounded linear map between Banach spaces, then $T^* : Y^* \to X^*$ is weak*-to-weak* continuous.*

Proof The proof is very similar to the proof that *(i)* implies *(iii)* in Proposition 5.42. Consider a weak* neighborhood in X^*, say

$$W_{X^*} = W_{X^*}(x_1, \ldots, x_n; \epsilon) = \{x^* : |x^*(x_j)| < \epsilon, \ 1 \leq j \leq n\},$$

for $\{x_1, \ldots, x_n\} \subseteq X$ and $\epsilon > 0$. Suppose $y^* \in Y^*$ is such that $T^* y^* \in W_{X^*}$. Then $|T^* y^*(x_j)| < \epsilon$ for all $j \in \{1, \ldots, n\}$. But $T^* y^*(x) = y^*(Tx)$ for all $x \in X$, and so $|y^*(Tx_j)| < \epsilon$ for all $j \in \{1, \ldots, n\}$. Then $y^* \in W_{Y^*} = W_{Y^*}(Tx_1, \ldots, Tx_n; \epsilon)$, which is a weak* neighborhood of Y^*. This implies that $(T^*)^{-1}(W_{X^*}) \subseteq W_{Y^*}$. Similarly, $W_{Y^*} \subseteq (T^*)^{-1}(W_{X^*})$, and so T^* is weak*-to-weak* continuous, as required.   □

5.6 Mazur's Theorem

In this section we explore the consequences of convexity in weak topologies.

Theorem 5.45 (Mazur's Theorem) *Let X be a locally convex topological vector space. A convex subset of X is closed if and only if it is weakly closed.*

Proof Without loss of generality, assume X is a real topological vector space. A weakly closed set is always strongly closed, regardless of convexity. Suppose K is closed in the original topology, and let $\overline{K}^{(w)}$ denote the closure of K in the weak topology. Assume $x_0 \in \overline{K}^{(w)} \backslash K$. Then, by the Hahn–Banach Separation Theorem (Theorem 5.20), there is an $x^* \in X^*$ such that

$$x^*(x_0) > \sup\{x^*(x) : x \in K\}.$$

This contradicts the assumption x_0 is in the weak closure of K, and so $\overline{K}^{(w)} = K$. □

Example 5.46 Consider the real sequence space ℓ_p, where $1 < p < \infty$. As usual, for each $n \in \mathbb{N}$, let e_n be the sequence with 1 in the nth coordinate, and 0 elsewhere. We know that $e_n \to 0$ weakly as $n \to \infty$ (see Example 5.28), and so 0 is in the weak closure of the set $E = \{e_n : n \in \mathbb{N}\}$. Let co$(E)$ denote the set of convex linear combinations of elements in E:

$$\mathrm{co}(E) = \left\{ \sum_{j=1}^{m} \lambda_j e_j \ : \ \lambda_j \geq 0, \ \sum_{j=1}^{m} \lambda_j = 1, \ m \in \mathbb{N} \right\}.$$

We denote the closure of co(E) in the norm topology by $\overline{\mathrm{co}}(E)$. This set is convex and closed in the norm topology, and hence weakly closed by Mazur's Theorem (Theorem 5.45). It follows that $\overline{\mathrm{co}}(E)$ contains 0, since 0 is a weak limit point of E. We conclude that 0 can be approximated (in norm) by convex linear combinations of elements in $E = \{e_n : n \in \mathbb{N}\}$. In fact, if $1 < p < \infty$, then

$$\left\| \frac{1}{n}(e_1 + \cdots + e_n) \right\|_{\ell_p} = \frac{1}{n} \left(\sum_{j=1}^{n} 1^p \right)^{1/p} = n^{\frac{1}{p}-1} \xrightarrow[n \to \infty]{} 0.$$

The same cannot be said of ℓ_1—in this case, there exists no convex combination of elements in $\{e_n : n \in \mathbb{N}\}$ that will approximate 0. To see this, let $\lambda_1 e_1 + \cdots + \lambda_n e_n$ be any convex combination of elements from $\{e_n : n \in \mathbb{N}\}$. Then

$$\|\lambda_1 e_1 + \cdots + \lambda_n e_n\|_{\ell_1} = \|(\lambda_1, \lambda_2, \cdots, \lambda_n, 0, \dots)\|_{\ell_1} = \sum_{j=1}^{n} \lambda_j = 1.$$

In the above example, we introduced the notation co(E) to denote the set of convex linear combinations of elements in the set E. This idea will prove important later, and so we give the following definition.

Definition 5.47 Let X be a topological vector space and let A be any subset of X. The *convex hull* of A is the smallest convex subset of X that contains A. The convex hull of A is denoted by co(A) and consists of all convex linear combinations of elements in A; that is,

$$\mathrm{co}(A) = \left\{ \sum_{j=1}^{m} \lambda_j a_j \ : \ a_j \in A, \ \lambda_j > 0, \ \sum_{j=1}^{m} \lambda_j = 1, \ m \in \mathbb{N} \right\}.$$

The *closed convex hull* of A is the closure of the convex hull and is denoted $\overline{\mathrm{co}}(A)$.

Example 5.48 Let $X = L_p(0, 1)$, where $0 < p < 1$. (See Example A in Sect. 5.3.) In Example 5.19, we saw that the only nonempty open convex subset of X is X itself. Consequently, if $B_X = \{f : \|f\|_p \le 1\}$, then $\text{co}(B_X) = X$.

Let W be a subset of a topological vector space X. If f is a linear functional on X such that $|f(x)| \le M$ for all $x \in W$, then, by the definition of the convex hull, it must be that $|f(x)| \le M$ for all $x \in \text{co}(W)$. In the case of $L_p(0, 1)$, where $0 < p < 1$, from the example above, the convex hull of the unit ball is the entire space. Since no nonzero linear functionals are bounded on $L_p(0, 1)$ (see Example A in Sect. 5.3), it follows that there are no nonzero linear functionals bounded on the unit ball of $L_p(0, 1)$. Indeed, there are no nonzero linear functionals bounded on any nonempty open subsets of $L_p(0, 1)$ if $0 < p < 1$.

We now give an example of how convexity can help to solve optimization problems. We begin by making a definition.

Definition 5.49 Let X be a vector space. A function $f : X \to \mathbb{R}$ is called *convex* if

$$f(tx + (1 - t)y) \le tf(x) + (1 - t)f(y),$$

for all $\{x, y\} \subseteq X$ and $t \in [0, 1]$.

Our problem is as follows: Suppose K is a closed bounded convex set in a Banach space X and let $f : X \to \mathbb{R}$ be a continuous convex function. Does there exist some $x_0 \in K$ such that

$$f(x_0) = \min\{f(x) : x \in K\}?$$

There is no reason we should assume this minimum exists, as there is no compactness assumption made on K.

Suppose for a moment that X is reflexive. Then B_X is weakly compact, by Theorem 5.41. Since K is closed in norm, K is weakly closed, by Mazur's Theorem. It follows that K is weakly compact. (Here, we use the fact that B_X is weakly compact and absorbent.) We now have continuity and compactness, but not in the same topology: f is continuous in the norm topology, and K is compact in the weak topology.

Despite this, we claim that if X is a reflexive Banach space and if f is a convex function, then f does attain its minimum value on K. Let $\alpha = \inf\{f(x) : x \in K\}$. Our goal is to show that $\alpha > -\infty$ and that $f(x_0) = \alpha$ for some $x_0 \in K$.

Suppose $\alpha = -\infty$ and, for each $n \in \mathbb{N}$, define $K_n = \{x \in K : f(x) \le -n\}$. For each $n \in \mathbb{N}$, the set K_n is closed (and hence weakly closed by Mazur's Theorem), convex, and nonempty (since $\alpha = -\infty$). Therefore, $(K_n)_{n=1}^{\infty}$ forms a nested sequence of weakly compact nonempty sets. By the Nested Interval Property (Corollary B.7), it must be that $\bigcap_{n=1}^{\infty} K_n \ne \emptyset$. But this implies that there is some $x_0 \in K$ such that $f(x_0) \le -n$ for all $n \in \mathbb{N}$, an impossibility. Consequently, $\alpha > -\infty$.

Define for $n \in \mathbb{N}$ a sequence of sets $K_n' = \{x \in K : f(x) \le \alpha + 1/n\}$. As before, $(K_n')_{n=1}^{\infty}$ is a nested sequence of weakly compact nonempty sets, and so $\bigcap_{n=1}^{\infty} K_n' \ne \emptyset$. If $x_0 \in \bigcap_{n=1}^{\infty} K_n'$, then $x_0 \in K$ and $f(x_0) = \alpha$, as required.

We summarize in the following proposition.

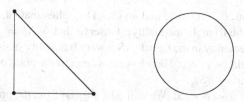

Fig. 5.2 Some elementary convex objects

Proposition 5.50 *Suppose K is a nonempty closed bounded convex set in a Banach space X and let $f : X \to \mathbb{R}$ be a continuous convex function. If X is reflexive, then there exists an $x_0 \in K$ such that $f(x_0) = \min\{f(x) : x \in K\}$.*

Proof See the discussion preceding the statement of the proposition. □

A special case of the above is $f(x) = \|u - x\|$, the function representing the distance between $x \in X$ and a fixed point $u \in X$. If X is a reflexive Banach space, and if K is a closed and bounded convex set in X (not containing u), then there exists a $x_0 \in K$ such that

$$\|u - x_0\| = \min_{x \in K} \|u - x\|.$$

That is, there exists some point $x_0 \in K$ which is closest to u. (Actually, the boundedness assumption on K is not needed for this statement to be true.)

5.7 Extreme Points

In this section, we consider sets K that are convex in some vector space X.

Definition 5.51 Let X be a vector space and suppose K is a convex subset of X. A point $x \in K$ is an *extreme point* of K if it does not lie on a line segment in K. That is, x is an extreme point of K provided that the following is true: If u and v are elements of K such that $x = (1 - t)u + tv$ for some $t \in (0, 1)$, then $x = u = v$.

The set of extreme points of K is denoted $\mathrm{ex}(K)$.

For example, a triangle has an extreme point at each vertex, while any boundary point of a circle is an extreme point. (See Fig. 5.2.)

Example 5.52 We now determine the extreme points for the unit ball B_X in several cases where X is a real Banach space. Note that no point of the interior of B_X can be extreme, and so we must consider only points on the boundary ∂B_X.

(i) $X = \ell_2$. Denote the inner product on ℓ_2 by $\langle \cdot, \cdot \rangle$. Suppose $x \in \ell_2$ is such that $\|x\| = 1$. Now let $\{u, v\} \subseteq B_{\ell_2}$ and suppose $x = (1 - t)u + tv$ for some $t \in (0, 1)$. By the triangle inequality, $\|u\| = \|v\| = 1$ (otherwise $\|x\| < 1$). Since $\|x\| = 1$, we have

$$1 = \langle x, x \rangle = (1 - t)\langle u, x \rangle + t\langle v, x \rangle. \tag{5.8}$$

By assumption, $0 < t < 1$, and so (5.7.1) implies that $\langle u, x \rangle = \langle v, x \rangle = 1$ (again using the triangle inequality). Observe that $\langle u, x \rangle = 1 = \|u\| \|x\|$, and thus we have equality in the Cauchy–Schwarz Inequality. This can only happen if $u = x$. Similarly, $v = x$. Therefore, x is an extreme point of B_{ℓ_2} whenever x is on the boundary ∂B_{ℓ_2}.

(ii) $X = L_p(0, 1)$, $1 < p < \infty$. We will take our cue from the preceding example. Suppose $f \in L_p(0, 1)$ is such that $\|f\|_p = 1$. Let $\{g, h\} \subseteq L_p(0, 1)$ and suppose that $f = (1 - t)g + th$ for some $t \in (0, 1)$. Then $\|g\|_p = \|h\|_p = 1$, by the triangle inequality (as in *(i)*). By the Hahn–Banach Theorem, there exists a linear functional ϕ in $L_p(0, 1)^*$ such that $\|\phi\| = 1$ and $\phi(f) = 1$. In fact, in this case we can write ϕ explicitly:

$$\phi(k) = \int_0^1 |f(x)|^{p-1}(\operatorname{sign} f(x))\, k(x)\, dx, \quad k \in L_p(0, 1).$$

Since $\phi(f) = \|f\|_p^p$, we have

$$1 = \phi(f) = (1 - t)\phi(g) + t\phi(h).$$

Again using the triangle inequality, we have that $\phi(g) = \phi(h) = 1$. By Hölder's Inequality,

$$|\phi(g)| \le \int |f(x)|^{p-1} |g(x)|\, dx \le \left(\int |f(x)|^{(p-1)q}\, dx \right)^{1/q} \|g\|_p = \|f\|_p^{p/q} \|g\|_p,$$

where $\frac{1}{p} + \frac{1}{q} = 1$ (and so $q = \frac{p}{p-1}$). Since the left and right sides of the above inequality are both equal to 1, we have equality in Hölder's Inequality. This happens only if there are positive constants a and b such that $a(|f|^{p-1})^q = b|g|^p$ (as members of $L_p(0, 1)$). Because $q = \frac{p}{p-1}$, this equality (which is valid almost everywhere) becomes $a|f|^p = b|g|^p$, which is equivalent to $a^{1/p}|f| = b^{1/p}|g|$. Since $\|f\|_p = \|g\|_p$, this can only happen if $|f| = |g|$ in $L_p(0, 1)$. A similar argument shows $|f| = |h|$ in $L_p(0, 1)$. From these equalities, together with the assumption that f is a convex combination of g and h, we conclude that $f = g$ and $f = h$. It follows that f is an extreme point of $B_{L_p(0,1)}$. The choice of f was arbitrary in $\partial B_{L_p(0,1)}$, and therefore any f on the boundary of the unit ball is an extreme point of the unit ball $B_{L_p(0,1)}$.

A similar argument shows that the extreme points of the unit ball in ℓ_p, where $1 < p < \infty$, are the elements of the boundary ∂B_{ℓ_p}.

(iii) $X = L_1(0, 1)$. If f is an extreme point of $B_{L_1(0,1)}$, then f is on the boundary of $B_{L_1(0,1)}$. Let f be a function in $L_1(0, 1)$ such that $\|f\|_1 = \int_0^1 |f(s)|\, ds = 1$. Define a new function F on $[0, 1]$ by

$$F(t) = \int_0^t |f(s)|\, ds, \quad t \in [0, 1].$$

The function F is continuous with $F(0) = 0$ and $F(1) = 1$. By the Intermediate Value Theorem, there exists a $\tau \in (0, 1)$ such that $F(\tau) = 1/2$.

Let $g = 2f \chi_{(0,\tau)}$ and $h = 2f \chi_{(\tau,1)}$. By the choice of τ, we have $\|g\|_1 = 1$ and $\|h\|_1 = 1$. We have found distinct functions g and h in $B_{L_1(0,1)}$ such that $f = \frac{1}{2}g + \frac{1}{2}h$. Therefore, f is not an extreme point of the unit ball of $L_1(0,1)$. Since f was an arbitrary element of $\partial B_{L_1(0,1)}$, we conclude that the unit ball in $L_1(0,1)$ has no extreme points.

(iv) $X = c_0$. We will show there are no extreme points in B_{c_0}. Suppose $x = (x_k)_{k=1}^{\infty}$ is a sequence in B_{c_0}. Then $\lim_{k\to\infty} x_k = 0$, and so there exists some $n \in N$ such that $|x_n| < 1/2$. We will define sequences $y = (y_k)_{k=1}^{\infty}$ and $z = (z_k)_{k=1}^{\infty}$ in c_0 so that $x = \frac{1}{2}y + \frac{1}{2}z$. If $x_n \neq 0$, then define y and z as follows:

$$y_k = \begin{cases} x_k & \text{if } k \neq n, \\ 0 & \text{if } k = n \end{cases} \quad \text{and} \quad z_k = \begin{cases} x_k & \text{if } k \neq n, \\ 2x_n & \text{if } k = n. \end{cases}$$

We have $x = \frac{1}{2}y + \frac{1}{2}z$ and, because $|x_n| < 1/2$, the sequences y and z are in B_{c_0}. The previous sequences work only if $x_n \neq 0$. If instead $x_n = 0$, then define y and z so that:

$$y_k = \begin{cases} x_k & \text{if } k \neq n, \\ \frac{1}{2} & \text{if } k = n \end{cases} \quad \text{and} \quad z_k = \begin{cases} x_k & \text{if } k \neq n, \\ -\frac{1}{2} & \text{if } k = n. \end{cases}$$

Once again, we have $x = \frac{1}{2}y + \frac{1}{2}z$, where the sequences y and z are in B_{c_0}. Therefore, x is not an extreme point of the set B_{c_0}.

(v) $X = C[0,1]$. In this case the extreme points of $B_{C[0,1]}$ are the two functions $\chi_{[0,1]}$ and $-\chi_{[0,1]}$. (That is, the constant functions 1 and -1.) If $f \in B_{C[0,1]}$ is a continuous function such that $|f(t)| < 1$ for some $t \in (0,1)$, then we may use an argument similar to the perturbation argument used in *(iv)*. (That is, we can put a small "wiggle" in the function.)

Similarly, if K is a compact Hausdorff space, then the extreme points of $B_{C(K)}$ are the two functions χ_K and $-\chi_K$, which are the constant functions 1 and -1 on K.

In the case that $C(K)$ is a *complex* Banach space, the extreme points of $B_{C(K)}$ are all functions $f \in C(K)$ for which $|f(s)| = 1$ for all $s \in K$.

(vi) $X = \ell_1$. The set of extreme points in B_{ℓ_1} is $\{\pm e_n : n \in \mathbb{N}\}$. First, let us show that each of these points is indeed extreme in B_{ℓ_1}. Fix some $n \in \mathbb{N}$. Let $y = (y_k)_{k=1}^{\infty}$ and $z = (z_k)_{k=1}^{\infty}$ be elements of B_{ℓ_1} such that $e_n = ay + bz$, where a and b are positive numbers such that $a + b = 1$. (Note that these conditions imply $y_n > 0$ and $z_n > 0$.) We have that $ay_n + bz_n = 1$ and $ay_k + bz_k = 0$ for all $k \neq n$. By assumption, we know that $a \neq 0$, and so $y_n = \frac{1}{a} - \frac{b}{a}z_n$ and $y_k = -\frac{b}{a}z_k$ for $k \neq n$. Once again making use of the triangle inequality, we see that $\|y\|_1 = 1$ and $\|z\|_1 = 1$. Computing $\|y\|_1$ directly, we have

$$\left(\sum_{k \neq n} |y_k|\right) + y_n = \left(\sum_{k \neq n} \frac{b}{a}|z_k|\right) + \left(\frac{1}{a} - \frac{b}{a}z_n\right) = \left(\frac{b}{a}\sum_{k=1}^{\infty}|z_k|\right) + \left(\frac{1}{a} - \frac{2b}{a}z_n\right).$$

Thus,

$$\|y\|_1 = \frac{b}{a}\|z\|_1 + \left(\frac{1}{a} - \frac{2b}{a}z_n\right).$$

But $\|y\|_1 = 1$ and $\|z\|_1 = 1$, and so

$$1 = \frac{b}{a}(1) + \left(\frac{1}{a} - \frac{2b}{a}z_n\right) = \frac{b+1-2bz_n}{a}.$$

A little arithmetic (and the fact that $a + b = 1$) reveals that $z_n = 1$, and so z must in fact be e_n. Therefore, e_n is an extreme point. A similar argument shows that $-e_n$ is an extreme point for each $n \in \mathbb{N}$.

Now we show that no other element of ∂B_{ℓ_1} is an extreme point. Suppose $x \in B_{\ell_1}$ with $\|x\|_1 = 1$, but $x \neq \pm e_n$ for any $n \in \mathbb{N}$. Then there must be at least two non-zero entries, say x_{m_1} and x_{m_2}. Without loss of generality, we may assume both terms are positive. Choose some constant $\epsilon > 0$ such that $\epsilon < \min\{x_{m_1}, x_{m_2}, 1 - x_{m_1}, 1 - x_{m_2}\}$. Define $y = (y_k)_{k=1}^{\infty}$ and $z = (z_k)_{k=1}^{\infty}$ in ℓ_1 as follows:

$$y_k = \begin{cases} x_k & \text{if } k \notin \{m_1, m_2\}, \\ x_{m_1} + \epsilon & \text{if } k = m_1, \\ x_{m_2} - \epsilon & \text{if } k = m_2 \end{cases} \quad \text{and} \quad z_k = \begin{cases} x_k & \text{if } k \notin \{m_1, m_2\}, \\ x_{m_1} - \epsilon & \text{if } k = m_1, \\ x_{m_2} + \epsilon & \text{if } k = m_2. \end{cases}$$

It follows that $x = \frac{1}{2}y + \frac{1}{2}z$, where $\{y, z\} \subseteq B_{\ell_1}$. Therefore, x is not an extreme point.

The next theorem describes a condition under which the set of extreme points is never empty.

Theorem 5.53 (**Krein–Milman Theorem**) *Suppose E is a locally convex Hausdorff topological vector space. If K is a nonempty compact convex subset of E, then $K = \overline{co}(exK)$. In particular, $ex(K) \neq \emptyset$.*

Before proving Theorem 5.53, let us observe a consequence.

Corollary 5.54 *If X is a Banach space, then $B_{X^*} = \overline{co}^{(w^*)}(exB_{X^*})$.*

Proof By Theorem 5.39, the set B_{X^*} is compact in the w^*-topology whenever X is a Banach space. The result then follows from the Krein–Milman Theorem. □

What makes this result so interesting is that we can now, courtesy of Example 5.52, conclude that neither c_0 nor $L_1(0, 1)$ is a dual space of a Banach space, since the unit balls in these spaces have no extreme points. While the unit ball of $C[0, 1]$ does have extreme points, it does not have enough (only two!) to construct the entire unit ball using only convex linear combinations. Therefore, $C[0, 1]$ cannot be the dual space of any Banach space, either.

In order to proceed with the proof of Theorem 5.53, we now introduce a definition and a lemma.

Definition 5.55 Let E be a topological vector space with nonempty subset K. A subset F of K is called *extremal* (in K) if F is a nonempty compact convex set such that the following holds: If $\{u, v\} \subseteq K$ and $(1 - t)u + tv \in F$ for all $t \in (0, 1)$, then $\{u, v\} \subseteq F$.

Lemma 5.56 *Suppose K is a nonempty compact convex subset of a locally convex topological vector space. Every extremal subset of K contains an extreme point.*

Proof Let K be a nonempty compact convex subset of the locally convex topological vector space E and suppose F is an extremal set in K. Consider the partially ordered set of all subsets of F that are extremal in K, where $G \geq H$ whenever $G \subseteq H$. We wish to find a *maximal* element of this partially ordered set (which in turn will be a *minimal* extremal subset of F).

Suppose $\mathcal{C} = (G_i)_{i \in I}$ is a chain of subsets of F that are extremal in K. Let G be the intersection of all sets in $\mathcal{C}$, so $G = \bigcap_{i \in I} G_i$. Then G is a compact and convex subset of F. For any finite collection of indices $i_1, \ldots, i_n$ in I, we have $\bigcap_{k=1}^{n} G_{i_k} = G_{i_j}$ for some $j \in \{1, \ldots, n\}$, because $\mathcal{C}$ is a chain of subsets. Therefore, by the Finite Intersection Property, G is nonempty.

We claim that G is extremal in K. Suppose $\{u, v\} \subseteq K$ and $(1 - t)u + tv \in G$ for all $t \in (0, 1)$. For each $i \in I$, we have $G \subseteq G_i$, and so $(1 - t)u + tv \in G_i$ for all $t \in (0, 1)$. But G_i is extremal in K, and hence $\{u, v\} \subseteq G_i$. It follows that $\{u, v\} \subseteq \bigcap_{i \in I} G_i = G$. Thus, G is extremal in K.

By Zorn's Lemma, there exists a maximal element in the partially ordered set of subsets of F that are extremal in K. Hence, there is a minimal subset of F that is extremal in K. Denote this minimal extremal set by F_0. We will show that F_0 consists of only one element. Assume to the contrary that there exist distinct elements u and v in F_0. The space E is Hausdorff, and so $\{u\}$ and $\{v\}$ are closed sets. By the Hahn–Banach Separation Theorem (Theorem 5.20 for real spaces and Theorem 5.22 for complex spaces), there exists a continuous linear functional ϕ on E such that $\phi(u) \neq \phi(v)$. In particular, ϕ is not constant on F_0.

Without loss of generality, we may assume ϕ is real-valued. The functional ϕ is continuous on E, and therefore attains its maximum on the compact set F_0. Let

$$G_0 = \left\{ x \in F_0 : \phi(x) = \max_{\xi \in F_0} \phi(\xi) \right\}.$$

Let M denote the maximum value of ϕ on F_0, so that $\phi(x) = M$ for all $x \in G_0$.

The set G_0 is nonempty (by the continuity of ϕ), compact, and convex. It is also a proper subset of F_0, because ϕ is not constant. We claim that G_0 is extremal in K. Suppose $\{u, v\} \subseteq K$ and $(1 - t)u + tv \in G_0$ for all $t \in (0, 1)$. We know that $G_0 \subseteq F_0$, and we know that F_0 is extremal; hence $\{u, v\} \subseteq F_0$. For each $t \in (0, 1)$, we have that $(1 - t)u + tv \in G_0$, and so

$$M = \phi((1 - t)u + tv) = (1 - t)\phi(u) + t\phi(v),$$

for all $t \in (0, 1)$. From this we conclude that $\phi(u) = \phi(v) = M$, and consequently $\{u, v\} \subseteq G_0$. This implies G_0 is extremal, but this violates the minimality of F_0. We have derived a contradiction, and so it must be the case that F_0 contains only one element. The set F_0 is a single-point set and an extremal set. Therefore, the one element of F_0 is an extreme point.

We are now prepared to prove the Krein–Milman Theorem.

Proof of Theorem 5.53   Without loss of generality, we may assume E is a real topological vector space. The set K is extremal in itself, and so must contain an extreme point, by Lemma 5.56. Let $K_0 = \overline{\text{co}}(\text{ex}\,K)$, the closed convex hull of the set of extreme points in K. Suppose $x \in K \setminus K_0$. By the Hahn–Banach Separation Theorem (Theorem 5.20), there exists a continuous linear functional f on E such that

$$f(x) > \max_{y \in K_0} f(y). \tag{5.9}$$

Let

$$G_0 = \left\{ z \in K : f(z) = \max_{y \in K} f(y) \right\}.$$

The set G_0 is nonempty because f is continuous and K is compact; it is also disjoint from K_0 because of (5.9). The set G_0 is extremal (see the argument in the proof of Lemma 5.56), and consequently contains an extreme point of K, by Lemma 5.56. This, however, is a contradiction, because K_0 contains all of the extreme points of K, and K_0 and G_0 are disjoint. Therefore, $K = K_0$, as required.

The Krein–Milman Theorem originally appeared in a work by Krein and Milman in 1940 [21]. The local convexity assumption on E was needed to invoke the Hahn–Banach Separation Theorem (Theorem 5.20). Local convexity is a necessary condition, a fact which was not shown until the 1970s [32]. The Krein–Milman Theorem has a deep relationship with the Axiom of Choice. (See [4].)

5.8   Milman's Theorem

Suppose K is a compact Hausdorff space. The Riesz Representation Theorem identifies the dual space of the space of continuous functions on K as the space of regular Borel measures on K; that is, $C(K)^* = M(K)$. (See Theorem A.35.) We recall that the norm on $M(K)$ is the total variation norm: $\|\mu\|_M = |\mu|(K)$ for all $\mu \in M(K)$. We define the *probability measures* on K to be elements in the set

$$\mathcal{P}(K) = \{\mu \in M(K) : \mu \geq 0, \ \|\mu\|_M = 1\}.$$

This set is convex. It is also closed in the w^*-topology, which can be seen from the equality $\mathcal{P}(K) = \{\mu \in B_{M(K)} : \int_K 1 \, d\mu = 1\}$. (See Exercise 5.27.)

By the Banach–Alaoglu Theorem (Theorem 5.39), the unit ball $B_{M(K)}$ is compact in the w^*-topology, and hence $\mathcal{P}(K)$ is w^*-compact as a w^*-closed subset. A simple computation shows that $\mathcal{P}(K)$ is an extremal set in the unit ball of $M(K)$. (Again, see Exercise 5.27.) Since $\mathcal{P}(K)$ is a w^*-compact convex extremal set, Lemma 5.56 assures us that $\mathcal{P}(K)$ must have at least one extreme point.

Proposition 5.57   *Let K be a compact Hausdorff space. A probability measure in $M(K)$ is an extreme point of $\mathcal{P}(K)$ if and only if it is a Dirac measure.*

Proof We first show that δ_s is an extreme point of $\mathcal{P}(K)$ for $s \in K$. Suppose there exist probability measures μ and ν such that $\delta_s = (1-t)\mu + t\nu$ for some $t \in (0,1)$. It follows that $\mu(\{s\}) = \nu(\{s\}) = 1$. Thus, $\mu = \nu = \delta_s$, as required.

Now suppose μ is an extreme point of $\mathcal{P}(K)$. We will show that $\mu = \delta_s$ for some $s \in K$. Let $\mathcal{U} = \{U \text{ open } : \mu(U) = 0\}$ and let $V = \bigcup_{U \in \mathcal{U}} U$. We claim $\mu(V) = 0$. Suppose E is a compact subset of V. The collection of sets $\mathcal{U}$ forms an open cover of E, and so by compactness there exists a finite subcover, say $E \subseteq U_1 \cup \cdots \cup U_n$. The measure μ is nonnegative, and so (by subadditivity)

$$\mu(E) \leq \mu(U_1) + \cdots + \mu(U_n) = 0.$$

Therefore, $\mu(E) = 0$ for all compact subsets of V. By the regularity of μ,

$$\mu(V) = \sup\{\mu(E) : E \text{ is a compact subset of } V\} = 0.$$

Let $F = K \backslash V$. Then $\mu(F) = 1$. We wish to show that F contains only one point. Assume to the contrary that F contains more than one point. Let $\{s,t\} \subseteq F$. By the Hausdorff property, there are open sets W_1 and W_2 in K such that $s \in W_1$, $t \in W_2$, and $W_1 \cap W_2 = \emptyset$. Since W_1 and W_2 are not subsets of V, they have non-zero μ-measure. Define measures μ_1 and μ_2 on K as follows:

$$\mu_1(B) = \frac{\mu(W_1 \cap B)}{\mu(W_1)} \quad \text{and} \quad \mu_2(B) = \frac{\mu((K \backslash W_1) \cap B)}{\mu(K \backslash W_1)},$$

where B is any measurable subset of K. We note that $\mu(K \backslash W_1) \neq 0$ since $\mu \geq 0$ and $W_2 \subseteq K \backslash W_1$.

Both μ_1 and μ_2 are probability measures, and

$$\mu(W_1)\,\mu_1 + \mu(K \backslash W_1)\,\mu_2 = \mu.$$

By assumption, μ is an extreme point of $\mathcal{P}(K)$. Thus, since $\mu(W_1) + \mu(K \backslash W_1) = 1$, it follows that $\mu = \mu_1 = \mu_2$. This is not possible, however, since $\mu_1(W_1) = 1$ and $\mu_2(W_1) = 0$. We have derived a contradiction. Therefore, there can be no more than one point in the set F, say s. Since $\mu(F) = 1$, it follows that $\mu = \delta_s$, as required.

In the above proof, the set V is the maximal open set of μ-measure zero. The entire μ-mass of K is contained in $K \backslash V$. This motivates the next definition.

Definition 5.58 Suppose μ is a positive nonzero measure on K. If V is the maximal open set of μ-measure zero, then $K \backslash V$ is called the *support of* μ. If μ is a signed (or complex) measure, the support of μ is defined to be the support of $|\mu|$.

In the proof of Proposition 5.57, observe that the measures μ_1 and μ_2 were defined in such a way that they had disjoint supports. As a result, it was certainly the case that $\mu_1 \neq \mu_2$. (They "live" on different sets, so to speak.)

Theorem 5.59 (Milman's Theorem) *Suppose E is a locally convex Hausdorff topological vector space and let K be a compact subset of E. If D is a closed subset of K such that $K = \overline{co}(D)$, then $ex(K) \subseteq D$. Furthermore, for every $x \in K$, there* *exists a $\mu_x \in \mathcal{P}(D)$ such that $f(x) = \int_D f(y)\mu_x(dy)$ for all linear functionals* $f \in E^*$.

Proof Observe that $E \subseteq \mathbb{K}^{E^*}$, the collection of all scalar-valued functions on E^*, where the superspace is equipped with the product topology. The inclusion is made explicit by the embedding $\rho : E \to \mathbb{K}^{E^*}$ defined by

$$\rho(e) = (f(e))_{f \in E^*}, \quad e \in E.$$

(We often make this identification implicitly, suppressing the letter ρ.)

Introduce a map $T : M(D) \to \mathbb{K}^{E^*}$, defined by

$$T(\mu) = \left(\int_D f(y)\,\mu(dy) \right)_{f \in E^*}, \quad \mu \in M(D).$$

Observe that T is continuous in the w^*-topology on $M(D)$.

For each $s \in D$, we have

$$T(\delta_s) = (f(s))_{f \in E^*} = s. \tag{5.10}$$

By the w^*-continuity of T, then, it follows that T maps $\overline{\mathrm{co}}^{(w^*)}(\{\delta_s\}_{s \in D})$ onto $\overline{\mathrm{co}}^{(w)}(D)$. We note that we have the weak closure of $\mathrm{co}(D)$ in E because the topology E inherits from $\mathbb{K}^{E^*}$ is the weak topology. (See the comments after Definition 5.27.)

By Theorem 5.53 (the Krein–Milman Theorem) and Proposition 5.57, we make the identification $\overline{\mathrm{co}}^{(w^*)}(\{\delta_s\}_{s \in D}) = \mathcal{P}(D)$. By assumption, $K = \overline{\mathrm{co}}(D)$, and so by Mazur's Theorem (Theorem 5.45), we have that $K = \overline{\mathrm{co}}^{(w)}(D)$. Therefore, the restriction $T|_{\mathcal{P}(D)} : \mathcal{P}(D) \to K$ is a surjection. This proves the second part of the theorem.

It remains to prove the first part of the theorem; that is, that the extreme points of K are in D. Suppose $x \in \mathrm{ex}(K)$. We claim that the set $T^{-1}(x) \subseteq \mathcal{P}(D)$ is an extremal set. Suppose $\{\mu, \nu\} \subseteq \mathcal{P}(D)$ and $(1-t)\mu + t\nu \in T^{-1}(x)$ for all $t \in (0,1)$. It follows that $(1-t)T(\mu) + tT(\nu) = x$ for all $t \in (0,1)$. By assumption, x is an extreme point in K, and consequently $T\mu = T\nu = x$. Therefore, $\{\mu, \nu\} \subseteq T^{-1}(x)$, and so $T^{-1}(x)$ is extremal.

By Lemma 5.56, $T^{-1}(x)$ contains an extreme point of $\mathcal{P}(D)$. Therefore, there exists some $s \in D$ such that $\delta_s \in T^{-1}(x)$, and consequently $T(\delta_s) = x$. By (5.10), however, $T(\delta_s) = s$, and so $x = s \in D$. The result follows.   □

5.9 Haar Measure on Compact Groups

We now turn our attention to topological groups. We saw in Sect. 3.4 that if G is a compact abelian metrizable group, then there exists a unique translation-invariant probability measure on G. We noted at the time that the metrizability assumption was not needed. Now, using the tools of the previous sections, we will extend this result to include compact topological groups that are not abelian. Let us review some definitions.

Definition 5.60 A group G is called a *topological group* if the set G is endowed with a topology for which the group operations (multiplication and inversion)

$$(s,t) \mapsto s \cdot t \quad \text{and} \quad s \mapsto s^{-1}, \quad (s,t) \in G \times G,$$

are continuous. If G is compact in the given topology, then G is called a *compact group*.

Classical examples of topological groups include $\mathbb{R}^n$ (where the group multiplication is given by addition) and the set of orthogonal $n \times n$ matrices $\mathcal{O}_n$ (where the group multiplication is matrix multiplication). Multiplication in a group is usually denoted either with a dot ($\cdot$) or by juxtaposition. When the group is abelian, however, it is traditional to use a plus symbol ($+$), provided it will not result in any confusion.

When G is a compact group, we denote the space of continuous functions on G by $C(G)$. The σ-algebra on G is implicitly taken to be the Borel σ-algebra generated by the open sets in G. We denote the Borel σ-algebra on G by $\mathcal{B}$.

Definition 5.61 Let G be a compact group with Borel algebra $\mathcal{B}$. A measure μ on G is called *left-invariant* if $\mu(gB) = \mu(B)$ for all $B \in \mathcal{B}$ and $g \in G$. Correspondingly, the measure μ is called *right-invariant* if $\mu(Bg) = \mu(B)$ for all $B \in \mathcal{B}$ and $g \in G$.

Theorem 5.62 (Existence of Haar Measure) *Suppose G is a compact group. There exists a unique left-invariant probability measure on the Borel sets of G. Furthermore, this measure is also the unique right-invariant probability measure on the Borel sets of G.*

Proof First, we assume we can find a left-invariant probability measure λ and a right-invariant probability measure μ. We will show that $\lambda = \mu$. (Notice that this will imply uniqueness.)

Let $f \in C(G)$. By Fubini's Theorem,

$$\int_G \left(\int_G f(s \cdot t) \lambda(dt) \right) \mu(ds) = \int_G \left(\int_G f(s \cdot t) \mu(ds) \right) \lambda(dt). \qquad (5.11)$$

By the left-invariance of λ,

$$\int_G \left(\int_G f(s \cdot t) \lambda(dt) \right) \mu(ds) = \int_G \left(\int_G f(t) \lambda(dt) \right) \mu(ds) = \int_G f(t) \lambda(dt),$$

since $\mu(G) = 1$. Similarly, by the right-invariance of μ,

$$\int_G \left(\int_G f(s \cdot t) \mu(ds) \right) \lambda(dt) = \int_G \left(\int_G f(s) \mu(ds) \right) \lambda(dt) = \int_G f(s) \mu(ds),$$

since $\lambda(G) = 1$. Substituting into (5.11), we obtain

$$\int_G f(t) \lambda(dt) = \int_G f(s) \mu(ds).$$

The choice of $f \in C(G)$ was arbitrary, and so by duality we conclude $\lambda = \mu$.

It remains to prove the existence of a left-invariant measure on G. (A similar argument will produce a right-invariant measure.) We begin by defining for each $s \in G$ a left-multiplication operator $L_s : C(G) \to C(G)$ by

$$(L_s f)(t) = f(s \cdot t), \quad t \in G.$$

Observe that $\|L_s\| = 1$, $L_s^{-1} = L_{s^{-1}}$, and $L_u \circ L_s = L_{u \cdot s}$, whenever u and s are in G.

Claim 1 *Let* $f \in C(G)$. *The map* $s \mapsto L_s f$ *is continuous from* G *into* $C(G)$.

We wish to estimate, for s and s' in G, the quantity

$$\|L_s f - L_{s'} f\|_\infty = \sup_{t \in G} |f(s \cdot t) - f(s' \cdot t)|.$$

Multiplication in the group is continuous, and so the map $(s, t) \mapsto f(s \cdot t)$ is continuous on $G \times G$. Since the group G is compact, we conclude the map $(s, t) \mapsto f(s \cdot t)$ is in fact uniformly continuous. Therefore, for any given $\epsilon > 0$, there exists an open neighborhood V_ϵ of the identity such that $|f(s \cdot t) - f(s' \cdot t')| < \epsilon$ whenever $s' \cdot s^{-1} \in V_\epsilon$ and $t' \cdot t^{-1} \in V_\epsilon$. In this case, we have $t' = t$, and so if $s' \cdot s^{-1} \in V_\epsilon$, then

$$\|L_s f - L_{s'} f\|_\infty = \sup_{t \in G} |f(s \cdot t) - f(s' \cdot t)| < \epsilon.$$

This proves Claim 1.

Claim 2 *Let* $\mu \in C(G)$. *The map* $s \mapsto L_s^* \mu$ *is* w^*-*continuous from* G *into* $M(G)$.

Observe that $L_s^* : M(G) \to M(G)$. Then, for $f \in C(G)$, by the definition of the adjoint, $\int f \, dL_s^* \mu = \int L_s f \, d\mu$. If s and s' are in G, then

$$\left| \int_G f \, dL_s^* \mu - \int_G f \, dL_{s'}^* \mu \right| = \left| \int_G L_s f \, d\mu - \int_G L_{s'} f \, d\mu \right| \le \|L_s f - L_{s'} f\|_\infty \|\mu\|_M.$$

The rest follows from Claim 1.

Claim 3 *If* $s \in G$, *then* $L_s^*(\mathcal{P}(G)) \subseteq \mathcal{P}(G)$.

If $f \ge 0$, then $\int f \, dL_s^* \mu = \int L_s f \, d\mu \ge 0$, whenever $\mu \ge 0$. Thus, $L_s^* \mu \ge 0$ for any $\mu \in \mathcal{P}(G)$. Furthermore,

$$L_s^* \mu(G) = \int_G 1 \, L_s^* \mu(dt) = \int_G L_s(1) \, \mu(dt) = \mu(G) = 1.$$

Thus, $L_s^* \mu$ is in $\mathcal{P}(G)$ whenever μ is a probability measure. This proves Claim 3.

The gist of Claim 3 is that the set $\mathcal{P}(G)$ is invariant under multiplication on the left; i.e., $\mathcal{P}(G)$ is *left-invariant*. We wish to find a set with this property that contains only one element. To that end, let $\mathcal{K}$ be the collection of all weak*-compact convex subsets of $\mathcal{P}(G)$ that are left-invariant; that is, all weak*-compact convex subsets K such that $L_s^* K \subseteq K$ for all $s \in G$. Define a partial order $\le$ on $\mathcal{K}$ so that $A \le B$ when $A \subseteq B$. We know that $\mathcal{K}$ is nonempty, because $\mathcal{P}(G) \in \mathcal{K}$. If $(C_i)_{i \in I}$ is a chain in $\mathcal{K}$, then $C = \bigcap_{i \in I} C_i$ is nonempty, by the Finite Intersection Property. Furthermore, C is a lower bound for the chain $(C_i)_{i \in I}$. Therefore, by Zorn's Lemma, there exists a minimal element of $\mathcal{K}$, say K.

We wish to show that K is a single-point set. Assume to the contrary that μ_1 and μ_2 are distinct elements in K. Let $\nu = \frac{1}{2}(\mu_1 + \mu_2)$. Then $\nu \in K$, by convexity. Define a new set $E = \{L_s^* \nu : s \in G\}$. By Claim 2, the set E is weak*-compact (as the image of the compact set G under a weak*-continuous mapping). Furthermore, $E \subseteq K$, by the left-invariance of K.

For all u and s in G,

$$L_u^*(L_s^* \nu) = L_u^* L_s^* \nu = L_{s \cdot u}^* \nu \in E.$$

Thus $L_u^*(E) \subseteq E$, and so E is left-invariant.

Let $K_0 = \overline{\text{co}}^{(w^*)}(E)$. The set K_0 is convex by construction. We also have that K_0 is weak*-compact, because it is weak*-closed in the weak*-compact set K. By construction, K_0 is left-invariant, and so $K_0 \in \mathcal{K}$. But $K_0 \subseteq K$ and K is minimal in $\mathcal{K}$. Therefore, $K = K_0$.

By the Krein–Milman Theorem (Theorem 5.53), there is some extreme point in K; and by Milman's Theorem (Theorem 5.59), every extreme point of K is in E. Therefore, there exists some $s \in G$ such that $L_s^* \nu$ is extreme in K. Recalling the definition of ν, we see that

$$L_s^* \nu = \frac{1}{2}(L_s^* \mu_1 + L_s^* \mu_2).$$

But $L_s^* \mu_1$ and $L_s^* \mu_2$ are in K, by left-invariance, and $L_s^* \nu$ is extreme in K. Therefore, $L_s^* \nu = L_s^* \mu_1 = L_s^* \mu_2$. If we multiply all sides of this equation by s^{-1} on the left (that is, apply $L_{s^{-1}}^*$ to all sides), we discover that $\nu = \mu_1 = \mu_2$. This violates the assumption that μ_1 and μ_2 are distinct. Thus, K contains only one element, say λ. The measure λ is the desired left-invariant probability measure on G.

Definition 5.63 Let G be a compact group. The unique left-invariant probability measure on the Borel subsets of G is called *Haar measure* on G.

5.10 The Banach–Stone Theorem

In this section, we prove a classical theorem about the structure of spaces of continuous functions. We recall that two Banach spaces X and Y are called *isometrically isomorphic* if there exists a continuous linear bijection that preserves norms. That is, if there exists some linear bijection $T : X \to Y$ such that $\|T\| = \|T^{-1}\| = 1$.

Theorem 5.64 (Banach–Stone Theorem) *Suppose K_1 and K_2 are compact Hausdorff spaces. If $C(K_1)$ and $C(K_2)$ are isometrically isomorphic, then K_1 and K_2 are homeomorphic. Furthermore, if $T : C(K_1) \to C(K_2)$ is an isometric isomorphism,* *then there exists some $u \in C(K_2)$ such that $|u(s)| = 1$ for all $s \in K_2$, and such that*

$$Tf(s) = u(s) f(\phi(s)), \quad s \in K_2,$$

where $\phi : K_2 \to K_1$ is a homeomorphism.

Before proving the Banach–Stone Theorem, we will provide a simple lemma that will not only help us now, but will come in handy later, too. In order to prove this lemma, however, we need to make use of another result from general topology.

Theorem 5.65 (Urysohn's Lemma) *A topological space X is normal if and only if any two disjoint closed subsets A and B can be separated by a continuous function. That is, if there exists a continuous function $f : X \to [0, 1]$ such that $f|_A = 0$ and $f|_B = 1$.*

We recall that a topological space X is *normal* if for disjoint closed sets E and F, there exist disjoint open sets U and V such that $E \subseteq U$ and $F \subseteq V$. We will not prove Urysohn's Lemma; however, we will observe that, as a consequence, if K is a compact Hausdorff space, then $C(K)$ *separates the points* of K. That is, if a and b are distinct points in K, then there is a function $f \in C(K)$ such that $f(a) = 0$ and $f(b) = 1$. (See Exercise 5.7.)

Lemma 5.66 *Let K be a compact Hausdorff space. If $\Delta = \{\delta_s : s \in K\}$, then K is homeomorphic to Δ with the subspace topology inherited from $(M(K), w^*)$.*

Proof We remind the reader that for each $s \in K$, the Dirac measure at s is a measure δ_s defined so that $\int_K f \, d\delta_s = f(s)$ for all $f \in C(K)$. The set Δ is closed in the w^* topology and $\Delta \subseteq B_{M(K)}$. Therefore, Δ is w^*-compact.

Define a map $\psi : K \to \Delta$ by $\psi(s) = \delta_s$ for every $s \in K$. Clearly, ψ is a surjection. Suppose that s and t are two distinct points in K such that $\psi(s) = \psi(t)$. Then $\delta_s = \delta_t$. This means that $\delta_s(f) = \delta_t(f)$ for all $f \in C(K)$. Thus, $f(s) = f(t)$ for all $f \in C(K)$. This contradicts the fact that $C(K)$ separates the points of K. (See the comments before the statement of Lemma 5.66.) Therefore, $\psi(s) = \psi(t)$ only if $s = t$, and so ψ is an injection as well as a surjection.

We next show that ψ is a homeomorphism by showing that it is a continuous closed map (so that it maps closed sets to closed sets). Certainly, ψ is continuous in the w^* topology on Δ, since for every $f \in C(K)$,

$$\psi(s)(f) = \int_K f \, d\delta_s = f(s), \quad s \in K,$$

and because f is continuous (by assumption). Now let F be a closed set in K. Then F is compact, because it is a closed subset of the compact set K. Since ψ is continuous for Δ with the w^* topology, it follows that $\psi(F)$ is w^*-compact in Δ. Since $\psi(F)$ is a compact set in a Hausdorff topology, it must be closed (in that topology). Therefore, ψ is a closed map, and it follows that ψ is a homeomorphism. (See Exercise 5.4.)

We are now prepared to prove the Banach–Stone Theorem.

Proof of Theorem 5.64 Let K_1 and K_2 be compact Hausdorff spaces and suppose $T : C(K_1) \to C(K_2)$ is an isometric isomorphism. We wish to show that K_1 and K_2

are homeomorphic. Observe that T maps extreme points of $B_{C(K_1)}$ to extreme points of $B_{C(K_2)}$. To see this, let f be an extreme point in $B_{C(K_1)}$ and suppose that g and h are functions in $B_{C(K_2)}$ such that $Tf = \frac{1}{2}(g + h)$. Then $f = \frac{1}{2}(T^{-1}g + T^{-1}h)$, and from this we deduce that $T^{-1}g = T^{-1}h = f$ (because f is an extreme point). It follows that $g = h = Tf$, and so Tf is an extreme point in $B_{C(K_2)}$ whenever f is an extreme point in $B_{C(K_1)}$. In particular, since χ_{K_1} (which is identically equal to 1 on K_1) is extreme in $B_{C(K_1)}$, its image $T(\chi_{K_1})$ is extreme in $B_{C(K_2)}$. Consequently, we have that $|T(\chi_{K_1})(s)| = 1$ for all $s \in K_2$.

If we define an operator $S : C(K_1) \to C(K_2)$ by $Sf = (Tf)/T(\chi_{K_1})$ for all functions $f \in C(K_1)$, then S is an isometry such that $S(\chi_{K_1}) = \chi_{K_2}$. Therefore, we may assume without loss of generality that $T(\chi_{K_1}) = \chi_{K_2}$.

The adjoint $T^* : M(K_2) \to M(K_1)$ is also an isometry because $(T^*)^{-1} = (T^{-1})^*$, and so $T^*(B_{M(K_2)}) = B_{M(K_1)}$. Suppose $\mu \in \mathcal{P}(K_2)$. Then

$$(T^*\mu)(K_1) = \int_{K_1} 1 \, d\, T^*\mu = \int_{K_2} T(1) \, d\mu = \int_{K_2} 1 \, d\mu = \mu(K_2) = 1.$$

(Here we use the fact that $\chi_{K_1} = 1$ on K_1 and $\chi_{K_2} = 1$ on K_2.) The measure $T^*\mu$ is then an element of $B_{M(K_1)}$ such that $T^*\mu(K_1) = 1$. It can be shown that these two facts imply that $T^*\mu \in \mathcal{P}(K_1)$. (See Exercise 5.28.)

As an isometry, T^* will map extreme points to extreme points. By Proposition 5.57, the extreme points in $M(K_1)$ and $M(K_2)$ are the Dirac masses, and so for each $s \in K_2$, there must be some $t_s \in K_1$ (depending on s) such that $T^*\delta_s = \delta_{t_s}$. Define a map $\phi : K_2 \to K_1$ by $\phi(s) = t_s$ for each $s \in K_2$. Then,

$$T^*\delta_s = \delta_{\phi(s)}, \quad s \in K_2.$$

The map ϕ is a bijection from K_2 onto K_1, which follows from the fact that T^* is a bijection from $M(K_2)$ onto $M(K_1)$. We wish to show that ϕ is a homeomorphism, and as such we must show that both ϕ and ϕ^{-1} are continuous.

Let $\psi : K_1 \to \{\delta_t : t \in K_1\}$ and $\varphi : K_2 \to \{\delta_s : s \in K_2\}$ be defined by $\psi(t) = \delta_t$ and $\varphi(s) = \delta_s$ for all $t \in K_1$ and $s \in K_2$. By Lemma 5.66, the maps ψ and φ are homeomorphisms. Observe that, for all $s \in K_2$,

$$\phi(s) = \psi^{-1}(\delta_{\phi(s)}) = \psi^{-1}(T^*\delta_s) = (\psi^{-1} \circ T^* \circ \varphi)(s).$$

We assumed T was continuous, and hence $T^* : M(K_2) \to M(K_1)$ is weak*-to-weak* continuous, by Proposition 5.44. Therefore, ϕ is continuous. Similarly, we can show that $\phi^{-1} = \varphi^{-1} \circ (T^{-1})^* \circ \psi$, and so ϕ^{-1} is continuous. Thus, ϕ is a homeomorphism.

Finally, we have

$$Tf(s) = \int_{K_2} (Tf) \, d\delta_s = \int_{K_1} f \, d(T^*\delta_s) = \int_{K_1} f \, d\delta_{\phi(s)} = f(\phi(s)).$$

The factor u appearing in the statement of the theorem does not appear now because of the normalization we made at the beginning of the proof. Had we not assumed

$T(\chi_{K_1}) = \chi_{K_2}$, then we would have $u = T(\chi_{K_1})$, which is a function with the property that $|T(\chi_{K_1})(s)| = 1$ for all $s \in K_2$.

Remark T here are other ways to prove that K_1 and K_2 are homeomorphic using properties of $C(K_1)$ and $C(K_2)$. For example, it is possible to prove that K_1 and K_2 are homeomorphic using only the fact that $C(K_1)$ and $C(K_2)$ are isomorphic as rings, so that no norm structure is required in the proof. (See [15] for more.)

Exercises

Exercise 5.1 Let a and b be real numbers such that $a < b$. Show explicitly that (a, b) is not a compact set in $\mathbb{R}$ by finding an open cover with no finite subcover. (Use the standard topology on $\mathbb{R}$.)

Exercise 5.2 Prove the following theorems:

(a) Every closed subset of a compact space is compact.
(b) The image of a compact space under a continuous map is compact.
(c) Every compact subset of a Hausdorff space is closed.
(d) Let X be a compact space and Y be a Hausdorff space. If $f : X \to Y$ is continuous, then f is a *closed map*. (That is, $f(C)$ is closed in Y whenever C is closed in X.)

Exercise 5.3 Let K be a compact topological space and suppose $\phi : K \to E$ is a continuous one-to-one map, where E is a Hausdorff topological space. Show that ϕ is a homeomorphism onto its image $\phi(K)$. (*Hint:* See Exercise 5.2.)

Exercise 5.4 Let X and Y be topological spaces. If $\phi : X \to Y$ is a continuous closed bijection, show that ϕ is a homeomorphism.

Exercise 5.5 Let X be a Hausdorff space. If A and B are disjoint compact subsets of X, then show there exist disjoint open sets U and V such that $A \subseteq U$ and $B \subseteq V$.

Exercise 5.6 Suppose X is a compact Hausdorff space. Show that X is a *normal* space. That is, if E and F are disjoint closed subsets of X, show there exist disjoint open sets U and V such that $E \subseteq U$ and $F \subseteq V$. (*Hint:* Use Exercise 5.5.)

Exercise 5.7 Suppose X is a compact Hausdorff space. Use Urysohn's Lemma (Theorem 5.65) and Exercise 5.6 to show that $C(X)$ separates the points of X. That is, if a and b are distinct points in X, show there is a function $f \in C(X)$ such that $f(a) = 0$ and $f(b) = 1$.

Exercise 5.8 Show that limits in a Hausdorff space are unique. That is, if X is a Hausdorff space, show that a sequence $(x_n)_{n=1}^{\infty}$ in X cannot converge to two distinct limits x and $\tilde{x}$.

Exercise 5.9 Prove a metric space is second countable if and only if it is separable.

Exercise 5.10 (a) Suppose (M, d) is a metric space and A is a set. If $f : A \to M$ is an injective function, show that $d_A(x, y) = d(f(x), f(y))$ for $(x, y) \in A \times A$ defines a metric on A.
(b) Show that $\rho(x, y) = |\log(y/x)|$ defines a metric on the set $\mathbb{R}^+ = (0, \infty)$.

Exercise 5.11 Let $d(x, y) = |\phi(x) - \phi(y)|$, where $\phi(x) = x/(1 + |x|)$. Show that d is a metric on $\mathbb{R}$ that is not complete.

Exercise 5.12 Show that the space $L_p(0, 1)$, where $0 < p < 1$, is a complete metric space with metric $d(f, g) = \|f - g\|_p^p$, where $\|f\|_p^p = \int_0^1 |f(t)|^p \, dt$. (*Hint:* Use Lemma 2..4, which still applies in this case, despite the fact that $\| \cdot \|_p^p$ is not a norm.)

Exercise 5.13 Let $L_0(0, 1)$ denote the space of all (equivalence classes of) Lebesgue measurable functions on $[0, 1]$. Define

$$d(f, g) = \int_0^1 \min\left(1, |f(s) - g(s)|\right) ds, \quad \{f, g\} \subseteq L_0(0, 1).$$

Prove that d is a metric on $L_0(0, 1)$. Furthermore, show that $d(f_n, f) \to 0$ if and only if $f \to 0$ in measure. Conclude that $L_0(0, 1)$ is a topological vector space (i.e., show that addition and scalar multiplication are continuous).

Exercise 5.14 Show that any continuous linear functional on $L_0(0, 1)$ is identically zero.

Exercise 5.15 Suppose (Ω, μ) is a positive measure space such that $\mu(\Omega) = 1$.

(a) If $0 < p < q \leq 1$, then show $\|f\|_p \leq \|f\|_q$ for all measurable functions f.
(b) Assume that f is a measurable function such that $\|f\|_r < \infty$ for some $r \leq 1$. Prove that

$$\lim_{p \to 0^+} \|f\|_p = \exp\left(\int_\Omega \log |f(\omega)| \, \mu(d\omega)\right),$$

where we adopt the convention that $e^{-\infty} = 0$.

(Compare to Exercise 2.13.)

Exercise 5.16 Suppose (Ω, μ) is a positive measure space such that $\mu(\Omega) = 1$. Let $L_0(\mu)$ denote the space of all (equivalence classes of) μ-measurable functions on Ω. For any measurable function f, define

$$\|f\|_0 = \exp\left(\int_\Omega \log |f(\omega)| \, \mu(d\omega)\right).$$

(See Exercise 5.15.) If $d(f, g) = \|f - g\|_0$ for all measurable functions f and g in $L_0(\mu)$, does d define a metric on $L_0(\mu)$?

Exercise 5.17 Let X be a locally convex topological vector space with η a base of absolutely convex neighborhoods of 0. Verify that the topology on X is generated by the family of Minkowski functionals $\{p_U\}_{U \in \eta}$. Deduce that $x_n \to 0$ in X if and only if $p_U(x_n) \to 0$ for all $U \in \eta$.

Exercise 5.18 Consider the set $\partial B_{\ell_2} = \{x \in \ell_2 : \|x\|_2 = 1\}$. Show that ∂B_{ℓ_2} is closed in the norm topology, but not the weak topology on ℓ_2. (This example shows that the convexity assumption cannot be omitted from Mazur's Theorem.)

Exercise 5.19 Let K be a compact subset of a Hausdorff topological vector space E, and suppose C is a closed subset of E. Show that $C - K = \{x - y : x \in C, y \in K\}$ is a closed subset of E.

Exercise 5.20 Let E be a locally convex topological vector space and suppose K is a closed linear subspace of E. If $x_0 \notin K$, show that there exists a continuous linear functional $f \in E^*$ such that $f(x_0) = 1$, but $f(x) = 0$ for all $x \in K$.

Exercise 5.21 Let E be a real locally convex topological vector space. Suppose K is a nonempty compact convex subset of E, and C is a nonempty closed convex subset of E, and that $K \cap C = \emptyset$. Show there is a continuous linear functional ϕ on E such that

$$\inf_{x \in C} \phi(x) > \sup_{y \in K} \phi(y).$$

(We say ϕ *separates* K and C.)

Exercise 5.22 Let X be a real Banach space and let E be a weak*-closed subspace of X^*. If ϕ is a weak* continuous linear functional on E with $\|\phi\| = 1$, show for any $\epsilon > 0$ there exists an $x \in X$ with $\|x\| < 1 + \epsilon$ such that $\phi(e^*) = e^*(x)$ for all $e^* \in E$. (*Hint:* Consider the sets $C = \{e^* \in E : \phi(e^*) = 1\}$ and $K = (1+\epsilon)^{-1} B_{X^*}$.)

Exercise 5.23 Let $(X, \|\cdot\|)$ be a real reflexive Banach space and let $\phi \in X^*$. Define a map $f : X \to \mathbb{R}$ by $f(x) = \frac{1}{2}\|x\|^2 - \phi(x)$ for all $x \in X$. Show that f attains a minimum value.

Exercise 5.24 Let $(f_n)_{n=1}^\infty$ be a bounded sequence in $C[0, 1]$. Show that $f_n(s) \to 0$ for every $s \in [0, 1]$ if and only if $f_n \to 0$ weakly.

Exercise 5.25 Let $p \in [1, \infty)$ and for each $n \in \mathbb{N}$ let e_n be the sequence with 1 in the nth coordinate, and 0 elsewhere. Show that the sequence $(ne_n)_{n=1}^\infty$ does not converge weakly to 0 in ℓ_p. (Compare to Example 5.28.)

Exercise 5.26 Let $p \in (1, \infty)$ and for each $n \in \mathbb{N}$ define a function $f_n : [0, 1] \to \mathbb{R}$ by $f_n(x) = n^{1/p} \chi_{[0,1/n]}(x)$ for all $x \in [0, 1]$. Show that $f_n \to 0$ weakly in $L_p(0, 1)$, but not in norm. (Recall that χ_A is the *characteristic function* of the measurable set A.)

Exercise 5.27 Suppose K is a real compact Hausdorff space. Show that the set $\mathcal{P}(K)$ of regular Borel probability measures on K is a convex and w^*-closed subset of $M(K)$, the set of regular Borel measures on K. Show that $\mathcal{P}(K)$ is an extremal set in the unit ball of $M(K)$. (See Sect. 5.8.)

Exercise 5.28 Let K be a compact Hausdorff space and let ν be a Borel measure on K so that $\|\nu\|_{M(K)} \leq 1$ and $\nu(K) = 1$. Show that ν is a probability measure.

Exercise 5.29 Let G be a group that is also a topological space. Show that G is a topological group if and only if the map $g : G \times G \to G$ defined by $g(x, y) = x^{-1}y$ is continuous.

Exercise 5.30 Let X be a real separable Banach space. Show that B_{X^*} is metrizable in the weak* topology. (*Hint:* Let $(x_n)_{n=1}^{\infty}$ be a countable dense subset in X and define $\phi(x^*) = (x^*(x_n))_{n=1}^{\infty} \in \mathbb{R}^{\mathbb{N}}$.)

Exercise 5.31 Let X be a Banach space. If $x \in X$, use the Banach–Alaoglu Theorem to prove that there exists an element $x^* \in X^*$ such that $\|x^*\| = 1$ and $x^*(x) = \|x\|$. (*Note:* We proved this in Proposition 3.29 using the Hahn–Banach Theorem.)

Exercise 5.32 A subset E of a topological vector space X is called *bounded* if for every open neighborhood V of 0, there exists an $n \in \mathbb{N}$ such that $E \subseteq nV$. Show that any compact subset of a topological vector space is bounded.

Exercise 5.33 A topological vector space X has the *Heine–Borel property* if every closed and bounded subset of X is compact. (See Exercise 5.32 for the definition of a bounded set in a topological vector space.)

(a) Show that a Banach space has the Heine–Borel property if and only if it is finite-dimensional. (*Hint:* Use Lemma 5.36.)
(b) Show that (X^*, w^*) has the Heine–Borel property if X is a Banach space.

Exercise 5.34 Show that $C[0, 1]$ is not reflexive by showing that $B_{C[0,1]}$ is not compact in the weak topology. (*Hint:* Find a $\Lambda \in C[0, 1]^*$ such that $\Lambda(B_{C[0,1]})$ is open.)

Exercise 5.35 Let X be an infinite-dimensional Banach space. Show that (X^*, w^*) is of the first category in itself.

Chapter 6
Compact Operators and Fredholm Theory

6.1 Compact Operators

Suppose X is a vector space (over $\mathbb{R}$ or $\mathbb{C}$) and let $T : X \to X$ be a linear operator. Let us recall some basic definitions from linear algebra. The *kernel* (or *nullspace*) of T is the subspace of X given by $\ker(T) = \{x : Tx = 0\}$. The *range* (or *image*) of T is given by $\mathrm{ran}(T) = \{Tx : x \in X\}$. We say that $x \in X$ is an *eigenvector* of T if $x \neq 0$ and there exists some scalar λ (called an *eigenvalue*) such that $Tx = \lambda x$.

The behavior of linear operators on finite-dimensional vector spaces has been studied for a long time and is well understood. Some of the most basic and important theorems of linear algebra rely heavily on the dimension of the underlying vector space X. Consider the following well-known theorems.

Theorem 6.1 *Let X be a finite-dimensional vector space and suppose $T : X \to X$ is a linear operator. Then:*

(i)   *The map T is one-to-one if and only if T maps X onto X.*
(ii) $\dim(X) = \dim(\ker T) + \dim(\mathrm{ran} T)$ *(Rank-Nullity Theorem).*
(iii)   *If X is a nontrivial complex vector space, then T has at least one eigenvector.*

Notice that *(ii)* implies *(i)*, and certainly *(ii)* is dependent upon the underlying dimension of X. To see how *(iii)* depends on the finite-dimensionality of X, recall that eigenvalues can be calculated by solving the equation $\det(\lambda I - A) = 0$ for λ, where I is the identity matrix and A is any matrix representation of T. Because X is finite-dimensional, the expression $\det(\lambda I - A)$ is a polynomial, and as such must have a root because $\mathbb{C}$ is algebraically closed (by the Fundamental Theorem of Algebra).

These theorems rely on the finite-dimensionality of X, and so any attempt to generalize them to infinite-dimensional vector spaces requires careful consideration. In fact, as stated, Theorem 6.1 is false in a general Banach space. To see this explicitly, we now consider some examples, where the Banach spaces can be either real or complex.

© Springer Science+Business Media, LLC 2014
A. Bowers, N. J. Kalton, *An Introductory Course in Functional Analysis*,
Universitext, DOI 10.1007/978-1-4939-1945-1_6

Example 6.2 Let $\{p, q\} \subseteq [1, \infty]$ and consider the shift operators $T : \ell_p \to \ell_p$ given by

$$T(\xi_1, \xi_2, \xi_3, \ldots) = (0, \xi_1, \xi_2, \xi_3, \ldots), \quad (\xi_k)_{k=1}^{\infty} \in \ell_p,$$

and $S : \ell_q \to \ell_q$ given by

$$S(\eta_1, \eta_2, \eta_3, \ldots) = (\eta_2, \eta_3, \eta_4, \ldots), \quad (\eta_k)_{k=1}^{\infty} \in \ell_q.$$

The map T is clearly one-to-one, but certainly is not onto. On the other hand, the map S is onto, but clearly not one-to-one. We therefore cannot extend Theorem 6.1(*i*) to infinite dimensions. Since the spaces ℓ_p and ℓ_q have infinite dimensions, it is not obvious if the statement of Theorem 6.1(*ii*) has any significant meaning in its current form.

To see that Theorem 6.1(*iii*) does not extend to infinite-dimensional vector spaces, suppose $(\xi_k)_{k=1}^{\infty}$ is an eigenvector for T. Then there exists some scalar λ such that

$$T(\xi_1, \xi_2, \xi_3, \ldots) = \lambda(\xi_1, \xi_2, \xi_3, \ldots),$$

or

$$(0, \xi_1, \xi_2, \xi_3, \ldots) = (\lambda \xi_1, \lambda \xi_2, \lambda \xi_3, \ldots).$$

The only way this can happen is if $\xi_k = 0$ for all $k \in \mathbb{N}$, contradicting the assumption that $(\xi_k)_{k=1}^{\infty}$ is an eigenvector. It follows that T does not have any eigenvectors. The map S, however, has many eigenvectors. (See Exercise 6.1.)

In the case $p = q$, notice that $S \circ T$ is the identity map on ℓ_p, but $T \circ S$ is not. It is also worth mentioning that if p and q are conjugate exponents ($1/p + 1/q = 1$), then $S = T^*$.

Example 6.3 Let $T : C[0, 1] \to C[0, 1]$ be defined by $Tf(x) = xf(x)$. It is clear that T is a bounded linear operator and that $\|T\| \leq 1$. If λ is an eigenvalue for T, then there exists some function $f \in C[0, 1]$ such that $xf(x) = \lambda f(x)$ for all $x \in [0, 1]$. It follows that $(x - \lambda)f(x) = 0$ for all $x \in [0, 1]$. This can happen only if $f = 0$. Therefore, T has no eigenvectors.

We see that our intuition from linear algebra can fail us in infinite dimensions. Indeed, for general linear operators on Banach spaces, much of what we know for finite-dimensional vector spaces fails to remain true. For this reason, we impose additional conditions on our operators. We now define one class of operators for which our intuition can serve as a guide.

Definition 6.4 Let X and Y be Banach spaces. A bounded linear operator $T : X \to Y$ is said to be a *finite rank* operator if $\dim(\operatorname{ran} T) < \infty$. In such a case, the number $\dim(\operatorname{ran} T)$ is called the *rank* of T and is denoted $\operatorname{rank}(T)$.

For finite rank operators, we can still use some tools from linear algebra. This class of operators is too small, however, and so we wish to introduce a less restrictive condition on our operators. To that end, we introduce a definition.

Definition 6.5 Let X be a topological space. A subset E of X is called *relatively compact* if it has compact closure; that is, if $\overline{E}$ is compact in X.

Example 6.6 Let X be a Banach space. By Goldstine's Theorem (Theorem 5.40), B_X is dense in $B_{X^{**}}$ in the weak* topology on X^{**}. By the Banach–Alaoglu Theorem (Theorem 5.39), $B_{X^{**}}$ is compact in the weak* topology on X^{**}. Therefore, B_X has weak*-compact closure, and so is relatively compact, in (X^{**}, w^*).

Definition 6.7 Let X and Y be Banach spaces. A bounded linear map $T : X \to Y$ is called a *compact operator* if $T(B_X)$ is a relatively compact set in Y.

It is important to note that the image of the unit ball under a compact operator need not be a compact set. (See Example 6.10.)

Example 6.8 Finite rank operators are compact, by the Heine–Borel Theorem.

Example 6.9 Suppose that X and Y are Banach spaces such that X is reflexive, and let $T : X \to Y$ be a bounded linear operator. By Proposition 5.42, the operator T is *weakly continuous*. (That is, T is continuous in the weak topologies on X and Y.) Since X is reflexive, the weak and weak* topologies on X coincide. Therefore, by the Banach–Alaoglu Theorem (Theorem 5.39), the unit ball B_X is weakly compact. We know that T is weakly continuous, and so $T(B_X)$ is weakly compact in Y. It follows that $T(B_X)$ is weakly closed in Y, and so $T(B_X)$ is closed in the norm topology on Y, by Mazur's Theorem (Theorem 5.45). Therefore, if T is a compact operator, then $T(B_X)$ is compact in the norm topology on Y.

In Example 6.9, the set $T(B_X)$ is compact if and only if T is a compact operator. The proof of this fact relies on the assumption that X is reflexive. If X is not reflexive, then $T(B_X)$ may not be closed (and so not compact) even if T is a compact operator. We illustrate this in the next example.

Example 6.10 Let $\lambda = (\lambda_n)_{n=1}^{\infty}$ be a sequence in ℓ_1 and define a function $T : c_0 \to \mathbb{R}$ by

$$T(\xi) = \sum_{n=1}^{\infty} \lambda_n \xi_n, \quad \xi = (\xi_n)_{n=1}^{\infty} \in c_0.$$

For simplicity, assume $\lambda_n > 0$ for each $n \in \mathbb{N}$. Observe that T is linear because the series defining $T(\xi)$ is absolutely convergent for all $\xi \in c_0$. For each $\xi \in c_0$,

$$|T(\xi)| = \Big| \sum_{n=1}^{\infty} \lambda_n \xi_n \Big| \leq \Big(\sup_{n \in \mathbb{N}} |\xi_n| \Big) \sum_{n=1}^{\infty} \lambda_n = \|\xi\|_{c_0} \|\lambda\|_{\ell_1}.$$

Thus, T is bounded and $\|T\| \leq \|\lambda\|_{\ell_1}$. In fact, $\|T\| = \|\lambda\|_{\ell_1}$ because

$$\|T\| = \sup_{\xi \in B_{c_0}} |T(\xi)| \geq |T(e_1 + \cdots + e_N)| = \sum_{n=1}^{N} \lambda_n, \tag{6.1}$$

for all $N \in \mathbb{N}$. (Recall that e_k is the sequence with a 1 in the kth coordinate, and zero elsewhere.)

We have shown that T is bounded and linear (and hence continuous); it is also compact because it is a rank 1 operator. The set $T(B_{c_0})$ is not closed, however, because

$\sup_{\xi \in B_{c_0}} |T(\xi)| = \|\lambda\|_{\ell_1}$, by (6.1), but there is no $\xi \in c_0$ for which this maximum is attained. In fact, $T(B_{c_0}) = (-\|\lambda\|_{\ell_1}, \|\lambda\|_{\ell_1})$, which is not closed, but has a compact closure.

For another example of a compact operator $T : X \to Y$ such that $T(B_X)$ is not closed in Y, see Example 5.43. (See also Exercise 5.34.)

In the case of a metric space, a notion closely related to compactness is that of *total boundedness*.

Definition 6.11 A subset of a metric space is called *totally bounded* if it can be covered by finitely many closed balls of radius ε for any $\varepsilon > 0$.

Certainly, if a subset E of a metric space is totally bounded, then any subset of E is also totally bounded.

Theorem 6.12 *A closed subset of a complete metric space is compact if and only if it is totally bounded.*

Proof Let M be a complete metric space with metric d. For ease of notation, we will denote the open ball of radius ε about x in M by $B_\varepsilon(x)$ and the closed ball of radius ε about x in M by $\overline{B}_\varepsilon(x)$. That is, $B_\varepsilon(x) = \{y \in M : d(x, y) < \varepsilon\}$ and $\overline{B}_\varepsilon(x) = \{y \in M : d(x, y) \leq \varepsilon\}$.

Suppose K is a compact (and hence closed) subset of M. Let $\varepsilon > 0$ be given. The collection of sets $\{B_\varepsilon(x) : x \in K\}$ forms an open cover of K, and so (by compactness) admits a finite subcover. Thus, there is a finite set $\{x_1, \dots, x_N\}$ in K such that

$$K \subseteq B_\varepsilon(x_1) \cup \cdots \cup B_\varepsilon(x_N) \subseteq \overline{B}_\varepsilon(x_1) \cup \cdots \cup \overline{B}_\varepsilon(x_N).$$

Thus, K is totally bounded.

Now suppose that K is a totally bounded closed subset of M. We wish to show that K is compact. We will assume K is not compact and derive a contradiction. Let $\mathcal{V}$ be a collection of open sets that cover K and suppose $\mathcal{V}$ does not contain a finite subcover of K. Since K is totally bounded, it can be covered by a finite collection of closed balls of radius 1, say

$$K \subseteq \overline{B}_1(x_{1,1}) \cup \cdots \cup \overline{B}_1(x_{1,n_1}).$$

Since K cannot be covered by finitely many sets in $\mathcal{V}$, there is some member of the collection $\{x_{1,1}, \dots, x_{1,n_1}\}$, call it x_1, such that $K \cap \overline{B}_1(x_1)$ cannot be covered by finitely many sets in $\mathcal{V}$. Let $K_1 = K \cap \overline{B}_1(x_1)$.

The set K_1 is totally bounded (because it is a subset of K). Thus, K_1 can be covered by a finite collection of closed balls of radius $1/2$, say

$$K_1 \subseteq \overline{B}_{\frac{1}{2}}(x_{2,1}) \cup \cdots \cup \overline{B}_{\frac{1}{2}}(x_{2,n_2}).$$

Since K_1 cannot be covered by finitely many sets in $\mathcal{V}$, there is some member of the collection $\{x_{2,1}, \dots, x_{2,n_2}\}$, call it x_2, such that $K_1 \cap \overline{B}_{\frac{1}{2}}(x_2)$ cannot be covered by finitely many sets in $\mathcal{V}$. Let $K_2 = K_1 \cap \overline{B}_{\frac{1}{2}}(x_2)$.

Continuing inductively, we find a sequence of points $(x_n)_{n=1}^\infty$ in M and construct a sequence of subsets $(K_n)_{n=1}^\infty$ of K such that

(i) $K_n \subseteq \overline{B}_{\frac{1}{n}}(x_n)$ for all $n \in \mathbb{N}$,
(ii) $K \supseteq K_1 \supseteq K_2 \supseteq K_3 \supseteq \cdots$, and
(iii) K_n cannot be covered by finitely many sets in $\mathcal{V}$ for any $n \in \mathbb{N}$.

For each $n \in \mathbb{N}$, choose $y_n \in K_n$. Then the sequence $(y_n)_{n=1}^\infty$ is a Cauchy sequence, by the properties in (i) and (ii), above. Since M is assumed to be a complete metric space, there is a point $y \in K$ (because K is closed) such that $y_n \to y$ as $n \to \infty$. We chose $y_n \in K_n$ for each $n \in \mathbb{N}$, and so it follows from (i) that $x_n \to y$ as $n \to \infty$.

By assumption, $\mathcal{V}$ is an open cover of K and $y \in K$. Hence, there is an open set $V \in \mathcal{V}$ such that $y \in V$. Since $x_n \to y$ as $n \to \infty$, there exists some $N \in \mathbb{N}$ such that $\overline{B}_{\frac{1}{n}}(x_n) \subseteq V$ for all $n \geq N$. This violates property (iii), because $K_n \subseteq \overline{B}_{\frac{1}{n}}(x_n)$ for all $n \in \mathbb{N}$, and so we have obtained a contradiction. Therefore, the cover $\mathcal{V}$ contains a finite subcover of K, and so K is compact.

Definition 6.13 We denote the collection of compact operators from X to Y by the symbol $\mathcal{K}(X, Y)$. We use $\mathcal{K}(X)$ to denote $\mathcal{K}(X, X)$.

The next theorem shows that the collection of compact operators is well-behaved.

Theorem 6.14 *Let X and Y be Banach spaces. The set $\mathcal{K}(X,Y)$ of compact operators from X to Y forms a closed linear subspace of $\mathcal{L}(X,Y)$.*

Proof First, we will show that $\mathcal{K}(X, Y)$ is a linear subspace. Suppose that $S : X \to Y$ and $T : X \to Y$ are compact operators and let α and β be scalars. We wish to show that the operator $\alpha S + \beta T$ is compact. Define a map $\phi : Y \times Y \to Y$ by

$$\phi(u, v) = \alpha u + \beta v, \quad (u, v) \in Y \times Y.$$

The map ϕ is continuous because Y is a topological vector space. By assumption, the set $\overline{S(B_X)} \times \overline{T(B_X)}$ is compact in $Y \times Y$. Therefore, the set $H = \phi(\overline{S(B_X)} \times \overline{T(B_X)})$ is compact in Y, as the continuous image of a compact set. Since $(\alpha S + \beta T)(B_X) \subseteq H$, and H is compact, it follows that $\overline{(\alpha S + \beta T)(B_X)}$ is compact, and so $\alpha S + \beta T$ is a compact operator.

It remains to show that $\mathcal{K}(X, Y)$ is a closed subspace. Suppose $T \in \mathcal{L}(X, Y)$ is a bounded linear operator and let $(T_n)_{n=1}^\infty$ be a sequence of compact operators in $\mathcal{K}(X, Y)$ such that $\|T_n - T\| \to 0$ as $n \to \infty$. We wish to show that $T(B_X)$ is relatively compact. By Theorem 6.12, it suffices to show that its closure is totally bounded. This will follow if we show, for any $\varepsilon > 0$, the set $T(B_X)$ is contained in a finite union of balls of radius ε in Y.

Fix $\varepsilon > 0$. There exists an $n \in \mathbb{N}$ such that $\|T_n - T\| < \varepsilon/2$. Because $T_n(B_X)$ is relatively compact, there exists a finite set of points $\{y_1, \ldots, y_N\}$ in Y such that

$$T_n(B_X) \subseteq \bigcup_{j=1}^N \left(y_j + \frac{\varepsilon}{2} B_Y\right).$$

If $x \in B_X$, then $\|T_n x - Tx\| < \varepsilon/2$, and so $Tx \in \bigcup_{j=1}^{N} (y_j + \varepsilon B_Y)$. Thus, $T(B_X)$ is contained in a finite union of ε-balls. We conclude that $\overline{T(B_X)}$ is totally bounded, and hence compact. Therefore, T is a compact operator, as required. $\qquad\square$

Remark 6.15 There is a more straightforward proof that $\mathcal{K}(X, Y)$ is a subspace of $\mathcal{L}(X, Y)$ that uses the fact that, in a metric space, compactness is equivalent to *sequential compactness* (i.e., every sequence has a convergent subsequence). We will prove this shortly (for complete metric spaces), but for now let us take this fact for granted.

Let $(x_n)_{n=1}^{\infty}$ be a sequence in B_X. There exists a subsequence $(x_{n_k})_{k=1}^{\infty}$ such that $(Sx_{n_k})_{k=1}^{\infty}$ converges (not necessarily in $S(B_X)$). There exists a further subsequence $(x_{n_{k_j}})_{j=1}^{\infty}$ such that $(Tx_{n_{k_j}})_{j=1}^{\infty}$ converges (again, not necessarily in $T(B_X)$). Then $((\alpha S + \beta T)(x_{n_{k_j}}))_{j=1}^{\infty}$ converges, and we obtain the desired result.

We will now show that compactness is equivalent to sequential compactness in a complete metric space. The theorem remains valid even without the completeness assumption, but we need it only for Banach spaces, and so we opt for a theorem with a simpler proof.

Theorem 6.16 *Let M be a complete metric space. A closed subset K of M is compact if and only if it is sequentially compact.*

Proof Let d be a complete metric on M. For notational simplicity, we will denote by $B_\delta(s)$ the open ball of radius δ about $s \in M$; that is, $B_\delta(s) = \{x \in M : d(s, x) < \delta\}$.

Suppose K is a compact set and let $(a_n)_{n=1}^{\infty}$ be a sequence in K. We will show that $(a_n)_{n=1}^{\infty}$ has a convergent subsequence. Since K is compact, we can cover K with a finite number of open balls of radius 1. At least one of these must contain a_n for infinitely many values of $n \in \mathbb{N}$. That is, there is a $x_1 \in K$ such that $a_n \in B_1(x_1)$ for infinitely many $n \in \mathbb{N}$. Let $N_1 = \{n : n \in \mathbb{N} \text{ and } a_n \in B_1(x_1)\}$. Denote the integers in N_1 by $(n_{1j})_{j=1}^{\infty}$, where $n_{1j} < n_{1k}$ whenever $j < k$. Observe that $(a_{n_{1j}})_{j=1}^{\infty}$ is a subsequence of $(a_n)_{n=1}^{\infty}$, all the terms of which are in $B_1(x_1)$.

As before, because K is compact, we can cover K using a finite number of open balls of radius $\frac{1}{2}$. At least one of these balls contains $a_{n_{1j}}$ for infinitely many $j \in \mathbb{N}$. That is, there exists a $x_2 \in K$ such that $a_{n_{1j}} \in B_{\frac{1}{2}}(x_2)$ for infinitely many $j \in \mathbb{N}$. Let $N_2 = \{n_{1j} : j \in \mathbb{N} \text{ and } a_{n_{1j}} \in B_{\frac{1}{2}}(x_2)\}$. Denote the integers in N_2 by $(n_{2j})_{j=1}^{\infty}$, where $n_{2j} < n_{2k}$ whenever $j < k$. Observe that $(a_{n_{2j}})_{j=1}^{\infty}$ is a subsequence of $(a_{n_{1j}})_{j=1}^{\infty}$, all the terms of which are in the ball $B_{\frac{1}{2}}(x_2)$.

Continuing inductively, for each $i \in \mathbb{N}$ we find $x_i \in K$ and a subsequence $(a_{n_{ij}})_{j=1}^{\infty}$ of $(a_{n_{(i-1)j}})_{j=1}^{\infty}$ such that $a_{n_{ij}} \in B_{\frac{1}{i}}(x_i)$.

We define a subsequence $(a_{n_i})_{i=1}^{\infty}$ of $(a_n)_{n=1}^{\infty}$ by letting $a_{n_i} = a_{n_{ii}}$ for each $i \in \mathbb{N}$. We claim that $(a_{n_i})_{i=1}^{\infty}$ is a Cauchy sequence. To see this, let $\varepsilon > 0$. Choose a positive integer $N \in \mathbb{N}$ such that $\frac{1}{N} < \frac{\varepsilon}{2}$. If $i > N$ and $j > N$, then a_{n_i} and a_{n_j} are members of the Nth subsequence, and consequently are in the open ball $B_{\frac{1}{N}}(x_N)$. By the triangle inequality,

$$d(a_{n_i}, a_{n_j}) \le d(a_{n_i}, x_N) + d(x_N, a_{n_j}) < \frac{\varepsilon}{2} + \frac{\varepsilon}{2} = \varepsilon.$$

We have established that the subsequence $(a_{n_i})_{i=1}^{\infty}$ is a Cauchy sequence. By assumption, M is a complete metric space, and so the Cauchy sequence $(a_{n_i})_{i=1}^{\infty}$ converges to some point in the closed set K. Since any sequence in K has a convergent subsequence, K is sequentially compact.

Now suppose K is sequentially compact. We will show that K is totally bounded. (Then K will be compact by Theorem 6.12.) Suppose to the contrary that K is not totally bounded. Then there exists some $\varepsilon > 0$ such that K cannot be covered by finitely many closed balls with radius ε. Pick any $a_1 \in K$. Since K cannot be covered by finitely many ε-balls, the set $K \setminus \overline{B_\varepsilon(a_1)}$ is nonempty. Choose some $a_2 \in K \setminus \overline{B_\varepsilon(a_1)}$. Similarly, $K \setminus (\overline{B_\varepsilon(a_1)} \cup \overline{B_\varepsilon(a_2)})$ cannot be empty, and must contain some element, say a_3.

Proceeding inductively, we construct a sequence of points $(a_n)_{n=1}^{\infty}$ in K such that $a_{n+1} \notin \overline{B_\varepsilon(a_1)} \cup \cdots \cup \overline{B_\varepsilon(a_n)}$ for all $n \in \mathbb{N}$. It follows that $d(a_n, a_m) \geq \varepsilon$ for all $m \neq n$, and so $(a_n)_{n=1}^{\infty}$ cannot have a convergent subsequence. This contradicts the assumption that K is sequentially compact. Therefore, K is totally bounded, and so is compact (by Theorem 6.12).

We return now to the space of compact operators between Banach spaces X and Y. In addition to being a closed subspace of $\mathcal{L}(X, Y)$, the collection of compact operators $\mathcal{K}(X, Y)$ also possesses an ideal structure.

Theorem 6.17 *Let W, X, Y, and Z be Banach spaces. If $T : X \to Y$ is a compact operator and the maps $A : W \to X$ and $B : Y \to Z$ are bounded linear operators, then the composition $B \circ T \circ A : W \to Z$ is compact. In particular, the space $\mathcal{K}(X)$ is a two-sided ideal in $\mathcal{L}(X)$.*

Proof The map A is bounded, and consequently $A(B_W) \subseteq \|A\| B_X$. We thus conclude that $T A(B_W) \subseteq \|A\| \overline{T(B_X)}$, the latter set being compact, since T is a compact operator. Finally, $B T A(B_W) \subseteq \|A\| B(\overline{T(B_X)})$. The set on the right of the previous inclusion is compact as the continuous image (under B) of a compact set. Therefore, $\overline{B T A(B_W)}$ is a closed subset of a compact set, and so is compact.

Example 6.18 Suppose X is a Banach space. The identity map $\mathrm{Id} : X \to X$ is a compact operator if and only if $\dim(X) < \infty$. Indeed, if $A : X \to X$ is any invertible bounded linear operator, then A is compact if and only if $\dim(X) < \infty$. This follows from Proposition 5.37.

Example 6.19 Let $p \in [1, \infty]$ be given and let $a = (a_j)_{j=1}^{\infty}$ be a bounded sequence (i.e., an element of ℓ_∞). Define a map $T_a : \ell_p \to \ell_p$ by

$$T_a(\xi_1, \xi_2, \xi_3, \dots) = (a_1\xi_1, a_2\xi_2, a_3\xi_3, \dots), \quad (\xi_j)_{j=1}^{\infty} \in \ell_p.$$

We claim this map is a bounded linear operator. Linearity is clear. To see it is bounded, let $\xi = (\xi_j)_{j=1}^{\infty} \in \ell_p$. Then

$$\|T_a\xi\|_p = \left(\sum_{j=1}^{\infty} |a_j \xi_j|^p \right)^{1/p} \leq \|a\|_\infty \|\xi\|_p.$$

This implies that $\|T_a\| \leq \|a\|_\infty$. In fact, $T_a(e_i) = a_i\, e_i$ for all $i \in \mathbb{N}$, and so it follows that $\|T_a\| = \|a\|_\infty$. (Note that e_i is an eigenvector with corresponding eigenvalue a_i for each $i \in \mathbb{N}$.)

Proposition 6.20 T_a is compact if and only if $a \in c_0$.

Proof If T_a is a compact operator, then $\overline{T_a(B_{\ell_p})}$ is a compact set, and consequently any sequence in $T_a(B_{\ell_p})$ must have a convergent subsequence. Suppose that $a \notin c_0$. Then there exists some $\delta > 0$ and a subsequence $(a_{n_j})_{j=1}^\infty$ such that $|a_{n_j}| > \delta$ for all $j \in \mathbb{N}$. Then, for all natural numbers j and k,

$$\|T_a e_{n_j} - T_a e_{n_k}\|_p = \|a_{n_j} e_{n_j} - a_{n_k} e_{n_k}\|_p = \left(|a_{n_j}|^p + |a_{n_k}|^p\right)^{1/p} > \delta.$$

It follows that $(T_a e_{n_j})_{j=1}^\infty$ has no convergent subsequence, and so T_a is not a compact operator.

Now suppose $a = (a_j)_{j=1}^\infty \in c_0$. For each $n \in \mathbb{N}$, define a sequence in c_0 by the rule $a^{(n)} = (a_1, \ldots, a_n, 0, \ldots)$. The sequence $a^{(n)}$ has only finitely many nonzero terms, and so $T_{a^{(n)}}$ is a finite rank operator, and hence is compact. For each $n \in \mathbb{N}$,

$$\|T_a - T_{a^{(n)}}\| = \|T_{a - a^{(n)}}\| = \sup_{k > n} |a_k|.$$

Since $a \in c_0$, this quantity tends to 0 as $n \to \infty$. Thus, T_a is the uniform limit of a sequence of finite rank operators. Therefore, T_a is compact, because $\mathcal{K}(\ell_p)$ is a closed subspace of $\mathcal{L}(\ell_p)$, by Theorem 6.14. $\square$

Example 6.21 Let K be a continuous scalar-valued function on the compact space $[0, 1] \times [0, 1]$. Define $T_K : C[0, 1] \to C[0, 1]$ by

$$T_K f(s) = \int_0^1 K(s, t)\, f(t)\, dt, \quad f \in C[0, 1], \;\; s \in [0, 1].$$

We must show that T_K is well-defined; that is, we must show that $T_K f$ is a continuous function on $[0, 1]$. If $s_n \to s$ as $n \to \infty$, then

$$\int_0^1 K(s_n, t)\, f(t)\, dt \longrightarrow \int_0^1 K(s, t)\, f(t)\, dt,$$

as $n \to \infty$, by the Dominated Convergence Theorem. Consequently, $T_K f$ is continuous on $[0, 1]$, by the sequential characterization of continuity.

It is clear that T_K is linear. To compute the bound, let $f \in C[0, 1]$. Then

$$|T_K f(s)| \leq \int_0^1 |K(s, t)\, f(t)|\, dt \leq \|K\|_\infty \|f\|_\infty.$$

This bound is uniform in $s \in [0, 1]$, and so $\|T_K f\|_\infty \leq \|K\|_\infty \|f\|_\infty$. Therefore, T_K is bounded and $\|T_K\| \leq \|K\|_\infty$.

Proposition 6.22 *If K is a continuous scalar-valued function on $[0, 1] \times [0, 1]$, then T_K is a compact operator.*

Proof Suppose K is a polynomial on $[0, 1] \times [0, 1]$, say $K(s, t) = \sum_{j=1}^{n} a_j s^{m_j} t^{n_j}$, where $(m_j)_{j=1}^{n}$ and $(n_j)_{j=1}^{n}$ are finite sequences in $\mathbb{N}$, and where $(a_j)_{j=1}^{n}$ is a finite sequence of scalars. Then

$$T_K f(s) = \sum_{j=1}^{n} a_j s^{m_j} \int_0^1 t^{n_j} f(t) \, dt.$$

Thus, $T_K f$ is a polynomial in the linear span of $\{s^{m_1}, \ldots, s^{m_n}\}$. This holds for all $f \in C[0, 1]$, and so T_K is a finite rank operator, and hence compact.

Now suppose $K \in C([0, 1] \times [0, 1])$ is a continuous function. By the Weierstrass Approximation Theorem, there exists a sequence of polynomials $(K_n)_{n=1}^{\infty}$ such that $\|K - K_n\|_{\infty} \to 0$ as $n \to \infty$. Then

$$\|T_K - T_{K_n}\| = \|T_{K-K_n}\| \le \|K - K_n\|_{\infty} \to 0,$$

as $n \to \infty$. Therefore, T is the limit of a sequence of finite rank operators, and so is compact (by Theorem 6.14).

Digression: Historical Comments

The linear operator T_K from Example 6.21 was considered by Fredholm in 1903, in what is considered by some to be the first paper on functional analysis [12]. The function K is called the *Fredholm kernel* (or simply the *kernel*) of the operator T_K (not to be confused with the nullspace $\ker T_K$).

Fredholm was interested in solving integral equations of the form

$$g(s) = \int_0^1 K(s, t) f(t) \, dt,$$

where f, g, and K had specified properties. This integral equation is an extension of the matrix equation

$$y_j = \sum_{k=1}^{n} a_{jk} x_k,$$

where $A = (a_{jk})_{j,k=1}^{n}$ is an $n \times n$ matrix (for some $n \in \mathbb{N}$), x and y are n-dimensional vectors, and $y = Ax$.

Fredholm's work was before the advent of measure theory, and so the focus was on the space of continuous functions on $[0, 1]$. The space $C[0, 1]$ is an infinite-dimensional vector space, and integral operators determine linear operators on this vector space. Fredholm set about trying to apply techniques of linear algebra to these kinds of operators.

With the advent of measure theory and Riesz's work on L_p-spaces, the focus on integral operators broadened to include these larger spaces of functions. For any p in the interval $[1, \infty]$, we define a *Fredholm operator* on $L_p(0, 1)$ to be any operator $T_K : L_p(0, 1) \to L_p(0, 1)$ of the form

$$T_K f(s) = \int_0^1 K(s, t) f(t) \, dt, \quad f \in L_p(0, 1), \ s \in [0, 1], \tag{6.2}$$

where K is a prescribed measurable function on $[0, 1] \times [0, 1]$. The function K is known as the *kernel* of T_K. (It is also variously known as the *nucleus* or the *Green's function* for T_K.) In order to insure that T_K is well-defined, the kernel K generally has some additional assumptions imposed upon it. For example, in the following proposition, which can be seen as an extension of Proposition 6.22, the kernel K is assumed to be an essentially bounded measurable function.

Proposition 6.23 *If $K \in L_\infty([0, 1] \times [0, 1])$, then T_K is a compact operator on $L_p(0, 1)$ for $1 < p < \infty$.*

Proof Let $p \in (1, \infty)$ and suppose $f \in L_p(0, 1)$. Then

$$\|T_K f\|_p = \left(\int_0^1 |T_K f(s)|^p \, ds \right)^{1/p} \leq \|T_K f\|_\infty = \sup_{s \in [0,1]} \left| \int_0^1 K(s,t) f(t) \, dt \right|.$$

Thus,

$$\|T_K f\|_p \leq \|K\|_\infty \|f\|_1 \leq \|K\|_\infty \|f\|_p.$$

(Note that $\| \cdot \|_1 \leq \| \cdot \|_p$ on a probability space, by Hölder's Inequality.) Therefore, $T_K f \in L_p(0, 1)$ whenever $f \in L_p(0, 1)$ and $\|T_K f\|_p \leq \|K\|_\infty \|f\|_p$. We have established that T_K is well-defined and $\|T_K\| \leq \|K\|_\infty$. To prove that T_K is compact, we use an argument similar to the one we used in Proposition 6.22. If $K = \chi_{A \times B}$, where A and B are measurable subsets of $[0, 1]$, then for all $f \in L_p(0, 1)$ and $s \in [0, 1]$,

$$T_K f(s) = \int_0^1 \chi_A(s) \chi_B(t) f(t) \, dt = \left(\int_B f(t) \, dt \right) \chi_A(s).$$

Thus, $T_K f \in \text{span}\{\chi_A\}$, and so T_K is a rank-one operator. Therefore, T_K is compact. Now suppose

$$K = \sum_{i=1}^n c_i \chi_{A_i \times B_i}, \tag{6.3}$$

where $n \in \mathbb{N}$ and, for each $i \in \{1, \ldots, n\}$, the sets A_i and B_i are measurable and c_i is a scalar. Then for all $f \in L_p(0, 1)$ and $s \in [0, 1]$,

$$T_K f(s) = \int_0^1 \sum_{i=1}^n c_i \chi_{A_i}(s) \chi_{B_i}(t) f(t) \, dt = \sum_{i=1}^n \left(c_i \int_{B_i} f(t) \, dt \right) \chi_{A_i}(s).$$

Consequently, $T_K f \in \text{span}\{\chi_{A_1}, \ldots, \chi_{A_n}\}$, and so T_K is an operator of rank at most n. Thus, T_K is compact.

If K is a bounded measurable function on $[0, 1] \times [0, 1]$, then the compactness of T_K follows from a density argument. Since K is a bounded function, it is also true that K is in $L_r([0,1] \times [0,1])$ for all $r \in (1, \infty)$. Observe that

$$\|T_K f\|_p = \left(\int_0^1 |T_K f(s)|^p \, ds \right)^{1/p} \leq \left(\int_0^1 \left(\int_0^1 |K(s,t) f(t)| \, dt \right)^p ds \right)^{1/p}.$$

Thus, applying Hölder's Inequality to the inner integral, we have

$$\|T_K f\|_p \leq \left(\int_0^1 \left(\int_0^1 |K(s,t)|^q \, dt \right)^{p/q} ds \right)^{1/p} \|f\|_p, \tag{6.4}$$

where $1/p + 1/q = 1$. We will show that the right side of (6.4) is bounded by $\|K\|_r \|f\|_p$, where $r = \max\{p,q\}$. Suppose that $q \leq p$, then for any $s \in [0,1]$, we have that $(\int_0^1 |K(s,t)|^q \, dt)^{1/q} \leq (\int_0^1 |K(s,t)|^p \, dt)^{1/p}$. Consequently,

$$\|T_K f\|_p \leq \left(\int_0^1 \int_0^1 |K(s,t)|^p \, dt \, ds \right)^{1/p} \|f\|_p = \|K\|_p \|f\|_p.$$

Now suppose $p \leq q$ and let $A(s) = (\int_0^1 |K(s,t)|^q \, dt)^{1/q}$. Then

$$\|T_K f\|_p \leq \left(\int_0^1 A(s)^p \, ds \right)^{1/p} \|f\|_p \leq \left(\int_0^1 A(s)^q \, ds \right)^{1/q} \|f\|_p$$

$$= \left(\int_0^1 \int_0^1 |K(s,t)|^q \, dt \, ds \right)^{1/q} \|f\|_p = \|K\|_q \|f\|_p.$$

Therefore, $\|T_K f\|_p \leq \|K\|_r \|f\|_p$, where $r = \max\{p,q\}$.

Functions of the type in (6.3) are dense in $L_r([0,1] \times [0,1])$. Thus, there exists a sequence of kernels $(K_n)_{n=1}^{\infty}$, all of the type in (6.3), such that $K = \lim_{n \to \infty} K_n$, where the limit is in the norm on $L_r([0,1] \times [0,1])$. Because $\|T_K f\|_p \leq \|K\|_r \|f\|_p$, it follows that $T_K f = \lim_{n \to \infty} T_{K_n} f$ for all $f \in L_p(0,1)$. For each $n \in \mathbb{N}$, the operator T_{K_n} is a finite rank operator on $L_p(0,1)$. Therefore, T_K is the limit of a sequence of finite rank operators, and so T_K is compact, by Theorem 6.14.

Example 6.24 Let us explore which kernels we can use to define a Fredholm operator as in (6.2). In order for $T_K : L_p(0,1) \to L_p(0,1)$ to be well-defined, we must ensure that $T_K f \in L_p(0,1)$ for all $f \in L_p(0,1)$. In the proof of Proposition 6.23, we saw that

$$\|T_K f\|_p \leq \left(\int_0^1 \left(\int_0^1 |K(s,t)|^q \, dt \right)^{p/q} ds \right)^{1/p} \|f\|_p.$$

(See (6.4).) If we choose K so that the above quantity is finite, then T_K will be a well-defined bounded linear operator.

An important case is when $p = 2$ (and thus $q = 2$). In this case, the inequality becomes

$$\|T_K f\|_2 \leq \left(\int_0^1 \int_0^1 |K(s,t)|^2 \, dt \, ds \right)^{1/2} \|f\|_2 = \|K\|_2 \|f\|_2. \tag{6.5}$$

Therefore, if $K \in L_2([0,1] \times [0,1])$, then T_K is a bounded linear operator from $L_2(0,1)$ to $L_2(0,1)$ and $\|T_K\| \leq \|K\|_2$. An operator of this type is known as a *Hilbert–Schmidt operator*. We will say more about these operators in Chapter 7.

In the examples given above, each operator was shown to be compact by demonstrating that it was the limit (in the operator norm) of a sequence of finite rank operators. This leads to a natural question: *Is it true that every compact operator is a limit (in the operator norm) of a sequence of finite rank operators?*

This question, known as the "Approximation Problem," motivated much of the early development of functional analysis. It was not until 1973 that Per Enflo settled the issue and proved that not every compact operator was the limit of a sequence of finite rank operators [9]. Famously, Per Enflo was awarded a live goose by Stanislaw Mazur for solving the Approximation Problem. (Many problems included in the *Scottish Book* came with the promise of a reward for a solution, especially those that were considered difficult or important. In 1936, when Mazur included a problem in the *Scottish Book* [Problem 153] that was equivalent to the Approximation Problem, a live goose was considered very valuable. For a complete list of problems and prizes, see [24].)

Much earlier, in 1955, Alexander Grothendieck studied compact operators in depth [17]. He developed several conditions which were equivalent to the statement that every compact operator is the limit of a sequence of finite rank operators. (In fact, it was one of Grothendieck's alternate formulations that Enflo disproved.) Perhaps one of the most unexpected equivalences is the following.

Proposition 6.25 *Each compact operator between arbitrary Banach spaces is the limit of a sequence of finite rank operators if and only if the following is true:*

$$\text{If } K \in C([0,1] \times [0,1]) \text{ and } \int_0^1 K(s,t)\, K(t,u)\, dt = 0 \text{ for all } \{s,u\} \subseteq [0,1],$$

$$\text{then } \int_0^1 K(s,s)\, ds = 0.$$

(6.6)

The motivation behind (6.1.6) comes from the finite-dimensional theory of linear algebra. Let $n \in \mathbb{N}$ and suppose $A = (a_{jk})_{j,k=1}^n$ is an $n \times n$ matrix. Recall that the trace of the matrix A is

$$\text{trace}(A) = \sum_{i=1}^n a_{ii}.$$

It is known that if $A^2 = 0$, then $\text{trace}(A) = 0$. That is,

$$\sum_{j=1}^n a_{ij} a_{jk} = 0 \text{ for all } \{i,k\} \subseteq \{1,\dots,n\} \quad \Longrightarrow \quad \sum_{i=1}^n a_{ii} = 0.$$

Proposition 6.25 suggests that a similar result might be true for the Fredholm operator T_K, when the kernel K is continuous. We define the *trace of the Fredholm operator*

T_K to be

$$\text{trace}(T_K) = \int_0^1 K(s,s)\,ds. \tag{6.7}$$

Proposition 6.25 states that if each compact operator between arbitrary Banach spaces is the limit of a sequence of finite rank operators, then if K is a continuous kernel,

$$\int_0^1 K(s,t)\,K(t,u)\,dt = 0 \text{ for all } \{s,u\} \subseteq [0,1] \quad \Longrightarrow \quad \int_0^1 K(s,s)\,ds = 0.$$

Since it is not true that each compact operator between arbitrary Banach spaces is the limit of a sequence of finite rank operators, this implication does not hold for all $K \in C([0,1] \times [0,1])$. It can be shown, however, that the implication does hold if K is Hölder continuous with exponent α for all $\alpha > 1/2$.

6.2   A Rank-Nullity Theorem for Compact Operators

We resume the study of compact operators with a theorem of Schauder that guarantees the compactness of the adjoint of a compact operator.

Theorem 6.26   (Schauder's Theorem) *If X and Y are Banach spaces and the operator $T : X \to Y$ is compact, then $T^* : Y^* \to X^*$ is also compact.*

Proof By the Banach–Alaoglu Theorem (Theorem 5.39), we know that B_{Y^*} is w^*-compact. Consequently, it suffices to show that $T^* : (B_{Y^*}, w^*) \to (X^*, \|\cdot\|)$ is a continuous map, because then $T^*(B_{Y^*})$ will be compact as the continuous image of a compact set.

Suppose that $y_0^* \in B_{Y^*}$ and consider the closed neighborhood of $T^* y_0^*$ given by $T^* y_0^* + \delta B_{X^*}$, where $\delta > 0$. Let V denote the preimage of this set in B_{Y^*}; that is,

$$V = (T^*)^{-1}(T^* y_0^* + \delta B_{X^*}) \cap B_{Y^*}.$$

We need to show that V is a w^*-neighborhood of y_0^* relative to B_{Y^*}; that is, we must find a w^*-open set W (open in the w^*-topology of Y^*) containing y_0^* such that $W \cap B_{Y^*} \subseteq V$.

By assumption, T is a compact operator, and so $T(B_X)$ is a relatively compact set in Y. By Theorem 6.12, a relatively compact set is totally bounded, and hence there exists a finite set $\{x_1, \dots, x_n\} \subseteq B_X$ such that

$$T(B_X) \subseteq \bigcup_{j=1}^n \left(T x_j + \frac{\delta}{3} B_Y \right). \tag{6.8}$$

Define a w^*-open neighborhood W of y_0^* by

$$W = \left\{ y^* \in Y^* : |y^*(Tx_j) - y_0^*(Tx_j)| < \frac{\delta}{3}, \; j \in \{1, \dots, n\} \right\}.$$

We claim that $W \cap B_{Y^*} \subseteq V$. To that end, we will show

$$T^*(W \cap B_{Y^*}) \subseteq T^* y_0^* + \delta B_{X^*}.$$

Pick $y^* \in W \cap B_{Y^*}$. We need to show that $\|T^* y^* - T^* y_0^*\| \le \delta$.

Let $x \in B_X$. By (6.8), there exists some j such that $\|Tx - Tx_j\| \le \delta/3$. By the triangle inequality,

$$|(T^* y^* - T^* y_0^*)(x)| = |(y^* - y_0^*)(Tx)| \le |(y^* - y_0^*)(Tx_j)| + |(y^* - y_0^*)(Tx - Tx_j)|.$$

Because $y^* \in W$, it follows that $|(y^* - y_0^*)(Tx_j)| < \delta/3$. And since y^* and y_0^* are in B_{Y^*}, we conclude that

$$|(y^* - y_0^*)(Tx - Tx_j)| \le \|y^* - y_0^*\| \, \|Tx - Tx_j\| \le \frac{2\delta}{3}.$$

Consequently, $|(T^* y^* - T^* y_0^*)(x)| \le \delta$ for all $x \in B_X$, and so $\|T^* y^* - T^* y_0^*\| \le \delta$, as required.

We conclude that $T^* : (B_{Y^*}, w^*) \to (X^*, \| \cdot \|)$ is continuous, and hence $T^*(B_{Y^*})$ is compact as the continuous image of the weak*-compact set B_{Y^*}.

Notice that in the proof of Theorem 6.26, we did not have to take the closure of $T^*(B_{Y^*})$ in X^* to get a compact set. This is a direct consequence of the Banach–Alaoglu Theorem, which assures us that B_{Y^*} is always compact in the weak* topology.

In the proof of Theorem 6.26, we demonstrated the continuity of T^* when viewed as a map from (B_{Y^*}, w^*) to $(X^*, \| \cdot \|)$. In general, however, we cannot extend this to the entire space Y^*. That is, we cannot say that $T^* : (Y^*, w^*) \to (X^*, \| \cdot \|)$ is continuous. (See Exercise 6.3.) On the other hand, by Proposition 5.44, we do know that $T^* : (Y^*, w^*) \to (X^*, w^*)$ is always continuous.

Example 6.27 (Volterra operator) Our purpose for adding structure to our linear operators was to recover some of the properties of linear operators on finite-dimensional vector spaces. We will now provide a compact operator on a complex Banach space with no eigenvalues. Let $L_2(0, 1)$ be the complex Banach space of square-integrable functions on $[0, 1]$. Define a map $V : L_2(0, 1) \to L_2(0, 1)$ by

$$V f(x) = \int_0^x f(t) \, dt, \quad f \in L_2(0, 1), \; x \in [0, 1].$$

This operator, known as the *Volterra operator* (see Example 3.41), is a Hilbert–Schmidt operator with kernel

$$K(x, t) = \begin{cases} 1 & \text{if } t \le x, \\ 0 & \text{if } t > x. \end{cases}$$

Since a Hilbert–Schmidt operator is always compact (see Example 6.24), we conclude that V is a compact operator.

Suppose that $\lambda \in \mathbb{C}$ is an eigenvalue. Then $Vf(x) = \lambda f(x)$ for almost every $x \in [0, 1]$. If $\lambda = 0$, then $Vf = 0$a.e., and so $\int_0^x f(t)\, dt = 0$ for almost every x. This implies that $f = 0$a.e., and so $\lambda = 0$ is not an eigenvalue.

Suppose $\lambda \neq 0$. Then $f(x) = \frac{1}{\lambda} \int_0^x f(t)\, dt$ for almost every x. Then f is differentiable and $f'(x) = \frac{1}{\lambda} f(x)$ for all $x \in [0, 1]$. The general solution to this differential equation is $f(x) = C\, e^{x/\lambda}$, for a constant $C \in \mathbb{C}$. However, $f(0) = \frac{1}{\lambda} Vf(0) = 0$, and so $C = 0$. Therefore, $f = 0$, and so $\lambda \neq 0$ is not an eigenvalue.

Despite the previous example, we can recover some theory from the finite-dimensional case. In particular, we will be able to prove a version of the Rank-Nullity Theorem. In order to articulate the theorem, we will need some preparation.

Definition 6.28 Suppose L is a bounded linear operator. If K is a compact operator, then the operator $L - K$ is called a *compact perturbation of L* or a *perturbation of L by a compact operator*.

For the moment, we will be interested in $L = \lambda I$, where λ is a *nonzero* scalar, and $I : X \to X$ is the identity operator on the Banach space X. We refer to the map $L = \lambda I$ as a *scaling of the identity*. In what follows, we let $K : X \to X$ be a compact operator and define $T = \lambda I - K$, so that T is a compact perturbation of a scaling of the identity. Observe that

$$\ker T = \{x : Kx = \lambda x\},$$

and so $x \in \ker T$ if and only if x is an eigenvector of K with eigenvalue λ (when $x \neq 0$).

Lemma 6.29 *Let $T = \lambda I - K$, where λ is a nonzero scalar and K is a compact operator on a Banach space X. The set $\ker T$ is finite-dimensional.*

Proof On $\ker T$, we have the equality $K = \lambda I$. This implies that $I|_{\ker T}$ is a compact operator, and so $\dim(\ker T) < \infty$. (See Example 6.18.)   □

Lemma 6.30 *Let $T = \lambda I - K$, where λ is a nonzero scalar and K is a compact operator on a Banach space X. The range of T is a closed subspace of X.*

Proof Recall that the range of T is denoted by $\operatorname{ran} T$. Let $y \in \overline{\operatorname{ran} T}$. Pick $(x_n)_{n=1}^{\infty}$ in X such that $Tx_n \to y$ as $n \to \infty$. For each $n \in \mathbb{N}$, let

$$\alpha_n = d(x_n, \ker T) = \inf\{\|x_n - z\| : z \in \ker T\}.$$

Then there exists for each $n \in \mathbb{N}$ some $z_n \in \ker T$ such that $\|x_n - z_n\| \leq 2\alpha_n$.

Let $x_n' = x_n - z_n$. Then $\|x_n'\| \leq 2\alpha_n$. By the definition of x_n', we have

$$d(x_n', \ker T) = d(x_n, \ker T) = \alpha_n.$$

Furthermore, $Tx_n' = Tx_n \to y$ as $n \to \infty$.

We claim the sequence $(\alpha_n)_{n=1}^{\infty}$ is bounded. To prove this, assume that $\alpha_n \to \infty$ as $n \to \infty$. Without loss of generality, we may assume that $\alpha_n > 0$ for all $n \in \mathbb{N}$. Then we may divide by α_n, and hence

$$\lim_{n \to \infty} \left\| \frac{Tx_n'}{\alpha_n} \right\| = 0 \quad \text{and} \quad d\left(\frac{x_n'}{\alpha_n}, \ker T\right) = 1. \tag{6.9}$$

Furthermore, $\|x'_n/\alpha_n\| \le 2$. Consequently, since K is a compact operator, there is a subsequence $(x'_{n_k}/\alpha_{n_k})_{k=1}^\infty$ such that $\lim_{k\to\infty} K(x'_{n_k}/\alpha_{n_k})$ exists. Call this limit v.

Recalling that $T = \lambda I - K$, we see that

$$\lambda \cdot \frac{x'_{n_k}}{\alpha_{n_k}} = T\left(\frac{x'_{n_k}}{\alpha_{n_k}}\right) + K\left(\frac{x'_{n_k}}{\alpha_{n_k}}\right).$$

Letting $k \to \infty$, and using the limit in (6.9), we see that $\lambda\, x'_{n_k}/\alpha_{n_k} \to v$ as $k \to \infty$. But this implies that $v \in \ker T$:

$$T(v) = \lim_{k\to\infty} \lambda\, T\left(\frac{x'_{n_k}}{\alpha_{n_k}}\right) = 0.$$

If $v \in \ker T$, then $\lim_{k\to\infty} d(x'_{n_k}/\alpha_{n_k}, \ker T) = 0$, which contradicts the second equality in (6.9). It follows that $\alpha_n \not\to \infty$ as $n \to \infty$. The same argument shows that no subsequence of $(\alpha_n)_{n=1}^\infty$ can tend to infinity, and so there must be some $C > 0$ such that $|\alpha_n| \le C$ for all $n \in \mathbb{N}$.

Once again, we invoke the compactness of the operator K. We know $\|x'_n\| \le 2C$ for all $n \in \mathbb{N}$. Thus, there is a convergent subsequence $(K(x'_{n_k}))_{k=1}^\infty$ of $(K(x'_n))_{n=1}^\infty$. Suppose $\lim_{k\to\infty} K(x'_{n_k}) = v$. Then

$$\lambda x'_{n_k} = T(x'_{n_k}) + K(x'_{n_k}) \xrightarrow[k\to\infty]{} y + v.$$

Miraculously,

$$T\left(\frac{y+v}{\lambda}\right) = \lim_{k\to\infty} T(x'_{n_k}) = y,$$

and hence $y \in \operatorname{ran} T$. Our assumption was that $y \in \overline{\operatorname{ran} T}$, and thus we conclude that the range of T is closed.

Lemma 6.31 *Let $T = \lambda I - K$, where λ is a nonzero scalar and K is a compact operator on a Banach space X. The quotient $X/\operatorname{ran} T$ is finite-dimensional.*

Proof By Proposition 3.51, we have that

$$(X/\operatorname{ran} T)^* = (\operatorname{ran} T)^\perp = \{x^* \in X^* : x^*(Tx) = 0 \text{ for all } x \in X\}.$$

The statement that $x^*(Tx) = 0$ for all $x \in X$ is equivalent to the statement that $(T^*x^*)(x) = 0$ for all $x \in X$. But this means $T^*x^* = 0$, and hence $x^* \in \ker(T^*)$. Thus,

$$(X/\operatorname{ran} T)^* = \ker(T^*).$$

Since $T = \lambda I - K$, it follows that $T^* = \lambda I^* - K^*$. Therefore, by Theorem 6.26 (Schauder's Theorem) and Lemma 6.29, the set $\ker(T^*)$ is finite-dimensional. Hence,

$$\dim(X/\operatorname{ran} T)^* = \dim(\ker(T^*)) < \infty.$$

The result follows, since $\dim(X/\operatorname{ran} T) = \dim(X/\operatorname{ran} T)^*$. (See Exercise 6.11.) □

From the preceding proof, we extract a corollary that is of independent interest.

Corollary 6.32 *If X is a Banach space and T is a compact perturbation of a scaling of the identity, then* $\dim(X/\mathrm{ran}\,T) = \dim(\ker(T^*))$.

We are now ready to state (although not prove) our version of the Rank-Nullity Theorem for compact operators. We will see that it is really a statement about compact perturbations of scalings of the identity.

Theorem 6.33   (Rank-Nullity Theorem) *Let X be a Banach space with identity operator I. If* $T = \lambda I - K$, *where* λ *is a nonzero scalar and K is a compact operator on X, then* $\dim(\ker T) = \dim(X/\mathrm{ran}\,T)$.

Before we can prove this theorem, we require a few more technical results. The first of these, known as Riesz's Lemma, asserts that any closed proper subspace of a Banach space is always a prescribed distance away from some portion of the unit sphere ∂B_X.

Lemma 6.34   (Riesz's Lemma) *Suppose X is a Banach space and E is a proper closed subspace of X. Given any* $\varepsilon > 0$, *there exists some* $x \in \partial B_X$ *such that* $d(x, E) > 1 - \varepsilon$. *Furthermore, if E is finite-dimensional, then* $x \in \partial B_X$ *can be chosen so that* $d(x, E) = 1$.

Proof   Denote the metric on X by d. By assumption, E is a proper closed subspace of X, and so there exists some $u \in X \backslash E$ with $d(u, E) > 0$. It follows that, for any $\delta > 0$, there exists some $e \in E$ such that $d(u, E) < d(u, e) < d(u, E) + \delta$. We may assume without loss of generality that $\varepsilon < 1$. Pick $\delta = \frac{\varepsilon}{1-\varepsilon} d(u, E)$. Then there exists an $e \in E$ such that

$$\|u - e\| = d(u, e) < \left(1 + \frac{\varepsilon}{1 - \varepsilon}\right) d(u, E) = \frac{1}{1 - \varepsilon} d(u, E).$$

Observe that $\|u - e\| > 0$. If we let $x = \frac{u-e}{\|u-e\|}$, then

$$d(x, E) = \frac{1}{\|u - e\|} d(u - e, E) = \frac{1}{\|u - e\|} d(u, E) > 1 - \varepsilon,$$

as required.

Now assume E is finite-dimensional. For each $n \in \mathbb{N}$, pick $e_n \in E$ such that

$$\|u - e_n\| < \left(1 + \frac{1}{n}\right) d(u, E).$$

Note that $0 \in E$ (because E is a subspace of X), and so $d(u, E) \leq \|u\|$. Thus, for each $n \in \mathbb{N}$, we have

$$\|e_n\| \leq \|u\| + 2\,d(u, E) \leq 3\,\|u\|.$$

Therefore, the set $\{e_n : n \in \mathbb{N}\}$ is bounded in norm. Since E is finite-dimensional, the Heine–Borel property asserts the existence of a convergent subsequence, say

$(e_{n_k})_{k=1}^{\infty}$. Then there is some $e \in E$ such that $\lim_{k \to \infty} e_{n_k} = e$. By construction, it follows that $\|u - e\| = d(u, E)$. If $x = \frac{u-e}{\|u-e\|}$, then $x \in \partial B_X$ and $d(x, E) = 1$.

Remark 6.35 If E is a closed proper subspace of a *reflexive* Banach space X, then we can always find an $x \in X$ such that $d(x, E) = 1$, even if E is infinite-dimensional. (See Exercise 6.6.)

Lemma 6.36 *Let X be a Banach space and suppose $K : X \to X$ is a compact operator. Let $\delta > 0$ be given and suppose $(\lambda_n)_{n=1}^{\infty}$ is a sequence of scalars (either real or complex) such that $|\lambda_n| \geq \delta > 0$ for all $n \in \mathbb{N}$. If*

$$E_n = \ker\big((\lambda_1 I - K)(\lambda_2 I - K) \cdots (\lambda_n I - K)\big), \quad n \in \mathbb{N},$$

then each E_n is finite-dimensional and there exists some $N \in \mathbb{N}$ such that $E_m = E_N$ for all $m \geq N$.

Proof Denote the metric on X by d. Fix an $n \in \mathbb{N}$. Expanding, we have

$$(\lambda_1 I - K)(\lambda_2 I - K) \cdots (\lambda_n I - K) = \lambda_1 \cdots \lambda_n I - \hat{K},$$

where $\hat{K}$ is some compact operator. Consequently, the product is a compact perturbation of a scaling of the identity. Therefore $\dim E_n < \infty$, by Lemma 6.29.

By definition, the sets $(E_n)_{n=1}^{\infty}$ form a nested sequence $E_1 \subseteq E_2 \subseteq \cdots$. We wish to show that the inclusions eventually become equality. Define a subset $\mathbb{A}$ of the natural numbers to be $\mathbb{A} = \{n : E_n \neq E_{n-1}\}$.

Let $n \in \mathbb{A}$. By Riesz's Lemma (Lemma 6.34), there exists some $x_n \in E_n$ such that $\|x_n\| = d(x_n, E_{n-1}) = 1$. Since $x_n \in E_n$, we have that

$$(\lambda_1 I - K)(\lambda_2 I - K) \cdots (\lambda_n I - K)(x_n) = 0,$$

and so $(\lambda_n I - K)(x_n) \in E_{n-1}$. Then $K x_n \in \lambda_n x_n + E_{n-1}$. Consequently,

$$d(K x_n, E_{n-1}) = |\lambda_n| \, d(x_n, E_{n-1}) \geq \delta.$$

If $m \in \mathbb{A}$ is chosen such that $m < n$, then we can pick an element $x_m \in E_m$ such that $K x_m \in \lambda_m x_m + E_{m-1} \subseteq E_m$. Thus, $K x_m \in E_m \subseteq E_{n-1}$, and so

$$\|K x_n - K x_m\| = d(K x_n, K x_m) \geq \delta.$$

If $\mathbb{A}$ is infinite, then we can find a sequence $(x_n)_{n=1}^{\infty}$ in B_X such that $\|K x_m - K x_n\| \geq \delta$ for all $m \neq n$. This contradicts the assumption that K is a compact operator, because we should be able to find a convergent subsequence, but all the terms are at least δ apart. Therefore, the set $\mathbb{A}$ must be finite. The result follows.

Theorem 6.37 *Let K be a compact operator and let Λ be the set of nonzero eigenvalues of K. Then one of the following three things must be true:*

(i) $\Lambda = \emptyset$,
(ii) Λ is finite, or
(iii) $\Lambda = (\lambda_n)_{n=1}^{\infty}$, where $\lim\limits_{n \to \infty} |\lambda_n| = 0$.

Proof We will show that the set $\Lambda \cap \{z \in \mathbb{C} : |z| \geq r\}$ is finite for any $r > 0$. Assume the set is infinite. Then we can find a sequence $(\mu_k)_{k=1}^{\infty}$ in $\Lambda \cap \{z \in \mathbb{C} : |z| \geq r\}$ such that each element in the sequence is distinct. For each $n \in \mathbb{N}$, define the nested sets

$$E_n = \ker((\mu_1 I - K) \cdots (\mu_n I - K)).$$

Since the μ_k are distinct eigenvalues, we conclude that $E_1 \subset E_2 \subset E_3 \subset \cdots$, where the inclusions are proper. But $|\mu_k| \geq r$ for all $k \in \mathbb{N}$, and so, by Lemma 6.36, there exists some $N \in \mathbb{N}$ such that $E_n = E_N$ for all $n \geq N$. We have arrived at a contradiction. Therefore, for each $r > 0$, there are only finitely many eigenvalues inside the set $\{z \in \mathbb{C} : |z| \geq r\}$.

It follows that either the set of eigenvalues is finite or, if there are infinitely many eigenvalues, they must accumulate to 0.

Finally, we are ready to prove the Rank-Nullity Theorem for compact operators.

Proof of Theorem 6.33   Let us begin by recalling the setup. We start with a Banach space X and a compact operator $K : X \to X$. Let $T = \lambda I - K$, where $\lambda \neq 0$.

By Lemma 6.36, there exists an $N_1 \in \mathbb{N}$ such that $\ker(T^m) = \ker(T^{N_1})$ for all $m \geq N_1$. (In the lemma, let $\lambda_n = \lambda$ for all $n \in \mathbb{N}$.) Similarly, there exists an $N_2 \in \mathbb{N}$ such that $\ker((T^*)^m) = \ker((T^*)^{N_2})$ for all $m \geq N_2$. From the latter equality, we deduce that $\text{ran}(T^m) = \text{ran}(T^{N_2})$ for all $m \geq N_2$. (See Exercise 6.12.)

Let $N = \max\{N_1, N_2\}$. For ease of notation, let $W = \ker(T^N)$ and $V = \text{ran}(T^N)$. Observe that $T^N = \lambda^N I - \hat{K}$, for some compact operator $\hat{K}$. Thus, W is finite-dimensional (by Lemma 6.29) and V is closed (by Lemma 6.30). Since V is a closed subspace of a Banach space, it too is a Banach space. Notice that $T(W) \subseteq W$ and $T(V) \subseteq V$. (That is, the spaces W and V are invariant under T).

We recall that $T|_V$ is the restriction of the map T to the subspace V.

Claim 1. *The map* $T|_V : V \to V$ *is one-to-one.*

Suppose $v \in V$ is such that $Tv = 0$. Since $V = \text{ran}(T^N)$, there exists some $u \in X$ such that $v = T^N u$. Then

$$0 = Tv = T(T^N u) = T^{N+1} u.$$

It follows that $u \in \ker(T^{N+1}) = \ker(T^N)$, and hence $T^N u = 0$. Our assumption was that $v = T^N u$, and consequently $v = 0$. Thus, the map $T|_V : V \to V$ is one-to-one.

Claim 2. *The map* $T|_V : V \to V$ *is onto.*

Let $v \in V$. Since $V = \text{ran}(T^N) = \text{ran}(T^{N+1})$, there is some $u \in X$ such that

$$v = T^{N+1} u = T(T^N u) \in T(V).$$

Thus, there is some $v' \in V$ such that $v = T(v')$. Consequently, the map $T|_V$ is onto.

We have demonstrated that $T|_V : V \to V$ is a bounded linear bijection on the Banach space V. Thus, by the Bounded Inverse Theorem (Corollary 4.30), there is a bounded linear operator $S : V \to V$ such that $ST|_V = T|_V S = I|_V$. (That is, T has a bounded linear inverse on V).

Now, define a map $P : X \to V$ by

$$Px = S^N T^N x, \quad x \in X.$$

If $x \in V$, then $Px = x$, and so P is a projection onto V. Observe also that $Px = 0$ if and only if $T^N x = 0$ (because S is an isomorphism on V). Consequently, we have that $Px = 0$ if and only if $x \in \ker(T^N) = W$. We conclude that $\ker P = W$. Therefore, by Theorem 4.42, $X = V \oplus W$.

Since $X = V \oplus W$ and $T|_V$ is invertible, we deduce that $\ker T = \ker T|_W$. Additionally, since $X = V \oplus W$, we have $X/\mathrm{ran} T = W/\mathrm{ran} T|_W$. We know that W is finite-dimensional, and so we may apply the finite-dimensional Rank-Nullity Theorem to $T|_W$. Hence,

$$\dim(\ker T|_W) + \dim(\mathrm{ran} T|_W) = \dim W.$$

Consequently,

$$\dim(\ker T|_W) = \dim(W/\mathrm{ran} T|_W).$$

Making the necessary substitutions, the result follows. $\square$

Exercises

Exercise 6.1 Let $1 \le q < \infty$ and let $S : \ell_q \to \ell_q$ be the left shift operator given by

$$S(\eta_1, \eta_2, \eta_3, \dots) = (\eta_2, \eta_3, \eta_4, \dots), \quad (\eta_k)_{k=1}^\infty \in \ell_q.$$

Show that λ is an eigenvalue for S if and only if $|\lambda| < 1$. What are the eigenvalues if $q = \infty$?

Exercise 6.2 Suppose X and Y are Banach spaces and $T : X \to Y$ is a bounded linear operator. If X is reflexive, show that $T(B_X)$ is closed in Y.

Exercise 6.3 Suppose X and Y are Banach spaces and $T : X \to Y$ is a bounded linear operator. Show that $T^* : (Y^*, w^*) \to (X^*, \| \cdot \|)$ is continuous only if T^* has finite rank.

Exercise 6.4 Suppose X and Y are Banach spaces and $T : X \to Y$ is a compact operator. If $(x_n)_{n=1}^\infty$ is a sequence in X such that $x_n \to 0$ weakly, then show that $\lim_{n\to\infty} \|T(x_n)\| = 0$.

Exercise 6.5 Show that the converse to Exercise 6.4 is true if X is a reflexive Banach space. That is, if X and Y are Banach spaces such that X is reflexive, and $\lim_{n\to\infty} \|T(x_n)\| = 0$ whenever $x_n \to 0$ weakly, show that T is a compact operator.

Exercise 6.6 Let X be a reflexive Banach space and let E be an infinite-dimensional closed proper subspace of X. Show there exists an $x \in X$ with $\|x\| = 1$ such that

$d(x, E) = \inf\{\|x - e\| : e \in E\} = 1$. (*Hint:* Find a weak limit point of the sequence $(e_n)_{n=1}^{\infty}$ in the proof of Lemma 6.34.)

Exercise 6.7 Let X be a reflexive Banach space. Show that if $T : X \to X$ is a compact operator, then there exists an element $x \in X$ with $\|x\| = 1$ such that $\|Tx\| = \|T\|$. (*Hint:* Use Theorem 5.39.)

Exercise 6.8 Suppose X and Y are Banach spaces and $T : X \to Y$ is a compact operator. If $T(X)$ is dense in Y, then show that Y is separable.

Exercise 6.9 Suppose X and Y are Banach spaces and $T : X \to Y$ is a compact operator. If $T(X) = Y$, then show that Y has finite dimension.

Exercise 6.10 Suppose X and Y are Banach spaces and $T : X \to Y$ is a bounded linear operator. If $T(X)$ is a closed infinite-dimensional subset of Y, is T a compact operator? Explain your answer.

Exercise 6.11 Let E be a Banach space. Show that E is infinite-dimensional if and only if E^* is infinite-dimensional. Also, show that $\dim(E) = \dim(E^*)$ if E is finite-dimensional. (*Hint:* Use the Hahn–Banach Theorem.)

Exercise 6.12 Let X be a Banach space and suppose $K : X \to X$ is a compact operator. Let $\lambda \neq 0$ and let $T = \lambda I - K$ be a compact perturbation of a scaling of the identity.

(a) Show there exists an $N \in \mathbb{N}$ such that $\ker((T^*)^m) = \ker((T^*)^N)$ for all $m \geq N$.
(b) Use (a) to show that $\operatorname{ran}(T^m) = \operatorname{ran}(T^N)$ for all $m \geq N$, where N is the natural number from (a). (*Hint:* Use Exercise 5.20.)

Exercise 6.13 Suppose that $T : C[0, 1] \to C[0, 1]$ is a compact operator. Show that there exists a sequence $(T_n)_{n=1}^{\infty}$ of finite rank operators such that $\lim_{n \to \infty} \|T - T_n\| = 0$.

Exercise 6.14 Define a map $T_K : C[0, 1] \to C[0, 1]$ by

$$T_K f(x) = \int_0^1 \sin((x - y)\pi) f(y)\, dy, \quad f \in C[0, 1], \ x \in [0, 1].$$

Show that T_K is a compact operator. Compute $\ker(T_K)$ and $\operatorname{ran}(T_K)$.

Exercise 6.15 Let X and Y be Banach spaces and suppose $T : X \to Y$ is a bounded linear operator. The operator T is called *weakly compact* if the set $T(B_X)$ is relatively compact in the weak topology on Y. Show that T is weakly compact if and only if $T^{**}(X^{**}) \subseteq Y$.

Exercise 6.16 (Gantmacher's Theorem) Prove the following: If T is weakly compact, then T^* is weakly compact.

Exercise 6.17 Let X and Y be Banach spaces. Show that if S and T in $\mathcal{L}(X, Y)$ are weakly compact operators, then $S + T$ is a weakly compact operator.

Exercise 6.18 Assume that W, X, Y, and Z are Banach spaces and suppose that the map $T : X \to Y$ is a weakly compact operator. Show that if $A : W \to X$ and $B : Y \to Z$ are bounded linear operators, then $BTA : W \to Z$ is weakly compact.

Exercise 6.19 Prove the following theorems of Pettis:

(a) If X is reflexive and $T : X \to \ell_1$ is a bounded linear operator, then T is compact.
(b) If X is reflexive and $S : c_0 \to X$ is a bounded linear operator, then S is compact.

Chapter 7
Hilbert Space Theory

In this chapter, we will consider the *spectral theory* for compact hermitian operators on a Hilbert space.

7.1 Basics of Hilbert Spaces

Before we begin our discussion of linear operators on a Hilbert space, we recall the basic definitions and the important theorems we will use.

Definition 7.1 Let H be a complex vector space. A *complex inner product* on H is a map $(\cdot, \cdot) : H \times H \to \mathbb{C}$ that satisfies the following properties:

(i) $(x, x) \geq 0$ for all $x \in H$, and $(x, x) = 0$ if and only if $x = 0$,
(ii) $(x, y) = \overline{(y, x)}$,
(iii) $(\alpha x_1 + \beta x_2, y) = \alpha(x_1, y) + \beta(x_2, y)$, and
(iv) $(x, \alpha y_1 + \beta y_2) = \overline{\alpha}(x, y_1) + \overline{\beta}(x, y_2)$,

where $\{x, x_1, x_2, y, y_1, y_2\} \subseteq H$ and $\{\alpha, \beta\} \subseteq \mathbb{C}$. When H is equipped with a complex inner product, it is called a *complex inner product space*.

A map that satisfies *(i)* is said to be *positive definite*, while a map satisfying *(ii)* is said to be *conjugate symmetric*. When a map satisfies *(iii)* and *(iv)*, it is called *sesquilinear*. (The term *sesqui* comes from the Latin for *one and a half*.)

A *real inner product space* is similarly defined, except the scalars are real. (See Definition 3.19.) When the underlying vector space is real, a map satisfying *(iii)* and *(iv)* is called *bilinear*. We will assume H is a complex inner product space, unless otherwise stated.

Let H be a complex inner product space. Define a norm on H by $\|x\| = \sqrt{(x, x)}$ for all $x \in H$. The positive-definiteness and homogeneity of $\| \cdot \|$ are clear (from the definition of the inner product). We will verify subadditivity in a moment, but first we need an important inequality.

Theorem 7.2 (Cauchy–Schwarz Inequality) *If x and y are elements of an inner product space, then $|(x, y)| \leq \|x\| \|y\|$.*

© Springer Science+Business Media, LLC 2014
A. Bowers, N. J. Kalton, *An Introductory Course in Functional Analysis*,
Universitext, DOI 10.1007/978-1-4939-1945-1_7

Proof Observe that $\|x+ty\|^2 \geq 0$ for all $t \in \mathbb{R}$. If $x+ty = 0$ for some $t \in \mathbb{R}$, then $\|x\| = |t|\,\|y\|$, and so

$$|(x,y)| = |(-ty,y)| = |t|\,\|y\|^2 = \|x\|\,\|y\|.$$

Suppose now that $\|x+ty\|^2 > 0$ for all $t \in \mathbb{R}$. Expanding the norm as an inner product, we obtain

$$(x,x) + t(x,y) + t(y,x) + t^2(y,y) > 0.$$

Since $(y,x) = \overline{(x,y)}$, it follows that $(x,y) + (y,x) = 2\,\Re((x,y))$, and so

$$\|x\|^2 + 2t\,\Re((x,y)) + t^2\,\|y\|^2 > 0, \tag{7.1}$$

for all $t \in \mathbb{R}$. The left side of (7.1) is a quadratic in t with real coefficients that has no real zeros, and consequently the discriminant must be negative. Therefore,

$$4\left[\Re((x,y))\right]^2 - 4\|x\|^2\,\|y\|^2 < 0.$$

It follows that $|\Re((x,y))| \leq \|x\|\,\|y\|$. This is true for all x and y in H.

By assumption, $(x,y) \in \mathbb{C}$. Choose $\theta \in [0,2\pi)$ so that $e^{i\theta}(x,y) \in \mathbb{R}$. Then

$$|(x,y)| = |e^{i\theta}(x,y)| = |\Re((e^{i\theta}x,y))| \leq \|e^{i\theta}x\|\,\|y\| = \|x\|\,\|y\|,$$

as required. $\qquad\square$

In the case of an inner product space, subadditivity of $\|\cdot\|$ follows from the next inequality, from which we conclude that $\|\cdot\|$ is a norm.

Theorem 7.3 (Minkowski's Inequality) *If x and y are elements of an inner product space, then* $\|x+y\| \leq \|x\| + \|y\|$.

Proof Observe that

$$\|x+y\|^2 = (x+y,x+y) = \|x\|^2 + (x,y) + (y,x) + \|y\|^2.$$

By the Cauchy–Schwarz Inequality, $|(x,y)| \leq \|x\|\,\|y\|$, and so

$$\|x+y\|^2 \leq \|x\|^2 + 2\,\|x\|\,\|y\| + \|y\|^2 = (\|x\| + \|y\|)^2.$$

The result follows by taking square roots. $\qquad\square$

Definition 7.4 A *Hilbert space* is an inner product space H such that $(H, \|\cdot\|)$ is a Banach space, where $\|x\| = \sqrt{(x,x)}$ for all $x \in H$.

We recall that the norm on H is said to be *induced* by the inner product on H. (See Definitions 3.19 and 3.20 and the comments in between.)

Example 7.5 The classical examples of Hilbert space are the sequence space ℓ_2 and the function space $L_2(0, 1)$. The inner product on ℓ_2 is given by

$$(x, y) = \sum_{i=1}^{\infty} x_i \, \overline{y_i}, \quad \{x, y\} \subseteq \ell_2,$$

where $x = (x_i)_{i=1}^{\infty}$ and $y = (y_i)_{i=1}^{\infty}$. The inner product on $L_2(0, 1)$ is given by

$$(f, g) = \int_0^1 f(s) \, \overline{g(s)} \, ds, \quad \{f, g\} \subseteq L_2(0, 1).$$

It is not hard to check that $(\cdot, \cdot)$ is a complex inner product in each of these cases. We will see that these two examples are (in some sense) typical.

Definition 7.6 Let H be an inner product space. Two elements x and y in H are said to be *orthogonal* if $(x, y) = 0$. If E is a closed subspace of H, then x is said to be *orthogonal to E* if $(x, e) = 0$ for all $e \in E$. We will use the notation $x \perp y$ to indicate x is orthogonal to y, and $x \perp E$ to indicate x is orthogonal to E.

Example 7.7 Suppose $H = \ell_2$, the space of all square-summable sequences. For each $n \in \mathbb{N}$, let e_n denote the sequence with 1 in the nth coordinate, and zero everywhere else. Then $e_n \perp e_m$ for all $n \neq m$. Furthermore, if $E_n = \overline{\text{span}}\{e_n, e_{n+1}, \dots\}$, then $e_m \perp E_n$ whenever $m < n$.

Example 7.8 Let H be a Hilbert space and suppose $y \in H$. If $y \perp H$, then $(x, y) = 0$ for all $x \in H$. In particular, it must be the case that $(y, y) = 0$, and so $y = 0$. As a consequence, if y and y' are elements of H such that $(x, y) = (x, y')$ for all $x \in H$, it must be the case that $y - y' \perp H$, and so $y = y'$. This is a fact we shall exploit repeatedly.

Theorem 7.9 (Pythagorean Theorem) *If x and y are elements of an inner product space, and if $x \perp y$, then $\|x + y\|^2 = \|x\|^2 + \|y\|^2$.*

Proof By assumption, $(x, y) = 0$, and so

$$\|x + y\|^2 = \|x\|^2 + 2 \, \Re((x, y)) + \|y\|^2 = \|x\|^2 + \|y\|^2,$$

as required. □

Theorem 7.10. (Parallelogram Law) *If x and y are elements of an inner product space, then*

$$\|x + y\|^2 + \|x - y\|^2 = 2 \left(\|x\|^2 + \|y\|^2 \right).$$

Proof We merely need to expand the expression on the left, and the result follows directly. (See Exercise 3.1.) □

In fact, the Parallelogram Law characterizes inner product spaces. That is, if H is a normed vector space which satisfies the Parallelogram Law, then there exists a unique inner product that gives rise to the norm. If H is a real normed space, the inner product is given by

$$(x, y) = \frac{1}{4} \left(\|x + y\|^2 - \|x - y\|^2 \right).$$

In the case of a complex normed space,

$$(x, y) = \frac{1}{4} \left(\|x + y\|^2 - \|x - y\|^2 + i \|x + iy\|^2 - i \|x - iy\|^2 \right).$$

These are known as the *polarization formulas*. We leave it to the interested reader to verify that these formulas, along with the Parallelogram Law, determine an inner product that agrees with the given norm. (See Exercise 7.25.)

One consequence of the Parallelogram Law is that any Hilbert space is *uniformly convex*. (See Exercise 7.25.)

Lemma 7.11 (Closest Point Lemma) *Let H be a Hilbert space and suppose C is a nonempty closed convex subset of H. Given $x \in H$, there is a unique point $y \in C$ such that*

$$\|x - y\| = d(x, C) = \inf_{z \in C} \|x - z\|.$$

Proof Let $\delta = d(x, C)$. For each $n \in \mathbb{N}$, pick $y_n \in C$ such that

$$\delta \leq \|x - y_n\| \leq \delta + \frac{1}{n}. \tag{7.2}$$

This we can do, because $\delta = d(x, C)$ is an infimum. By the Parallelogram Law,

$$\|x - y_n\|^2 + \|x - y_m\|^2 = 2 \left(\left\| x - \frac{y_m + y_n}{2} \right\|^2 + \left\| \frac{y_m - y_n}{2} \right\|^2 \right).$$

Because C is convex, $\frac{y_m + y_n}{2}$ is in C. Therefore, $\left\| x - \frac{y_m + y_n}{2} \right\| \geq \delta$, and so

$$\|x - y_n\|^2 + \|x - y_m\|^2 \geq 2 \left(\delta^2 + \frac{1}{4} \|y_m - y_n\|^2 \right).$$

On the other hand, by (7.2),

$$\|x - y_n\|^2 + \|x - y_m\|^2 \leq \left(\delta^2 + \frac{2\delta}{n} + \frac{1}{n^2} \right) + \left(\delta^2 + \frac{2\delta}{m} + \frac{1}{m^2} \right).$$

Combining these inequalities, we conclude

$$\|y_m - y_n\|^2 \leq 4\delta \left(\frac{1}{n} + \frac{1}{m} \right) + 2 \left(\frac{1}{n^2} + \frac{1}{m^2} \right).$$

The right side of this inequality tends to zero as m and n approach ∞. Consequently, $(y_n)_{n=1}^{\infty}$ is a Cauchy sequence. Thus, there exists some $y \in C$ (because C is closed) such that $y = \lim_{n \to \infty} y_n$. It then follows from (7.2) that $\|x - y\| = \delta$.

To show uniqueness, suppose y and y' are in C such that $\|x - y\| = \|x - y'\| = \delta$. Again using the Parallelogram Law,

$$\|x - y\|^2 + \|x - y'\|^2 = 2\left(\left\|x - \frac{y + y'}{2}\right\|^2 + \left\|\frac{y - y'}{2}\right\|^2\right).$$

And so, by convexity (and the definition of δ),

$$\|x - y\|^2 + \|x - y'\|^2 \geq 2\left(\delta^2 + \frac{1}{4}\|y - y'\|^2\right).$$

Therefore,

$$2\delta^2 \geq 2\delta^2 + \frac{1}{2}\|y - y'\|^2,$$

and so $\|y - y'\| = 0$. Consequently, $y = y'$, as required.   □

Remark 7.12   In a reflexive space, a nonempty closed convex subset has a closest point, but it might not be unique.

Proposition 7.13 *Let H be a Hilbert space with closed subspace E. If $x \in H$, then there exists a unique $y \in E$ such that $x - y \perp E$.*

Proof Let $x \in H$ be given. Pick $y \in E$ to be the closest point to x, which exists (and is unique) by the Closest Point Lemma (Lemma 7.11). We will verify that $x - y \perp E$.

Suppose $e \in E$. Because y was chosen to be the point in E closest to x, it follows that $\|x - y + te\| \geq \|x - y\|$ for all $t \in \mathbb{R}$. Furthermore, since y is the *unique* element of E that is closest to x, we can conclude that equality holds only when $t = 0$. Squaring both sides of the inequality, we obtain:

$$\|x - y\|^2 + 2t\,\Re((x - y, e)) + t^2\,\|e\|^2 \geq \|x - y\|^2, \quad t \in \mathbb{R}.$$

Consequently,

$$2t\,\Re((x - y, e)) + t^2\,\|e\|^2 \geq 0, \quad t \in \mathbb{R}.$$

If $t > 0$, then $\Re((x - y, e)) > -t\,\|e\|^2/2$. Taking the limit as $t \to 0^+$, we conclude that $\Re((x - y, e)) \geq 0$. Similarly, if $t < 0$, we have $\Re((x - y, e)) < |t|\,\|e\|^2/2$. Taking the limit as $t \to 0^-$, we see that $\Re((x - y, e)) \leq 0$. Therefore, $\Re((x - y, e)) = 0$.

We have established that $\Re((x - y, e)) = 0$ for all $e \in E$. Because E is a linear subspace, it follows that $ie \in E$. Hence, $\Re((x - y, ie)) = 0$ for all $e \in E$. However,

$$\Re((x - y, ie)) = -\Im((x - y, e)),$$

and so $\Im((x - y, e)) = 0$ for all $e \in E$. We have thus demonstrated that $(x - y, e)$ has zero real and imaginary parts for all $e \in E$. Consequently, $(x - y, e) = 0$ for all $e \in E$. Therefore, $x - y \perp E$, as required.

It remains to show that y is the unique element of E such that $x - y \perp E$. Suppose $y' \in E$ also has the property that $x - y' \perp E$. Since $y - y' = (x - y') - (x - y)$, it follows that $y - y' \perp E$. However, $y - y' \in E$, and so $\|y - y'\|^2 = (y - y', y - y') = 0$. Thus, $y = y'$, as required.   □

We have established that for any $x \in H$, and any closed subspace E of H, there exists a unique $y \in E$ such that $x - y \perp E$. This motivates the next definition.

Definition 7.14 Let H be a Hilbert space and let E be a closed subspace of H. If $x \in H$, then the symbol $P_E x$ denotes the unique element of E with the property $x - P_E x \perp E$.

Lemma 7.15 *Let H be a Hilbert space with closed subspace E. If $P_E : H \to H$ is defined by $P_E(x) = P_E x$ for all $x \in H$, then P_E is a bounded linear projection with $\| P_E \| = 1$.*

Proof First we show linearity. Let x and y be elements of H and let α and β be complex numbers. If $e \in E$, then

$$(\alpha x + \beta y - (\alpha P_E x + \beta P_E y), e) = \alpha(x - P_E x, e) + \beta(y - P_E y, e) = 0.$$

This remains true for all $e \in E$, and thus $\alpha P_E x + \beta P_E y$ is an element of E such that $\alpha x + \beta y - (\alpha P_E x + \beta P_E y) \perp E$. However, $P_E(\alpha x + \beta y)$ is the unique element of E with this property. Therefore, $P_E(\alpha x + \beta y) = \alpha P_E x + \beta P_E y$, and so P_E is linear.

Now suppose $x \in H$. By definition, $P_E x \in E$ and so $x - P_E x \perp P_E x$. Thence, by the Pythagorean Theorem,

$$\| x \|^2 = \| P_E x \|^2 + \| x - P_E x \|^2.$$

Consequently, we have that $\| P_E x \| \leq \| x \|$. Certainly, if $x \in E$, then $P_E x = x$, and so $\| P_E \| = 1$. The result follows.   □

The next theorem, known as the Riesz–Fréchet Theorem (and sometimes the Riesz Representation Theorem for Hilbert Spaces), is one of the key results in Hilbert space theory.

Theorem 7.16 (Riesz–Fréchet Theorem) *Let H be a Hilbert space.*

(i) *If $v \in H$, and $\phi_v(x) = (x, v)$ for all $x \in H$, then $\phi_v \in H^*$ and $\| \phi_v \| = \| v \|$.*
(ii) *If $\phi \in H^*$, then there exists a unique $v \in H$ such that $\phi(x) = (x, v)$ for all $x \in H$, and $\| \phi \| = \| v \|$.*

Proof We prove *(i)* first. Certainly ϕ_v is linear. By the Cauchy-Schwarz Inequality, $\| \phi_v(x) \| \leq \| x \| \| v \|$, and so $\| \phi_v \| \leq \| v \|$. Since $\phi_v(v) = (v, v) = \| v \|^2$, we conclude that $\| \phi_v \| = \| v \|$.

Now we turn our attention to *(ii)*. By assumption, ϕ is a continuous linear functional, and consequently $E = \ker \phi$ is a closed subspace of H. If $\phi = 0$, then $v = 0$ satisfies the conclusions of *(ii)*, and so we may assume $\phi \neq 0$. Pick $u \notin E$. Then $\phi(u) \neq 0$. Without loss of generality, assume $\phi(u) = 1$.

Let $w = u - P_E u$ and suppose $x \in H$. The element u was chosen so that $\phi(u) = 1$. It follows that $\phi(x - \phi(x)u) = 0$, and so $x - \phi(x)u \in E$. By the definition of w, we know that $w \perp E$, and hence w is orthogonal to $x - \phi(x)u$. Therefore,

$$0 = (x - \phi(x)u, w) = (x, w) - \phi(x)(u, w),$$

and so $(x, w) = \phi(x)(u, w)$. Since $u = w + P_E u$ and $w \perp P_E u$, we have

$$(u, w) = (w, w) + (P_E u, w) = \| w \|^2.$$

Thus, $(x, w) = \phi(x)(u, w) = \phi(x)\|w\|^2$.

We assumed that $u \notin E$, and so $w \neq 0$. Let $v = w/\|w\|^2$. Then $\phi(x) = (x, v)$ for all $x \in X$. In the notation of (i), we have found a $v \in H$ such that $\phi = \phi_v$. Therefore, using the conclusion of (i), we have $\|\phi\| = \|v\|$. $\square$

The significance of Theorem 7.16 is the identification of H^* with H. When H is a real Hilbert space, the identification with H^* is an isometric isomorphism. When H is complex, it remains an isometry, but the correspondence is given by what is sometimes called a *conjugate isomorphism*. Specifically, if ϕ and ψ in H^* correspond to u and v in H (respectively), then $\alpha\phi + \beta\psi$ corresponds to $\overline{\alpha}u + \overline{\beta}v$. Sometimes this is called *conjugate linearity* or *anti-isomorphism*.

Regardless of the underlying scalar field (real or complex), a consequence of the identification of H with H^* is that any Hilbert space is reflexive.

7.2   Operators on Hilbert Space

In this section, we discuss operators on Hilbert space. We focus our attention not on general operators, but on a special type of operator, called *hermitian*. Before giving the definition of a hermitian operator, we consider the following existence lemma.

Lemma 7.17 *Let H be a Hilbert space. If $T : H \to H$ is a bounded linear operator, then there exists a unique linear operator $T^* : H \to H$ such that*

$$(Tx, y) = (x, T^*y), \quad \{x, y\} \subseteq H,$$

and such that $\|T^*\| = \|T\|$.

Proof Provided we can show such a map exists, the uniqueness of T^* is clear. If there is another such map $S : H \to H$ that satisfies the conclusion of the lemma, then

$$(x, Sy) = (x, T^*y), \quad \{x, y\} \subseteq H.$$

It follows that $Sy = T^*y$ for all $y \in H$, and so $S = T^*$. (See Example 7.8.)

We now show the existence of T^*. Let $y \in H$ and define $\phi_y : H \to \mathbb{C}$ by

$$\phi_y(x) = (Tx, y), \quad x \in H.$$

Certainly, ϕ_y is a linear functional and $\|\phi_y\| \leq \|T\| \|y\|$, by the Cauchy-Schwarz Inequality. By the Riesz–Fréchet Theorem, there exists a unique element in H (depending on y), which we call T^*y, such that $\phi_y(x) = (x, T^*y)$ for all $x \in H$, and such that $\|\phi_y\| = \|T^*y\|$. Define a map $T^* : H \to H$ by $T^*(y) = T^*y$ for all $y \in H$. By the uniqueness of T^*y in H, the map T^* is well defined, and (by construction) we have $(Tx, y) = (x, T^*y)$ for all x and y in H.

We now show that T^* is linear. Let y and y' be elements in H and let α and β be complex numbers. Then for all $x \in H$,

$$\left(x, T^*(\alpha y + \beta y')\right) = \left(Tx, \alpha y + \beta y'\right) = \overline{\alpha}\left(Tx, y\right) + \overline{\beta}\left(Tx, y'\right).$$

Then, since $(Tx, y) = (x, T^*y)$ for all x and y in H,

$$\left(x, T^*(\alpha y + \beta y')\right) = \overline{\alpha}\left(x, T^*y\right) + \overline{\beta}\left(x, T^*y'\right) = \left(x, \alpha T^*y + \beta T^*y'\right).$$

This is true for all $x \in H$, and so $T^*(\alpha y + \beta y') = \alpha T^*y + \beta T^*y'$. Therefore, T^* is linear.

To see that T^* is bounded, let $y \in H$. Then, invoking the Riesz–Fréchet Theorem,

$$\|T^*y\| = \sup_{x \in B_H} |(x, T^*y)| = \sup_{x \in B_H} |(Tx, y)| \leq \|T\|\,\|y\|,$$

where the last inequality comes from the Cauchy-Schwarz Inequality. Therefore, $\|T^*\| \leq \|T\|$, and so T^* is bounded.

It remains to show that $\|T^*\| = \|T\|$. We have already established $\|T^*\| \leq \|T\|$, and so we need only show the reverse inequality. To that end, we repeat the above construction on T^* to define an operator $T^{**} : H \to H$ with the property that

$$(T^*x, y) = (x, T^{**}y), \quad \{x, y\} \subseteq H,$$

and such that $\|T^{**}\| \leq \|T^*\|$. It suffices then to show that $T^{**} = T$.

Let x and y be in H. Then

$$(x, T^{**}y) = (T^*x, y) = \overline{(y, T^*x)} = \overline{(Ty, x)} = (x, Ty).$$

Therefore, $(x, T^{**}y) = (x, Ty)$ for all x and y in H, and so $T = T^{**}$. This completes the proof. $\square$

Definition 7.18 Let H be a Hilbert space and suppose $T : H \to H$ is a bounded linear operator. The unique linear operator $T^* : H \to H$ such that $(Tx, y) = (x, T^*y)$ for all x and y in H is called the *Hilbert space adjoint of T*.

The Hilbert space adjoint is analogous to the matrix adjoint from linear algebra, and retains many features of its finite-dimensional cousin. (See Exercise 7.2.)

We have now defined two adjoints for the operator $T : H \to H$, both denoted T^*. (See Definitions 3.36 and 7.18.) Recall that the *(operator) adjoint* of T is the bounded linear map $T^* : H^* \to H^*$ defined by $(T^*x^*)(x) = x^*(Tx)$ for all $x \in H$ and $x^* \in H^*$. It may seem as though there is a risk of confusion, but there is a natural correspondence between the two, via the Riesz–Fréchet Theorem. (See Exercise 7.11.)

Definition 7.19 Let H be a Hilbert space and suppose $T : H \to H$ is a bounded linear operator. If $T = T^*$, then T is called a *hermitian* operator. A real hermitian operator is also known as a *symmetric* operator.

Example 7.20 Suppose $T : L_2(0, 1) \to L_2(0, 1)$ is a bounded linear operator given by the formula

$$Tf(x) = \int_0^1 K(x, y) f(y) \, dy, \quad f \in L_2(0, 1), \quad x \in [0, 1],$$

where $K \in L_2([0, 1] \times [0, 1])$. We established that T is well-defined in Section 6.1. Let f and g be functions in $L_2(0, 1)$. Then

$$(Tf, g) = \int_0^1 \left(\int_0^1 K(x, y) f(y) \, dy \right) \overline{g(x)} \, dx.$$

By Fubini's Theorem, this equals

$$\int_0^1 \left(\int_0^1 K(x, y) \overline{g(x)} \, dx \right) f(y) \, dy.$$

Since T^* is the unique operator with the property that $(Tf, g) = (f, T^*g)$, we conclude that

$$\overline{T^*g(y)} = \int_0^1 K(x, y) \overline{g(x)} \, dx.$$

Taking complex conjugates, and swapping the roles of x and y, we obtain

$$T^*g(x) = \int_0^1 \overline{K(y, x)} \, g(y) \, dy.$$

Now that we have obtained a formula for the Hilbert space adjoint T^*, we see that T is hermitian if $K(x, y) = \overline{K(y, x)}$ for almost every x and y. This is reminiscent of the n-dimensional case ($n < \infty$), where a matrix $(a_{jk})_{j,k=1}^n$ is called *hermitian* if $a_{jk} = \overline{a_{kj}}$ for each j and k in the set $\{1, \ldots, n\}$.

Definition 7.21 Let I be a (possibly uncountable) index set. A subset $(e_j)_{j \in I}$ of H is said to be *orthonormal* if $\|e_j\| = 1$ for all $j \in I$, and if $(e_j, e_k) = 0$ for all $j \neq k$.

We recall that the *Kronecker delta* is defined to be

$$\delta_{jk} = \begin{cases} 1 & \text{if } j = k, \\ 0 & \text{if } j \neq k. \end{cases}$$

Using this notation, the subset $(e_j)_{j \in I}$ is called orthonormal if $(e_j, e_k) = \delta_{jk}$ for all indices j and k in I.

Observe that if $(e_j)_{j \in I}$ is an orthonormal set in H, then $\|e_j - e_k\| = \sqrt{2}$ whenever $j \neq k$ (by the Pythagorean Theorem). From this we conclude that any orthonormal subset of a separable Hilbert space is necessarily countable.

Lemma 7.22 *Every orthonormal set is contained in a maximal orthonormal set.*

Proof This follows from Zorn's Lemma. □

In what follows, we will restrict our attention to countable orthonormal sets, but much of what we say will remain true for uncountable sets, as well.

Theorem 7.23 (Bessel's Inequality) *Suppose $(e_j)_{j=1}^{N}$ is a countable orthonormal set, where $N \in \mathbb{N} \cup \{\infty\}$.*

(i) If $x \in H$, then $\sum_{j=1}^{N} |(x, e_j)|^2 \leq \|x\|^2$.

(ii) If $x \in H$ and $\sum_{j=1}^{N} |(x, e_j)|^2 = \|x\|^2$, then $\sum_{j=1}^{N} (x, e_j) e_j = x$.

Proof (i) Let $x \in H$. Choose $m \in \mathbb{N}$ such that $m \leq N$. Let $y = \sum_{j=1}^{m} (x, e_j) e_j$. For each $k \in \{1, \ldots, m\}$,

$$(y, e_k) = \sum_{j=1}^{m} (x, e_j)(e_j, e_k) = \sum_{j=1}^{m} (x, e_j) \delta_{jk} = (x, e_k).$$

Therefore, $(y - x, e_k) = 0$ for all $k \in \{1, \ldots, m\}$. It follows that $(y - x, y) = 0$, and so $x - y \perp y$. Consequently,

$$\|x\|^2 = \|x - y\|^2 + \|y\|^2, \tag{7.3}$$

by the Pythagorean Theorem. From this we conclude that $\|y\| \leq \|x\|$.

Computing the norm of y:

$$\|y\|^2 = (y, y) = \sum_{j=1}^{m} \sum_{k=1}^{m} (x, e_j) \overline{(x, e_k)} (e_j, e_k) = \sum_{j=1}^{m} |(x, e_j)|^2, \tag{7.4}$$

recalling that $(e_j, e_k) = \delta_{jk}$ for all indices j and k. Therefore, $\sum_{j=1}^{m} |(x, e_j)|^2 \leq \|x\|^2$ for all $m \leq N$, when m is finite.

In order to complete the proof of the first part of the theorem, we consider two cases. If $N \in \mathbb{N}$, then let $m = N$ and the proof is complete. If $N = \infty$, then

$$\sum_{j=1}^{\infty} |(x, e_j)|^2 = \lim_{m \to \infty} \sum_{j=1}^{m} |(x, e_j)|^2 \leq \lim_{m \to \infty} \|x\|^2 = \|x\|^2.$$

This proves part *(i)* of the theorem.

(ii) If $N < \infty$, then the assumption, together with (7.4), implies that $\|y\| = \|x\|$. Thus, because of (7.3), we have $\|x - y\| = 0$, and hence $x = y$.

Now suppose that $N = \infty$. For any positive integers m and n such that $m < n$, arguing as in (7.4), we have

$$\Big\| \sum_{j=m+1}^{n} (x, e_j) e_j \Big\|^2 = \sum_{j=m+1}^{n} |(x, e_j)|^2.$$

By assumption, the series $\sum_{j=1}^{\infty} |(x, e_j)|^2$ converges. Consequently, the sequence

$$\left(\sum_{j=1}^{n} (x, e_j) e_j \right)_{n=1}^{\infty}$$

is a Cauchy sequence in the norm on H. By completeness, there exists a $u \in H$ such that $u = \sum_{j=1}^{\infty} (x, e_j) e_j$. By direct computation, we see that $(u, e_j) = (x, e_j)$ for all $j \in \mathbb{N}$. Thus, we have that $x - u \perp u$. Therefore, by the Pythagorean Theorem, we have that $\|x\|^2 = \|x - u\|^2 + \|u\|^2$. By assumption, $\|x\| = \|u\|$. It follows that $\|x - u\| = 0$, and so we conclude that $x = u$, as required. $\qquad \square$

Definition 7.24 Let H be a Hilbert space. An orthonormal sequence $(e_n)_{n=1}^{\infty}$ is called an *orthonormal basis* for H if $x = \sum_{j=1}^{\infty} (x, e_j) e_j$ for every $x \in H$.

Theorem 7.25 *Let H be a Hilbert space and suppose $(e_j)_{j=1}^{\infty}$ is an orthonormal sequence in H. The following are equivalent:*

(i) The sequence $(e_j)_{j=1}^{\infty}$ is an orthonormal basis for H.

(ii) If $x \in H$, then $\|x\|^2 = \sum_{j=1}^{\infty} |(x, e_j)|^2$. (Parseval's Identity.)

(iii) $(e_j)_{j=1}^{\infty}$ is a maximal orthonormal sequence.

Proof We begin by assuming *(i)*. Let $(e_j)_{j=1}^{\infty}$ be an orthonormal basis for H and let $x \in H$. In the proof of Bessel's Inequality (Theorem 7.23 (Bessel's Inequality)), we showed that the series $\sum_{j=1}^{\infty} (x, e_j) e_j$ converges to x and that $\|x\|^2 = \sum_{j=1}^{\infty} |(x, e_j)|^2$. (See (7.4) and the text following it.) Therefore, *(i)* implies *(ii)*.

Assume now that *(ii)* is true, but that $(e_j)_{j=1}^{\infty}$ is not maximal. Then there exists some $x \in H$ with $\|x\| = 1$ such that $(x, e_n) = 0$ for all $n \in \mathbb{N}$. This is a violation of *(ii)*, and so $(e_j)_{j=1}^{\infty}$ must be maximal. This proves that *(ii)* implies *(iii)*.

Now assume *(iii)*, but suppose that $(e_j)_{j=1}^{\infty}$ is not a basis for H. Let $x \in H$ be such that $\sum_{j=1}^{\infty} (x, e_j) e_j$ is not x. The series $\sum_{j=1}^{\infty} (x, e_j) e_j$ converges to some $u \in H$. (See the proof of Theorem 7.23 (Bessel's Inequality).) By construction, $x - u \perp e_j$ for all $j \in \mathbb{N}$. Since $x \neq u$ (by assumption), the element $\frac{x-u}{\|x-u\|}$ has norm one and is orthogonal to e_j for each $j \in \mathbb{N}$. This contradicts the maximality of $(e_j)_{j=1}^{\infty}$. Thus, $(e_j)_{j=1}^{\infty}$ is an orthonormal basis for H, and so *(iii)* implies *(i)*, as required. $\qquad \square$

Corollary 7.26 *Every nonzero separable Hilbert space has an orthonormal basis.*

Proof Let H be a nonzero separable Hilbert space. Then H has a maximal orthonormal set, by Lemma 7.22. This set is countable, because H is separable. (See the comments preceding Lemma 7.22.) By Theorem 7.25, a countable maximal orthonormal set is an orthonormal basis. $\qquad \square$

Example 7.27 (Identification of separable Hilbert spaces). Let H be an infinite-dimensional separable Hilbert space. Let $(e_i)_{i \in I}$ be an orthonormal basis for H. (Such a basis is known to exist, by Corollary 7.26.) By the Pythagorean Theorem, we have $\|e_i - e_j\| = \sqrt{2}$ whenever $i \neq j$. Since H is separable, it must be the case that I is a countable set. Therefore, *every infinite-dimensional separable Hilbert space has a countable orthonormal basis.*

Without loss of generality, we may assume $I = \mathbb{N}$. Since $(e_i)_{i \in \mathbb{N}}$ is an orthonormal basis for H, we can uniquely express every $x \in H$ as $x = \sum_{j=1}^{\infty} (x, e_j) e_j$. We define a map $T : H \to \ell_2$ by

$$Tx = ((x, e_j))_{j=1}^{\infty}, \quad x \in H.$$

By Parseval's Identity, this map is well-defined, and indeed $\|Tx\| = \|x\|$. Consequently, T is an isometry. It follows that *all infinite-dimensional separable Hilbert spaces are isometrically isomorphic to ℓ_2.* Therefore, as far as Banach spaces are concerned, there is only one infinite-dimensional separable Hilbert space. Choosing an orthonormal basis for H is essentially representing H as ℓ_2.

To emphasize this result, we place it in a theorem.

Theorem 7.28 *Every infinite-dimensional separable Hilbert space is isometrically isomorphic to ℓ_2.*

Proof See the discussion preceding the statement of the theorem. □

The next example can be seen as a special case of the previous.

Example 7.29 (The Fourier transform and separability of L_2) Let $L_2(\mathbb{T})$ denote the space of (equivalence classes of) complex-valued square-integrable functions on the probability space $(\mathbb{T}, \frac{d\theta}{2\pi})$, where $\mathbb{T} = [0, 2\pi)$. The space $L_2(\mathbb{T})$ is a Hilbert space with inner product given by

$$(f, g) = \frac{1}{2\pi} \int_0^{2\pi} f(\theta) \overline{g(\theta)} \, d\theta, \quad \{f, g\} \subseteq L_2(\mathbb{T}).$$

(Compare to Example 7.5.)

For each $n \in \mathbb{Z}$, define $e_n : \mathbb{T} \to \mathbb{C}$ by

$$e_n(\theta) = e^{in\theta}, \quad \theta \in \mathbb{T}.$$

We claim that $(e_n)_{n \in \mathbb{Z}}$ is an orthonormal basis for $L_2(\mathbb{T})$. The sequence $(e_n)_{n \in \mathbb{Z}}$ is an orthonormal set, because

$$(e_n, e_m) = \frac{1}{2\pi} \int_0^{2\pi} e^{in\theta} e^{-im\theta} \, d\theta = \begin{cases} 1 & \text{if } n = m, \\ 0 & \text{if } n \neq m. \end{cases}$$

The maximality of $(e_n)_{n \in \mathbb{Z}}$ follows from the density of trigonometric polynomials: The Weierstrass Approximation Theorem states that the set of trigonometric polynomials is dense in $C(\mathbb{T})$, and Lusin's Theorem (Theorem A.36) states that the set of continuous functions is dense in $L_2(\mathbb{T})$.

If $f \in L_2(\mathbb{T})$, then the *Fourier series* of f is given by the series

$$\sum_{n \in \mathbb{Z}} (f, e_n) \, e_n.$$

Since the sequence $(e_n)_{n \in \mathbb{Z}}$ is an orthonormal basis for $L_2(\mathbb{T})$, we see that the Fourier series of $f \in L_2(\mathbb{T})$ will always converge to f in the $L_2(\mathbb{T})$-norm. (Compare this to Example 4.17.)

We recall that the *Fourier transform of* f is given by the formula $\hat{f}(n) = (f, e_n)$ for $n \in \mathbb{Z}$. The Fourier transform represents $L_2(\mathbb{T})$ as $\ell_2(\mathbb{Z})$. To see this precisely, define a map $\mathcal{F} : L_2(\mathbb{T}) \to \ell_2(\mathbb{Z})$, also called the *Fourier transform*, by

$$\mathcal{F}(f) = (\hat{f}(n))_{n \in \mathbb{Z}}, \quad f \in L_2(\mathbb{T}).$$

It is not hard to show that $\mathcal{F}$ is an isomorphism. By Parseval's Identity (Theorem 7.25), we see that

$$\|f\|_{L_2(\mathbb{T})} = \|\mathcal{F}(f)\|_{\ell_2(\mathbb{Z})}, \quad f \in L_2(\mathbb{T}).$$

Therefore, $\mathcal{F}$ determines an isometric isomorphism, and so $L_2(\mathbb{T})$ and $\ell_2(\mathbb{Z})$ are identical as Banach spaces. The identification of $L_2(\mathbb{T})$ and $\ell_2(\mathbb{Z})$ is sometimes known as the *Riesz–Fischer Theorem*, after F. Riesz and E.S. Fischer, each of whom proved it (independently) in 1907 [11,30].

We are now ready to state the main result of this section, which nicely mirrors the matrix theory of finite-dimensional vector spaces. While we state (and prove) the result for separable Hilbert spaces, it remains true for general Hilbert spaces.

Theorem 7.30 *Let H be an infinite-dimensional separable Hilbert space. Suppose $T : H \to H$ is a compact hermitian operator. There exists an orthonormal basis $(e_n)_{n=1}^{\infty}$ of H such that e_n is an eigenvector of T for each $n \in \mathbb{N}$; that is, there exists a sequence $(\lambda_n)_{n=1}^{\infty}$ such that $T e_n = \lambda_n \, e_n$ for all $n \in \mathbb{N}$.*

Remark 7.31 If T is a compact hermitian operator, then any eigenvalue λ of T must be real. To see this, suppose $x \in H$ is an eigenvector with $\|x\| = 1$ such that $Tx = \lambda x$. Using the properties of inner products, and keeping in mind that x has norm one, we see that

$$\lambda = (\lambda x, x) = (Tx, x) = \overline{(x, Tx)} = \overline{(x, \lambda x)} = \overline{\lambda}.$$

Also, observe that the sequence of eigenvalues $(\lambda_n)_{n=1}^{\infty}$ promised in Theorem 7.30 must converge to 0. (We proved this for a general compact operator in Theorem 6.37.)

Before proceeding with the proof of Theorem 7.30, we require some preliminary results.

Lemma 7.32 *Let H be a nonzero Hilbert space. If $T : H \to H$ is a compact hermitian operator, there exists an $x \in H$ with $\|x\| = 1$ such that $\|Tx\| = \|T\|$.*

Proof First, we claim that $\|T^2\| = \|T\|^2$. Certainly, for any $x \in H$, we have

$$\|T^2(x)\| = \|T(Tx)\| \le \|T\| \, \|Tx\| \le \|T\|^2 \, \|x\|.$$

It follows that $\|T^2\| \le \|T\|^2$.

By the hermitian assumption on T, for any $x \in H$,

$$\|Tx\|^2 = (Tx, Tx) = (T^2x, x).$$

Therefore, by the Cauchy-Schwarz Inequality, it follows that $\|Tx\|^2 \leq \|T^2x\| \|x\|$. Taking the supremum over all $x \in B_H$, we conclude that $\|T\|^2 \leq \|T^2\|$, and hence $\|T\|^2 = \|T^2\|$, as claimed.

Now, pick a sequence $(x_n)_{n=1}^{\infty}$ in H such that $\|x_n\| \leq 1$ for all $n \in \mathbb{N}$, and such that

$$\lim_{n \to \infty} \|T^2x_n\| = \|T^2\| = \|T\|^2. \tag{7.5}$$

We assumed T was compact, and so $T(B_H)$ is relatively compact in H. Thus, passing to a subsequence if necessary, we may assume (without loss of generality) that $(Tx_n)_{n=1}^{\infty}$ converges to some $y \in H$. Since $\|Tx_n\| \leq \|T\|$ for all $n \in \mathbb{N}$, it must be the case that $\|y\| \leq \|T\|$. Let $x = y/\|T\|$. Then $\|x\| \leq 1$. Furthermore,

$$Tx = \frac{1}{\|T\|}Ty = \lim_{n \to \infty} \frac{1}{\|T\|}T^2x_n.$$

Therefore, by (7.5), $\|Tx\| = \|T\|$, as required. (Note this implies $\|x\| = 1$.) $\quad\square$

Remark 7.33 The hermitian assumption is not required in Lemma 7.32. In fact, the conclusion of Lemma 7.32 holds for any compact operator $T : X \to X$, whenever X is a reflexive Banach space. (See Exercise 6.7.)

We require one more lemma.

Lemma 7.34 *Let H be a nonzero Hilbert space. If $T : H \to H$ is a compact hermitian operator, then either $\|T\|$ or $-\|T\|$ is an eigenvalue.*

Proof We may assume $T \neq 0$. By Lemma 7.32, there exists some $x \in H$ with $\|x\| = 1$ such that $\|Tx\| = \|T\|$. We make the observation that, since T is a hermitian operator,

$$(T^2x, x) = (Tx, Tx) = \|Tx\|^2 = \|T\|^2. \tag{7.6}$$

Define $u = T^2x - (T^2x, x)x$. Notice that $u = T^2x - \|T\|^2 x$, by (7.6).

By the definition of u, we have

$$(u, x) = (T^2x, x) - (T^2x, x)(x, x) = (T^2x, x) - (T^2x, x)\|x\|^2 = 0.$$

Consequently, we have $u \perp x$, and in particular $u \perp (T^2x, x)x$. By the Pythagorean Theorem, it follows that

$$\|u\|^2 + \|(T^2x, x)x\|^2 = \|u + (T^2x, x)x\|^2. \tag{7.7}$$

On the left side of (7.7), we observe

$$\|(T^2x,x)x\|^2 = |(T^2x,x)|^2 \|x\|^2 = \|T\|^4,$$

by (7.6). On the right side of (7.7), recalling the definition of u, we have

$$\|u + (T^2x,x)x\|^2 = \|T^2x\|^2 \le \|T\|^4.$$

It follows that

$$\|u\|^2 + \|T\|^4 \le \|T\|^4.$$

Therefore, $\|u\| = 0$, and so $T^2x = \|T\|^2 x$.

If $Tx = \|T\|x$, then x is an eigenvector for T with eigenvalue $\|T\|$. Suppose instead that $Tx \ne \|T\|x$. Then $y = Tx - \|T\|x$ is a nonzero element of H. Computing:

$$Ty + \|T\|y = (T^2x - \|T\|Tx) + (\|T\|Tx - \|T\|^2 x) = 0.$$

Consequently, $Ty = -\|T\|y$, and so $y \ne 0$ is an eigenvector for T with eigenvalue $-\|T\|$. We have established that either $\|T\|$ or $-\|T\|$ is an eigenvalue for T, and we are done. $\qquad\square$

We are now prepared to prove Theorem 7.30.

Proof of Theorem 7.30 By Lemma 7.34, there exists a nonempty set of orthonormal eigenvectors, and so there exists a maximal set, by Zorn's Lemma. Let $(e_n)_{n=1}^N$ be a maximal set of orthonormal eigenvectors, where $N \in \mathbb{N} \cup \{\infty\}$. If $N < \infty$, let $\mathbb{A} = \{1, \dots, N\}$; otherwise let $\mathbb{A} = \mathbb{N}$. For each $n \in \mathbb{A}$, let λ_n denote the eigenvalue corresponding to the eigenvector e_n.

Let $H_0 = \{x : (x, e_n) = 0 \text{ for all } n \in \mathbb{A}\}$. We will show that $H_0 = \{0\}$. We begin by claiming that $T(H_0) \subseteq H_0$. Suppose that $x \in H_0$. Then for all $n \in \mathbb{A}$,

$$(Tx, e_n) = (x, Te_n) = (x, \lambda_n e_n) = 0.$$

Therefore, $Tx \in H_0$, and consequently $T|_{H_0} : H_0 \to H_0$ is a well-defined compact hermitian operator.

Observe that H_0 is a Hilbert space. Thus, if $H_0 \ne \{0\}$, then $T|_{H_0} : H_0 \to H_0$ has an eigenvector x_0, by Lemma 7.34. But then $(e_n)_{n \in \mathbb{A}} \cup \{x_0/\|x_0\|\}$ forms an orthonormal set of eigenvectors for $T : H \to H$. This contradicts the maximality of $(e_n)_{n \in \mathbb{A}}$, and so $H_0 = \{0\}$.

We have established that no nonzero element of H is orthogonal to the orthonormal set $(e_n)_{n \in \mathbb{A}}$. It follows that $(e_n)_{n \in \mathbb{A}}$ is maximal amongst all orthonormal sets, and thus it is an orthonormal basis for H, by Theorem 7.25. This completes the proof. $\qquad\square$

7.3 Hilbert–Schmidt Operators

Let a and b be real numbers such that $a < b$. Recall that $L_2(a, b)$ denotes the set of (equivalence classes of) Borel measurable functions $f : [a, b] \to \mathbb{C}$ such that

$$\|f\|_{L_2(a,b)} = \left(\int_a^b |f(x)|^2 \, dx \right)^{1/2} < \infty.$$

This space is a complex Hilbert space with the inner product $(f, g) = \int_a^b f(x) \overline{g(x)} \, dx$ for f and g in $L_2(a, b)$.

Suppose that $K : [a, b] \times [a, b] \to \mathbb{C}$ is in $L_2([a, b] \times [a, b])$, so that

$$\|K\|_{L_2}^2 = \int_a^b \left(\int_a^b |K(x, y)|^2 \, dy \right) dx < \infty. \tag{7.8}$$

A consequence of (7.8) is that $\int_a^b |K(x, y)|^2 \, dy < \infty$ for almost every $x \in [a, b]$.

We use K to define a map $T : L_2(a, b) \to L_2(a, b)$ by

$$Tf(x) = \int_a^b K(x, y) f(y) \, dy, \quad f \in L_2(a, b), \ x \in [a, b]. \tag{7.9}$$

We will show that T is a well-defined bounded linear operator on $L_2(a, b)$.

By Hölder's Inequality,

$$\int_a^b |K(x, y) f(y)| \, dy \le \left(\int_a^b |K(x, y)|^2 \, dy \right)^{1/2} \left(\int_a^b |f(y)|^2 \, dy \right)^{1/2}. \tag{7.10}$$

This quantity is finite for almost every x in $[a, b]$ whenever $f \in L_2(a, b)$. Therefore, the integral in (7.9) exists for almost every x in $[a, b]$, provided that the function f is an element of $L_2(a, b)$. Linearity of T is now evident, and so T is a well-defined linear operator.

It remains to show that T is bounded. From (7.10), we see that

$$|Tf(x)| \le \left(\int_a^b |K(x, y)|^2 \, dy \right)^{1/2} \|f\|_{L_2(a,b)}.$$

By squaring, and then integrating with respect to x, we discover

$$\int_a^b |Tf(x)|^2 \, dx \le \left(\int_a^b \int_a^b |K(x, y)|^2 \, dy \, dx \right) \|f\|_{L_2(a,b)}^2.$$

Therefore, $\|Tf\|_{L_2(a,b)} \le \|K\|_{L_2} \|f\|_{L_2(a,b)}$. Consequently, the map T is a well-defined bounded linear operator and $\|T\| \le \|K\|_{L_2}$.

Definition 7.35 A map $T : L_2(a, b) \to L_2(a, b)$ is a *Hilbert-Schmidt operator* if

$$Tf(x) = \int_a^b K(x, y) f(y) \, dy, \quad f \in L_2(a, b), \ x \in [a, b],$$

for some $K \in L_2([a, b] \times [a, b])$. The function K is the *kernel* associated with T, or simply the kernel of T.

We wish to identify T^*, the Hilbert space adjoint of T. To that end, suppose f and g are in $L_2(a,b)$. By Fubini's Theorem,

$$(Tf, g) = \int_a^b \left(\int_a^b K(x, y) \, f(y) \, dy \right) \overline{g(x)} \, dx = \int_a^b \left(\int_a^b K(x, y) \, \overline{g(x)} \, dx \right) f(y) \, dy.$$

The Hilbert space adjoint T^* is the unique operator that satisfies the equation $(Tf, g) = (f, T^*g)$ for all f and g in $L_2(a,b)$, and therefore

$$T^*g(x) = \int_a^b \overline{K(y, x)} \, g(y) \, dy. \tag{7.11}$$

(See Example 7.20.) Comparing (7.3.4) with the definition of T in (7.9), we see that $T^* = T$ precisely when $K(x, y) = \overline{K(y, x)}$ for almost every x and y in $[a, b]$. This motivates the next definition.

Definition 7.36   Suppose T is a Hilbert-Schmidt operator on $L_2(a, b)$ with kernel K. We say that K is a *hermitian kernel*, or simply that K is *hermitian*, if $K(x, y) = \overline{K(y, x)}$ for almost every x and y in $[a, b]$. A real-valued hermitian kernel is also known as a *symmetric* kernel.

Throughout the remainder of this section, $(e_n)_{n=1}^\infty$ will denote an orthonormal basis for $L_2(a, b)$. (We know an orthonormal basis exists by Corollary 7.26. The basis is countable by an argument similar to that used in Example 7.29.)

Proposition 7.37   *Let $(e_n)_{n=1}^\infty$ be an orthonormal basis for $L_2(a, b)$. For each m and n in $\mathbb{N}$, define a function $f_{mn} : [a, b] \times [a, b] \to \mathbb{C}$ by*

$$f_{mn}(x, y) = e_m(x) \overline{e_n(y)}, \quad \{x, y\} \subseteq [a, b].$$

The set $(f_{mn})_{m,n=1}^\infty$ is an orthonormal basis for $L_2([a, b] \times [a, b])$.

Proof   For ease of notation, let $L_2 = L_2([a, b] \times [a, b])$. The inner product on L_2 is given by

$$(g, f) = \int_a^b \int_a^b g(x, y) \, \overline{f(x, y)} \, dx \, dy,$$

where g and f are functions in L_2. We wish to show that $(f_{mn})_{m,n=1}^\infty$ is an orthonormal basis for L_2. It suffices to show that if $g \in L_2$ and $(g, f_{mn}) = 0$ for all natural numbers m and n, then $g = 0$ in L_2.

Let $g \in L_2$ and suppose for all m and n in $\mathbb{N}$,

$$(g, f_{mn}) = \int_a^b \int_a^b g(x, y) \, \overline{e_m(x)} \, e_n(y) \, dx \, dy = 0. \tag{7.12}$$

By Hölder's Inequality,

$$\int_a^b |g(x, y) \overline{e_m(x)}| \, dx \leq \left(\int_a^b |g(x, y)|^2 \, dx \right)^{1/2} < \infty, \tag{7.13}$$

for almost every y. For each $m \in \mathbb{N}$, define a function h_m on $[a,b]$ by

$$h_m(y) = \int_a^b g(x,y)\,\overline{e_m(x)}\,dx, \quad y \in [a,b].$$

The function h_m is well-defined by (7.13), and furthermore,

$$\|h_m\|^2_{L_2(a,b)} = \int_a^b |h_m(y)|^2\,dy \le \int_a^b \left(\int_a^b |g(x,y)|^2\,dx \right) dy = \|g\|^2_{L_2}.$$

It follows that $h_m \in L_2(a,b)$ for each $m \in \mathbb{N}$.

For every $n \in \mathbb{N}$, by (7.12), we have

$$(h_m, \overline{e_n}) = \int_a^b h_m(y)\,e_n(y)\,dy = 0. \tag{7.14}$$

Since $(e_n)_{n=1}^\infty$ is an orthonormal basis for $L_2(a,b)$, so too is $(\overline{e_n})_{n=1}^\infty$, and thus the equality in (7.14) implies for each $m \in \mathbb{N}$ that $h_m = 0\,\text{a.e.}(y)$. Thus, for each $m \in \mathbb{N}$,

$$\int_a^b g(x,y)\,\overline{e_m(x)}\,dx = 0 \ \text{ a.e.}(y).$$

A countable collection of measure zero sets is still measure zero, and so for almost every y,

$$\int_a^b g(x,y)\,\overline{e_m(x)}\,dx = 0,$$

for every $m \in \mathbb{N}$. Again invoking the fact that $(e_n)_{n=1}^\infty$ is an orthonormal basis, we conclude that $g(x,y) = 0\,\text{a.e.}(x)$ for almost every y. Therefore, $g = 0$ in L_2, as required. $\qquad\qquad\square$

Proposition 7.38 *A Hilbert-Schmidt operator on $L_2(a,b)$ is a compact operator.*

Proof Let $(e_n)_{n=1}^\infty$ be an orthonormal basis for the Hilbert space $L_2(a,b)$ and let $(f_{mn})_{m,n=1}^\infty$ be the orthonormal basis for $L_2([a,b] \times [a,b])$ given in Proposition 7.37.

Suppose that T_K is a Hilbert-Schmidt operator with kernel K. Because $(f_{mn})_{m,n=1}^\infty$ is an orthonormal basis for $L_2([a,b] \times [a,b])$, we have that

$$K = \sum_{n=1}^\infty \sum_{m=1}^\infty (K, f_{mn})\,f_{mn}, \tag{7.15}$$

where the series converges in the norm on $L_2([a,b] \times [a,b])$. To be precise, if for each $N \in \mathbb{N}$,

$$K_N = \sum_{n=1}^N \sum_{m=1}^N (K, f_{mn})\,f_{mn},$$

then $\|K - K_N\|_{L_2} \to 0$ as $N \to \infty$.

The Hilbert-Schmidt operator with kernel K_N is given by

$$T_{K_N} g(x) = \int_a^b \sum_{n=1}^N \sum_{m=1}^N (K, f_{mn}) f_{mn}(x, y) g(y) \, dy.$$

Recalling the definition of $f_{mn}(x, y)$, this becomes

$$\sum_{n=1}^N \sum_{m=1}^N (K, f_{mn}) \int_a^b e_m(x) \overline{e_n(y)} \, g(y) \, dy = \sum_{n=1}^N \sum_{m=1}^N (K, f_{mn}) (g, e_n) e_m(x).$$

Consequently,

$$T_{K_N} g(x) = \sum_{m=1}^N \left(\sum_{n=1}^N (K, f_{mn}) (g, e_n) \right) e_m(x).$$

Therefore, the range of T_{K_N} is at most N-dimensional, and so the rank of T_{K_N} is at most N. (In particular, the rank of T_{K_N} is finite.)

Observe that $T_K - T_{K_N} = T_{K-K_N}$. We know that $\|T_{K-K_N}\| \leq \|K - K_N\|_{L_2}$ (see the comments at the start of this section) and $\|K - K_N\|_{L_2} \to 0$ as $N \to \infty$ (by construction). It follows that $\|T_K - T_{K_N}\| \to 0$ as $N \to \infty$. Therefore, T_K is a compact operator, as the limit of a sequence of finite rank operators.   □

Now assume, in addition to being square-integrable, that K is a hermitian kernel, so that $K(x, y) = \overline{K(y, x)}$ for almost every x and y. By Proposition 7.38, the map T_K is a compact hermitian operator. From Theorem 7.30, we know that there exists an orthonormal basis for $L_2(a, b)$ composed of eigenvectors for T_K. Denote this set of orthonormal eigenvectors by $(e_n)_{n=1}^\infty$ and let the corresponding sequence of eigenvalues be $(\lambda_n)_{n=1}^\infty$. Let $(f_{mn})_{m,n=1}^\infty$ be the orthonormal basis for $L_2([a, b] \times [a, b])$ given in Proposition 7.37.

Using the fact that $\overline{f_{mn}(x, y)} = \overline{e_m(x)} e_n(y)$ for all x and y in $[a, b]$, we compute:

$$(K, f_{mn}) = \int_a^b \left(\int_a^b K(x, y) e_n(y) \, dy \right) \overline{e_m(x)} \, dx = \int_a^b T e_n(x) \cdot \overline{e_m(x)} \, dx. \quad (7.16)$$

By assumption, $T e_n = \lambda_n e_n$ for all $n \in \mathbb{N}$, and hence

$$(K, f_{mn}) = \int_a^b \lambda_n e_n(x) \cdot \overline{e_m(x)} \, dx = \lambda_n (e_n, e_m) = \lambda_n \delta_{mn}.$$

Thus, by Parseval's Identity (Theorem 7.25),

$$\|K\|_{L_2}^2 = \sum_{n=1}^\infty \sum_{m=1}^\infty |(K, f_{mn})|^2 = \sum_{n=1}^\infty \sum_{m=1}^\infty |\lambda_n \delta_{mn}|^2 = \sum_{n=1}^\infty |\lambda_n|^2.$$

We summarize in the following theorem.

Theorem 7.39 *Let T_K be a Hilbert-Schmidt operator with hermitian kernel K. If $(\lambda_n)_{n=1}^\infty$ is the sequence of eigenvalues of T_K, then*

$$\int_a^b \int_a^b |K(x,y)|^2 \, dx \, dy = \sum_{n=1}^\infty |\lambda_n|^2 < \infty,$$

and $\|T_K\| = \sup_{n\in\mathbb{N}} |\lambda_n|.$

Proof The proof of the integral equation can be found in the discussion preceding the statement of the theorem. The equality $\|T_K\| = \sup_{n\in\mathbb{N}} |\lambda_n|$ follows from the fact that $T e_n = \lambda_n e_n$ for all e_n in the orthonormal basis $(e_n)_{n=1}^\infty$. $\square$

7.4   Sturm–Liouville Systems

In this section, we will see an example of how to apply the theory of compact hermitian operators to solve differential equations. For simplicity, we will suppose the scalar field is $\mathbb{R}$. We will consider a special case of a system of differential equations known as a *Sturm-Liouville system*. The system we will consider is

$$\begin{cases} y'' + q(x)y = f(x), & \{f,q\} \subseteq C[a,b], & (DE) \\ y(a) = y(b) = 0. & & (BC) \end{cases} \tag{7.17}$$

We wish to find $y \in C^{(2)}[a,b]$ (a twice continuously differentiable function) that satisfies the *differential equation* (DE) subject to the *boundary conditions* (BC). While we start with the assumption $f \in C[a,b]$, we will later extend to $f \in L_2(a,b)$.

Throughout what follows, we will make the following assumption on the *homogeneous system* (i.e., the system with $f = 0$):

Assumption 1 *The only solution to the homogeneous system is $y = 0$.*

Example 7.40 Our basic model of a Sturm-Liouville system comes from a vibrating string. Suppose that $a = 0$ and $b = \pi$. Consider the following differential system:

$$\begin{cases} y'' = f(x), & f \in C[0,\pi], & (DE') \\ y(0) = y(\pi) = 0. & & (BC') \end{cases}$$

For the differential equation (DE'), the homogeneous equation $y'' = 0$ has general solution $y(t) = \alpha + \beta t$. The only way that this can satisfy (BC') is if $\alpha = \beta = 0$, and so the vibrating string model satisfies our basic assumption.

Let us return our attention to the Sturm-Liouville system in (7.17). Suppose u is a solution to the initial value problem

$$\begin{cases} y'' + q(x)y = 0, & q \in C[a,b], & (DE_0) \\ y(a) = 0, \ y'(a) = 1. & & (IC_1) \end{cases}$$

We know that a solution u to this system exists by the general theory of ordinary differential equations. If $u(b) = 0$, then u is a solution to the homogeneous system. It would follow that $u = 0$, by Assumption 1. This violates the initial conditions in (IC_1), and so $u(b) \neq 0$.

Next, let v be a solution to the initial value problem

$$\begin{cases} y'' + q(x)y = 0, & q \in C[a,b], & (DE_0) \\ y(b) = 0, \ y'(b) = 1. & & (IC_2) \end{cases}$$

As before, such a solution is know to exist by general theory. Furthermore, $v(a) \neq 0$, or else it would be trivial, violating the initial conditions in (IC_2).

Consider the function

$$y(x) = \phi(x)u(x) + \psi(x)v(x), \quad x \in [a,b],$$

where ϕ and ψ are differentiable functions to be determined at a later time. We will insist only that ϕ and ψ satisfy the following assumption:

Assumption 2 ϕ and ψ are differentiable functions such that $\phi'u + \psi'v = 0$.

Keeping Assumption 2 in mind, let us differentiate y:

$$y' = \phi'u + \phi u' + \psi'v + \psi v' = \phi u' + \psi v'.$$

Continue by computing the second derivative of y:

$$y'' = \phi'u' + \phi u'' + \psi'v' + \psi v''.$$

It follows that

$$y'' + qy = \phi'u' + \psi'v' + (u'' + qu)\phi + (v'' + qv)\psi.$$

By assumption, u and v satisfy (DE_0), and so we conclude that

$$y'' + qy = \phi'u' + \psi'v'.$$

Our goal is to find a solution to the differential system in (7.17). From the preceding calculations, we see that this will be accomplished if we can solve the following differential system:

$$\begin{cases} \phi'u + \psi'v = 0, \\ \phi'u' + \psi'v' = f, & \psi(a) = 0, \ \phi(b) = 0. \end{cases}$$

A little algebraic manipulation reveals:

$$(u'v - uv')\phi' = fv \quad \text{and} \quad (uv' - u'v)\psi' = fu. \tag{7.18}$$

Let the *Wronskian* of u and v be given by the formula

$$W = uv' - u'v = \det \begin{pmatrix} u & v \\ u' & v' \end{pmatrix}. \tag{7.19}$$

Then, (7.18) becomes

$$\phi' = -\frac{fv}{W} \quad \text{and} \quad \psi' = \frac{fu}{W}, \tag{7.20}$$

provided that $W(x) \neq 0$ for any $x \in [a, b]$.

We will show that W is a nonzero constant function. To that end, we compute the derivative:

$$W' = u'v' + uv'' - u''v - u'v' = uv'' - u''v.$$

By assumption, $u'' + qu = 0$ and $v'' + qv = 0$, and so

$$W' = u(-qv) - (-qu)v = 0.$$

Therefore, W is a constant function. To verify that W is nonzero, we recall that u and v satisfy the initial conditions (IC_1) and (IC_2), respectively. Observe that

$$W(a) = u(a)v'(a) - u'(a)v(a) = -v(a)$$

and

$$W(b) = u(b)v'(b) - u'(b)v(b) = u(b).$$

We know that $v(a) \neq 0$ and $u(b) \neq 0$, and so W is a nonzero constant function. Let $\alpha = W(a)$ be the value of this constant.

To solve for ϕ and ψ, we now integrate the equations in (7.20). The results are

$$\psi(x) = \int_a^x \frac{f(t)u(t)}{W(t)}\, dt = \frac{1}{\alpha} \int_a^x f(t)u(t)\, dt,$$

and

$$\phi(x) = -\int_b^x \frac{f(t)v(t)}{W(t)}\, dt = \frac{1}{\alpha} \int_x^b f(t)v(t)\, dt.$$

We now have an integral formula for y:

$$y(x) = \frac{1}{\alpha} \left(v(x) \int_a^x f(t)u(t)\, dt + u(x) \int_x^b f(t)v(t)\, dt \right).$$

Define a map $K : [a, b] \times [a, b] \to \mathbb{R}$ by

$$K(x,t) = \begin{cases} \frac{1}{\alpha} v(x)u(t) & \text{if } t \leq x, \\ \frac{1}{\alpha} u(x)v(t) & \text{if } t \geq x. \end{cases} \tag{7.21}$$

Then

$$y(x) = \int_a^b K(x,t)f(t)\,dt.$$

The map K is continuous on $[a,b] \times [a,b]$ and is symmetric; i.e., $K(x,y) = K(y,x)$ for all x and y in $[a,b]$. It follows that K defines a compact symmetric (hermitian) operator

$$T_K f(x) = \int_a^b K(x,t)f(t)\,dt, \quad f \in L_2(a,b), \ x \in [a,b]. \qquad (7.22)$$

Furthermore, for any $f \in L_2(a,b)$ we have that $y = T_K f$ is a solution to the system in (7.17).

Since K is a symmetric (hermitian) kernel, we know that T_K defines a bounded linear map into $L_2(a,b)$. In this case, since K is continuous, we can actually improve this to the statement $T_K : L_2(a,b) \to C[a,b]$.

We first show that $T_K f$ is continuous for all $f \in L_2(a,b)$. To see this, let $\epsilon > 0$ be given and observe that for all x and y in $[a,b]$,

$$|T_K f(x) - T_K f(y)| = \Big| \int_a^b \Big(K(x,t) - K(y,t) \Big) f(t)\,dt \Big|.$$

By Hölder's Inequality, this is bounded by

$$\left(\int_a^b |K(x,t) - K(y,t)|^2\,dt \right)^{1/2} \|f\|_{L_2(a,b)}.$$

By assumption, K is continuous on a compact set, and so there exists a $\delta > 0$ such that

$$|K(x,t) - K(y,t)| < \frac{\epsilon}{\|f\|_{L_2(a,b)}\sqrt{b-a}},$$

for all $t \in [a,b]$, whenever $|x-y| < \delta$. Therefore, $|T_K f(x) - T_K f(y)| \le \epsilon$ whenever $|x - y| < \delta$, and so $T_K f$ is continuous on $[a,b]$.

The argument to show T_K is bounded on $L_2(a,b)$ is similar. For each $x \in [a,b]$, by Hölder's Inequality, we have

$$|T_K f(x)| = \Big| \int_a^b K(x,t)f(t)\,dt \Big| \le \left(\int_a^b |K(x,t)|^2\,dt \right)^{1/2} \|f\|_{L_2(a,b)}.$$

The function K is continuous on the compact set $[a,b] \times [a,b]$, and hence it is bounded on $[a,b] \times [a,b]$. Consequently,

$$\|T_K f\|_{C[a,b]} \le \sqrt{b-a}\,\|K\|_{C([a,b]\times[a,b])}\|f\|_{L_2(a,b)}.$$

Therefore, $T_K : L_2(a,b) \to C[a,b]$ is bounded and $\|T_K\| \le \sqrt{b-a}\,\|K\|_{C([a,b]\times[a,b])}$.

Suppose now that e is an eigenvector for T_K (an *eigenfunction* in this case). Then there exists a nonzero scalar λ such that $T_K e = \lambda e$. (Note that $\lambda \neq 0$ because

$y = T_K e$ is a solution to $y'' + qy = e$. If $y = T_K e = 0$, the differential equation implies that $e = 0$, which contradicts the assumption that e is an eigenvector.)

We know that $y = T_K e$ is a solution to $y'' + qy = e$ (by construction) and that $T_K e = \lambda e$, where $\lambda \neq 0$. Substituting $y = \lambda e$ into the differential equation, we have $(\lambda e'') + q(\lambda e) = e$, or

$$e'' + qe = \frac{1}{\lambda} e.$$

This leads us to the following theorem.

Theorem 7.41 *Let a and b be real numbers such that $a < b$ and let $q \in C[a,b]$. There is a sequence of scalars $(\alpha_n)_{n=1}^\infty$ such that $|\alpha_n| \to \infty$ as $n \to \infty$ and such that, for each $n \in \mathbb{N}$, there exists a twice-differentiable function $e_n \in C[a,b]$ that satisfies*

$$e_n'' + qe_n = \alpha_n e_n, \quad e_n(a) = 0, \quad e_n(b) = 0,$$

and $(e_n)_{n=1}^\infty$ forms an orthonormal basis for $L_2(a,b)$.

Furthermore, the solution to the system of differential equations

$$\begin{cases} y'' + q(x)y = f(x), & f \in L_2(a,b), \\ y(a) = y(b) = 0, \end{cases}$$

is given by

$$y = \sum_{n=1}^\infty \frac{1}{\alpha_n} (f, e_n) e_n.$$

Proof Let T_K be the operator defined by (7.21) and (7.22). By Theorem 7.30, the operator T_K has a sequence of eigenvectors $(e_n)_{n=1}^\infty$ that form an orthonormal basis for $L_2(a,b)$. For each $n \in \mathbb{N}$, let λ_n be the eigenvalue associated with e_n, and define $\alpha_n = 1/\lambda_n$. Since $\lambda_n \to 0$ as $n \to \infty$ (by Theorem 6.37, or even Theorem 7.39), it follows that $|\alpha_n| \to \infty$ as $n \to \infty$. We know that the eigenvectors satisfy the differential equation because of the discussion prior to the statement of the theorem.

Finally, suppose $f \in L_2(a,b)$. We have demonstrated that $y = T_K f$ is a solution to the given differential system. Since $(e_n)_{n=1}^\infty$ is an orthonormal basis for $L_2(a,b)$, we conclude that

$$y = T_K f = \sum_{n=1}^\infty (T_K f, e_n) e_n.$$

Since T_K is symmetric (hermitian), we have

$$(T_K f, e_n) = (f, T_K e_n) = \lambda_n (f, e_n),$$

for all $n \in \mathbb{N}$. The result follows because $\lambda_n = 1/\alpha_n$. $\square$

Example 7.42 We return now to the example of the vibrating string, which is described by the following system of differential equations:

$$\begin{cases} y'' = f(x), & f \in L_2(0, \pi), & (DE') \\ y(0) = y(\pi) = 0. & & (BC') \end{cases}$$

We will parallel the argument used in the general case to see how it works in this example. First consider the initial value problem at the left endpoint:

$$y'' = 0, \quad y(0) = 0, \ y'(0) = 1.$$

The solution to this initial value problem is $u(x) = x$ for all $x \in [0, \pi]$.

Next, we consider the initial value problem at the right endpoint:

$$y'' = 0, \quad y(\pi) = 0, \ y'(\pi) = 1.$$

The solution to this initial value problem is $v(x) = x - \pi$ for all $x \in [0, \pi]$.

From the general case, we know that the value of W in (7.19) is a constant. In this example, we can calculate it explicitly:

$$W = u v' - u' v = \pi.$$

We now define the kernel K as in (7.21):

$$K(x, t) = \begin{cases} \frac{1}{\pi} (x - \pi) t, & \text{if } t \leq x, \\ \frac{1}{\pi} x (t - \pi), & \text{if } t \geq x. \end{cases} \tag{7.23}$$

As usual, we let T_K be the Hilbert-Schmidt operator with kernel K:

$$T_K f(x) = \int_0^\pi K(x, t) f(t) \, dt, \quad f \in L_2(0, \pi), \ x \in [0, \pi].$$

The next step is to calculate the eigenvectors and eigenvalues for T_K. According to Theorem 7.41, we can consider solutions to the equation

$$y'' - \alpha y = 0, \quad y(0) = 0, \ y(\pi) = 0. \tag{7.24}$$

The differential equation in (7.24) is a homogeneous linear second order ordinary differential equation, and the solution depends on the sign of α. We know that $\alpha \neq 0$, because the only solution to the homogeneous equation is $y = 0$. Suppose $\alpha > 0$. Then $\alpha = \beta^2$ for some $\beta > 0$. The differential equation becomes $y'' - \beta^2 y = 0$, and the general solution to this differential equation is

$$y(x) = A e^{\beta x} + B e^{-\beta x},$$

where A and B are real numbers. The boundary conditions imply that $A = 0$ and $B = 0$, and so there are no positive eigenvalues.

Now let $\alpha = -\beta^2$, where $\beta > 0$. The differential equation becomes $y'' + \beta^2 y = 0$, and this has general solution

$$y(x) = A \cos(\beta x) + B \sin(\beta x),$$

where A and B are real numbers. The boundary condition $y(0) = 0$ implies that $A = 0$. At the other endpoint (and setting $A = 0$), the condition $y(\pi) = 0$ implies that $\sin(\beta \pi) = 0$. This happens whenever $\beta \in \mathbb{N}$ (since $\beta > 0$).

We have established that the solutions to (7.24) are:

$$y = \sin(nx), \quad \alpha = -n^2, \quad n \in \mathbb{N}.$$

We wish to normalize y. Computing the $L_2(0, \pi)$-norm of y:

$$\|y\|_{L_2(0,\pi)} = \left(\int_0^\pi \sin^2(nx) \, dx \right)^{1/2} = \sqrt{\frac{\pi}{2}}.$$

Therefore, for each $n \in \mathbb{N}$, we let

$$\alpha_n = -n^2 \quad \text{and} \quad e_n = \sqrt{\frac{2}{\pi}} \sin(nx).$$

We now have the following interesting result.

Theorem 7.43 $(\sqrt{2/\pi} \sin(nx))_{n=1}^\infty$ *is an orthonormal basis for* $L_2(0, \pi)$.

Proof See Theorem 7.41 and Example 7.42. $\square$

The eigenvalues of T_K are the reciprocals of the α_n values. Consequently, the eigenvalues of T_K are

$$\lambda_n = -\frac{1}{n^2}, \quad n \in \mathbb{N}, \tag{7.25}$$

where λ_n is the eigenvalue corresponding to e_n. Since we have an explicit formula for the kernel K of T_K (see (7.23)), we also get an interesting summation formula via Theorem 7.39:

$$\sum_{n=1}^\infty \frac{1}{n^4} = \int_0^\pi \int_0^\pi |K(x,t)|^2 \, dt \, dx \tag{7.26}$$

$$= \int_0^\pi \int_0^x \frac{1}{\pi^2} (x - \pi)^2 \, t^2 \, dt \, dx + \int_0^\pi \int_x^\pi \frac{1}{\pi^2} x^2 (t - \pi)^2 \, dt \, dx.$$

We leave it to the reader to verify this equality. (See Exercise 7.21.)

Exercises

Exercise 7.1 Let H be a Hilbert space and suppose x and y are nonzero elements of H. Show that $\|x + y\| = \|x\| + \|y\|$ if and only if $y = cx$, where $c > 0$.

Exercise 7.2 Let H be a Hilbert space. Suppose T and S are operators on H and let $\alpha \in \mathbb{C}$. Show that $(T + S)^* = T^* + S^*$, $(\alpha T)^* = \overline{\alpha} T^*$, and $(ST)^* = T^* S^*$.

Exercise 7.3 Suppose T and S are hermitian operators on a Hilbert space. Show that TS is hermitian if and only if $TS = ST$.

Exercise 7.4 Let H be a normed space that satisfies the Parallelogram Law (Theorem 7.10.). Show that H is an inner product space. (*Hint:* Use the polarization formulas given after the statement of Theorem 7.10..)

Exercise 7.5 Let H be a complex inner product space and assume $A : H \to H$ is a bounded linear operator. Show that (Ax, y) can be written as

$$\frac{1}{4}\left[\big(A(x+y), x+y\big) - \big(A(x-y), x-y\big) - i\big(A(x+iy), x+iy\big) + i\big(A(x-iy), x-iy\big) \right].$$

Exercise 7.6 Let H be an inner product space and assume A and B are bounded linear operators on H.

(a) Show that if $(Ax, y) = (Bx, y)$ for all x and y in H, then $A = B$.
(b) Show that if $(Ax, x) = (Bx, x)$ for all $x \in H$ and H is a *complex* inner product space, then $A = B$.
(c) What assumptions need to be added to A and B in order for (b) to hold when H is a *real* inner product space?

Exercise 7.7 Suppose A is a bounded linear operator on the complex inner product space H. Show that

$$\|A\| = \sup_{x \in B_H} |(Ax, x)|.$$

Show that the same formula holds in a real inner product space if A is hermitian.

Exercise 7.8 Let H be a complex Hilbert space and let $T : H \to H$ be a bounded linear operator. Show that $T = T^*$ if and only if $(Tx, x) \in \mathbb{R}$ for all $x \in H$. (This equivalence cannot hold in a real Hilbert space because it is necessarily true that $(Tx, y) \in \mathbb{R}$ for all x and y in H when H is a real Hilbert space.)

Exercise 7.9 Let H be an inner product space and assume $T : H \to H$ is a bounded linear operator. Show that $\|T^* T\| = \|T\|^2$.

Exercise 7.10 Let $T : H \to H$ be such that $T(0) = 0$ and $\|T(x) - T(y)\| = \|x - y\|$ for all x and y in H. Show that T is a linear isometry from H to itself. (*Hint:* Show first that $(T(x), T(y)) = (x, y)$ for all x and y in H).

Exercise 7.11 Let H be a Hilbert space. For each $\phi \in H^*$, let $v_\phi \in H$ be the unique element (from the Riesz–Fréchet Theorem) satisfying $\phi(x) = (x, v_\phi)$ for all $x \in H$.

If $T_O^* : H^* \to H^*$ is the operator adjoint of T (see Definition 3.36) and $T_A^* : H \to H$ is the Hilbert space adjoint of T (see Definition 7.18), show that $v_{T_O^*(\phi)} = T_A^* v_\phi$ for all $\phi \in H^*$.

Exercise 7.12 If V is a closed subspace of a Hilbert space H, show $H = V \oplus V^\perp$.

Exercise 7.13 If V is a closed subspace of a Hilbert space H, show $(V^\perp)^\perp = V$. What is $(V^\perp)^\perp$ if V is not closed?

Exercise 7.14 Let H be a Hilbert space and suppose $(x_n)_{n=1}^\infty$ and $(y_n)_{n=1}^\infty$ are sequences in H. If $(x_n)_{n=1}^\infty$ and $(y_n)_{n=1}^\infty$ converge (in norm) to x and y, respectively, show that $\lim_{n\to\infty} (x_n, y_n) = (x, y)$.

Exercise 7.15 (Hellinger–Toeplitz Theorem) Let H be a Hilbert space. Prove the following: If $T : H \to H$ is a linear map that satisfies the equation $(Tx, y) = (x, Ty)$ for all x and y in H, then T is continuous.

Exercise 7.16 Let H be a Hilbert space with inner product $(\cdot, \cdot)$ and norm $\| \cdot \|$. If $(x_n)_{n=1}^\infty$ and $(y_n)_{n=1}^\infty$ are sequences in B_H, and $\lim_{n\to\infty} (x_n, y_n) = 1$, show that $\lim_{n\to\infty} \|x_n - y_n\| = 0$.

Exercise 7.17 Let H be an infinite-dimensional Hilbert space with inner product $(\cdot, \cdot)$. Show that the function $(\cdot, \cdot) : (H, w) \times (H, w) \to \mathbb{C}$ is continuous in each argument separately, but that it is not continuous on the product $(H, w) \times (H, w)$. (In this problem, (H, w) denotes the Hilbert space H endowed with the weak topology.)

Exercise 7.18 Solve the system of differential equations
$$\begin{cases} y'' + \lambda^2 y = 0, & \lambda \in \mathbb{R}, \\ y(0) = 1, \ y(2\pi) = 1. \end{cases}$$

Use your answer to show that $(e_n)_{n\in\mathbb{Z}}$ is an orthonormal basis for $L_2\left(\mathbb{T}, \frac{d\theta}{2\pi}\right)$, where $e_n(\theta) = e^{in\theta}$ and $\theta \in \mathbb{T} = [0, 2\pi)$.

Exercise 7.19 Use Parseval's Identity (Theorem 7.25) to show that $\displaystyle\sum_{n=1}^\infty \frac{1}{n^2} = \frac{\pi^2}{6}$:

(a) Use the function $f(x) = x$ for all $x \in [0, \pi)$ and Theorem 7.43.
(b) Use the function $f(\theta) = \theta$ for all $\theta \in [0, 2\pi)$ and Exercise 7.18.

Exercise 7.20 A theorem from linear algebra states that the trace of a square matrix equals the sum of its eigenvalues. If $T_K : L_2(a, b) \to L_2(a, b)$ is a Hilbert–Schmidt operator defined by the formula $T_K f(x) = \int_a^b K(x, y) f(y) \, dy$, then the *trace of the Hilbert–Schmidt operator* T_K is defined to be

$$\text{trace}(T_K) = \int_a^b K(x, x) \, dx,$$

whenever it exists. Let K be the kernel in (7.23) and show $\text{trace}(T_K) = \sum_{n=1}^\infty \lambda_n$, where $(\lambda_n)_{n=1}^\infty$ is the sequence of eigenvalues for T_K given in (7.25). (Compare to (6.1.7).)

Exercise 7.21 Compute the integral in (7.26) to show that $\sum_{n=1}^{\infty} \dfrac{1}{n^4} = \dfrac{\pi^4}{90}$.

Exercise 7.22 Suppose μ and ν are probability measures on a measure space (Ω, Σ). Assume that $\mu \ll \nu$ and ϕ is the Radon–Nikodým derivative of μ with respect to ν. Define a map $V : L_2(\mu) \to L_2(\nu)$ by $V(f) = \sqrt{\phi} f$ for all $f \in L_2(\mu)$. Show that this map is a well-defined isometry. Show that V is an isomorphism if and only if $\nu \ll \mu$.

Exercise 7.23 Recall that a function $f : [0, 1] \to \mathbb{R}$ is called *absolutely continuous* on $[0, 1]$ if f is differentiable almost everywhere (with respect to Lebesgue measure), and if $f' \in L_1(0, 1)$ satisfies the equation $f(x) - f(0) = \int_0^x f'(t) \, dt$ for all $x \in [0, 1]$.

(a) Let H denote the collection of all (real-valued) absolutely continuous functions on $[0, 1]$ such that $f(0) = 0$ and $f' \in L_2(0, 1)$. Show that

$$(f, g) = \int_0^1 f'(t) g'(t) \, dt, \quad \{f, g\} \subseteq H,$$

defines a complete inner product on H.

(b) Show that the map $T : H \to L_2(0, 1)$, defined by $Tf = f'$ for all $f \in H$, is an isomorphism. Find T^{-1}.

(c) Fix $a \in (0, 1)$ and define a map $\Lambda_a : H \to \mathbb{R}$ by $\Lambda_a(f) = f(a)$ for all $f \in H$. Show that Λ_a is a bounded linear functional. Find the element $\phi_a \in H$ such that $\Lambda_a(f) = (f, \phi_a)$ for all $f \in H$.

Exercise 7.24 Let H be a Hilbert space and suppose $T : H \to H$ is a compact operator. Show that T is the limit (in operator norm) of a sequence of finite-rank operators.

Exercise 7.25 A Banach space X is called *uniformly convex* if given $\epsilon > 0$ there exists a $\delta > 0$ such that $\|x - y\| < \epsilon$ whenever $\|x\| \le 1$, $\|y\| \le 1$, and $\|x + y\| > 2 - \delta$. Show that a Hilbert space is uniformly convex.

Exercise 7.26 Suppose X is a non-reflexive Banach space and let $\epsilon > 0$ be given.

(a) Show there exists an $x^{**} \in X^{**}$ such that $\|x^{**}\| = 1$ and

$$d(x^{**}, X) = \inf\{d(x^{**}, x) : x \in X\} > 1 - \epsilon.$$

(Here, d is the metric induced by the norm on X^{**}.)

(b) Let x^{**} be as found in (a). Show that there exists $x^* \in X^*$ with $\|x^*\| = 1$ and $x^{**}(x^*) > 1 - \epsilon/2$. Pick $x \in X$ with $\|x\| \le 1$ such that $x^*(x) > 1 - \epsilon/2$. Show that there exists $y^* \in X^*$ with $\|y^*\| = 1$ and $y^*(x^{**} - x) > 1 - \epsilon$.

(c) Let x and x^* be as found in (b). Use Goldstine's Theorem to show that there exists a $y \in X$ with $\|y\| \le 1$ such that $x^*(y) > 1 - \epsilon/2$ and $y^*(y - x) > 1 - \epsilon$. Deduce that $\|x + y\| > 2 - \epsilon$ and $\|x - y\| > 1 - \epsilon$.

(d) Deduce that every uniformly convex space is reflexive.

Exercise 7.27 For the following questions, assume $2 < p < \infty$.

(a) Show that there is a constant $c > 0$ such that

$$\frac{1}{2}\left(|1+t|^p + |1-t|^p\right) \geq 1 + c^p\,|t|^p, \quad t \in \mathbb{R}.$$

(*Hint:* Show that the function $\frac{|1+t|^p+|1-t|^p-2}{|t|^p}$ is bounded below.)

(b) Deduce from (a) that if f and g are functions in $L_p(0,1)$, then

$$\frac{1}{2}\left(\|f+g\|^p + \|f-g\|^p\right) \geq \|f\|^p + c^p\,\|g\|^p.$$

(c) Conclude that $L_p(0,1)$ is uniformly convex. (This is also true if $1 < p < 2$, but it is a little more tricky to show.)

Chapter 8
Banach Algebras

In this chapter, we will study Banach spaces with a multiplication. These spaces are called Banach algebras. Throughout this section, we will confine ourselves to considering complex Banach spaces, which will allow us to make use of powerful theorems from complex analysis. (For a brief review of results from complex analysis, see Sect. B.2 in the appendix.)

8.1 The Spectral Radius

We start with some basic definitions and properties.

Definition 8.1 The set A is called an *associative algebra* (over the scalar field $\mathbb{K}$) if it is a vector space over $\mathbb{K}$ together with an operation called *multiplication* (often denoted either by $\cdot$ or juxtaposition) that satisfies the following operations:

   (i) $a \cdot (b \cdot c) = (a \cdot b) \cdot c$ *(associativity)*,
  (ii) $(a + b) \cdot c = a \cdot c + b \cdot c$ *(right-distribution)*,
 (iii) $a \cdot (b + c) = a \cdot b + a \cdot c$ *(left-distribution)*,
 (iv) $\lambda(a \cdot b) = (\lambda a) \cdot b = a \cdot (\lambda b)$ *(bilinearity of scalar multiplication)*,

where a, b, and c are elements of the set A and λ is a scalar in $\mathbb{K}$.

We are interested in a type of associative algebra that is also a Banach space, and one in which the norm is what is called submultiplicative. A norm on an associative algebra is *submultiplicative* if

$$\|a \cdot b\| \le \|a\|\,\|b\|,$$

for all elements a and b in the Banach space.

Definition 8.2 A *Banach algebra* is a Banach space that is also an associative algebra with a submultiplicative norm.

Example 8.3 Suppose X is a Banach space. Then the space $\mathcal{L}(X)$ of linear operators on X forms a Banach algebra with multiplication given by composition.

© Springer Science+Business Media, LLC 2014
A. Bowers, N. J. Kalton, *An Introductory Course in Functional Analysis*,
Universitext, DOI 10.1007/978-1-4939-1945-1_8

In the space $\mathcal{L}(X)$ we distinguish a particular operator I that has the property $I(x) = x$ for all $x \in X$. This operator, known as the *identity operator*, has norm one and is such that $I \circ T = T = T \circ I$ for all $T \in \mathcal{L}(X)$. Elements of a Banach algebra with this property play a special role.

Definition 8.4 An element 1 in a Banach algebra A is called an *identity element* if

$$1 \cdot a = a = a \cdot 1, \quad a \in A.$$

We call an algebra *unital* if there is an identity 1 such that $\|1\| = 1$.

If there is a risk of confusing the identity element 1 in A with the number 1 in the scalar field, we will denote the identity element in A by 1_A.

We will assume that all Banach algebras with an identity are unital; that is, we will assume $\|1\| = 1$. We can make this assumption without any loss of generality. To see this, suppose A is a Banach algebra with identity 1_A, but $\|1_A\| \neq 1$. For each $a \in A$, let L_a denote left-multiplication:

$$L_a(b) = ab, \quad b \in A.$$

Observe that $\|L_a\| \leq \|a\|$, because the norm is submultiplicative. Additionally,

$$\|a\| = \|L_a(1_A)\| \leq \|L_a\| \, \|1_A\|.$$

Thus, if we let $k = \frac{1}{\|1_A\|}$, then

$$k\|a\| \leq \|L_a\| \leq \|a\|.$$

Notice also that if $L_a = L_b$ for a and b in A, then $L_a(1_A) = L_b(1_A)$, and so $a = b$. Therefore, the map $a \to L_a$ determines an embedding $A \to \mathcal{L}(A)$. Since $\|L_{1_A}\| = 1$, we can then think of A as a closed subalgebra of a unital Banach algebra.

Definition 8.5 A Banach algebra is *commutative* if $ab = ba$ for all elements a and b in the algebra.

Example 8.6 We now consider several examples of commutative Banach algebras.

(a) If K is a compact Hausdorff space, then $C(K)$ forms a commutative Banach algebra under pointwise multiplication:

$$(f \cdot g)(s) = f(s)\,g(s), \quad \{f, g\} \subseteq C(K), \; s \in K.$$

The identity in $C(K)$ is χ_K, the function that is constantly 1 on K.

(b) Let $\mathbb{D} = \{z \in \mathbb{C} : |z| < 1\}$. We denote by $A(\mathbb{D})$ the collection of all analytic functions $f : \mathbb{D} \to \mathbb{C}$ that are extendable to elements of $C(\overline{\mathbb{D}})$. When equipped with the supremum norm and pointwise multiplication, the space $A(\mathbb{D})$ is a Banach algebra, known as the *disk algebra*. (Completeness follows from Morera's Theorem (Theorem B.13).) We remark that $A(\mathbb{D})$ can be viewed as a subalgebra of both $C(\overline{\mathbb{D}})$ and $C(\mathbb{T})$. In the context of the complex plane, we use $\mathbb{T}$ to denote the unit circle

$$\mathbb{T} = \partial\mathbb{D} = \{z \in \mathbb{C} : |z| = 1\}.$$

The inclusion of $A(\mathbb{D})$ in $C(\mathbb{T})$ follows from the Maximum Modulus Theorem (Theorem B.11).

(c) Consider the sequence space $\ell_1 = \ell_1(\mathbb{Z}_+)$, where $\mathbb{Z}_+ = \{0, 1, 2, \ldots\}$. We will define a product $*$ on ℓ_1. Suppose $a = (a_k)_{k=0}^{\infty}$ and $b = (b_k)_{k=0}^{\infty}$ are sequences in ℓ_1. Define the product $a * b$ to be the sequence of coefficients of the formal power series

$$\left(\sum_{k=0}^{\infty} a_k t^k\right)\left(\sum_{k=0}^{\infty} b_k t^k\right). \tag{8.1}$$

Then, $(a * b)_k = \sum_{j=0}^{k} a_{k-j} b_j$ for $k \in \mathbb{Z}_+$, and so

$$a * b = (a_0 b_0,\ a_1 b_0 + a_0 b_1,\ a_2 b_0 + a_1 b_1 + a_0 b_2,\ \ldots).$$

From (8.1), we deduce that $\|a * b\|_1 \le \|a\|_1 \|b\|_1$, and so ℓ_1 becomes a Banach algebra with this multiplication. This example is known as a *convolution algebra* because the product is known as a *convolution*.

(d) Now consider the function space $L_1(\mathbb{R})$. We will turn this space into a convolution algebra. If f and g are functions in $L_1(\mathbb{R})$, we define the *convolution* of f with g to be the function

$$(f * g)(s) = \int_{-\infty}^{\infty} f(s - t)\, g(t)\, dt, \quad s \in \mathbb{R}.$$

To compute a bound for the L_1-norm of $f * g$, we observe that

$$\int_{-\infty}^{\infty} |(f * g)(s)|\, ds \le \int_{-\infty}^{\infty}\int_{-\infty}^{\infty} |f(s - t)\, g(t)|\, dt\, ds.$$

By Fubini's Theorem,

$$\int_{-\infty}^{\infty}\int_{-\infty}^{\infty} |f(s - t)\, g(t)|\, dt\, ds = \int_{-\infty}^{\infty}\left(\int_{-\infty}^{\infty} |f(s - t)|\, ds\right) |g(t)|\, dt.$$

By the translation-invariance of Lebesgue measure,

$$\int_{-\infty}^{\infty} |f(s - t)|\, ds = \int_{-\infty}^{\infty} |f(s)|\, ds = \|f\|_1,$$

and so we have $\|f * g\|_1 \le \|f\|_1 \|g\|_1$. The space $L_1(\mathbb{R})$ becomes a Banach algebra with multiplication given by convolution. It is worth noting that this Banach algebra lacks an identity element.

All of the Banach algebras listed in Example 8.6 are commutative, but not all Banach algebras are commutative. An example of a *noncommutative* Banach algebra is $\mathcal{L}(X)$, where X is a Banach space of dimension greater than one.

In some cases (such as Example 8.6(d)) the Banach algebra A will lack an identity element. In such a circumstance, the algebra can always be embedded in an algebra with an identity. Let

$$A' = A \oplus \mathbb{C},$$

and define a multiplication on A' by

$$(a, \lambda_1) \cdot (b, \lambda_2) = (ab + \lambda_1 b + \lambda_2 a, \; \lambda_1 \lambda_2),$$

where $\{a, b\} \subseteq A$ and $\{\lambda_1, \lambda_2\} \subseteq \mathbb{C}$. Then A' is a unital Banach algebra with identity element $(0, 1)$.

Equivalently, we may simply add an identity element to A. To do this, let 1 be an identity element and let

$$A'' = A \oplus 1\mathbb{C}.$$

This time, define a multiplication by

$$(a + 1\lambda_1) \cdot (b + 1\lambda_2) = ab + \lambda_1 b + \lambda_2 a + 1\lambda_1 \lambda_2,$$

where (again) $\{a, b\} \subseteq A$ and $\{\lambda_1, \lambda_2\} \subseteq \mathbb{C}$.

The above discussion allows us to consider, without loss of generality, only unital Banach algebras. This is a convention we will adopt throughout the remainder of this chapter.

Definition 8.7 Let A be a unital Banach algebra. An element $a \in A$ is said to be *invertible* if there exists some $a^{-1} \in A$ such that $a\,a^{-1} = 1 = a^{-1}\,a$. If it exists, a^{-1} is called the *inverse* of a.

Example 8.8 Consider the algebra $C(K)$ of continuous functions on the compact Hausdorff space K. A function $f \in C(K)$ is invertible (in the algebraic sense) if $f(s) \neq 0$ for any $s \in K$. In this case, the inverse of f is given by the function $1/f$.

If an inverse exists, it is necessarily unique, but not every element of a Banach algebra is invertible. To illustrate this point, consider the following example.

Example 8.9 Consider the Banach algebra $\mathcal{L}(\ell_2)$ of operators on ℓ_2 with identity operator I. Recall the left shift operator L and the right shift operator R on ℓ_2:

$$L(\xi_1, \xi_2, \xi_3, \ldots) = (\xi_2, \xi_3, \xi_4, \ldots) \quad \text{and} \quad R(\xi_1, \xi_2, \xi_3, \ldots) = (0, \xi_1, \xi_2, \ldots),$$

where $(\xi_n)_{n=1}^{\infty}$ is an element of ℓ_2. Neither R nor L has an inverse: R is not onto, and L has a zero eigenvalue. Observe that L and R do satisfy the relationship $LR = I$; however, they are not inverses of one another because $RL \neq I$.

In a Banach algebra A, we call b the *left-inverse* of a whenever $b \cdot a = 1$. Correspondingly, when this equation is satisfied, we call a the *right-inverse* of b. In the preceding example, L is the left-inverse of R and R is the right-inverse of L.

Simple algebra shows that if a has both a left-inverse and a right-inverse, then they must coincide (and hence a is invertible). (See Exercise 8.1.)

The next proposition, first due to the mathematician Carl Neumann (in the context of operators), will prove central to all that follows.

Proposition 8.10 (Neumann Series) *Let A be a unital Banach algebra and let $a \in A$. If $\|a\| < 1$, then $1 - a$ is invertible and the inverse is given by the Neumann series:*

$$(1 - a)^{-1} = \sum_{n=0}^{\infty} a^n = 1 + a + a^2 + a^3 + \cdots .$$

Proof For each $n \in \mathbb{N}$, let $S_n = 1 + a + \cdots + a^n$. Observe that the series $\sum_{n=0}^{\infty} \|a^n\|$ converges, since $\|a^n\| \leq \|a\|^n$ and $\|a\| < 1$. Consequently, $(S_n)_{n=1}^{\infty}$ is a Cauchy sequence, and hence converges to some $S \in A$.

Observe that $1 - a$ and S_n commute for each $n \in \mathbb{N}$, and

$$(1 - a) S_n = S_n (1 - a) = 1 - a^{n+1}.$$

Taking limits, we have
$$(1 - a) S = S (1 - a) = 1,$$

and so $S = \sum_{n=0}^{\infty} a^n$ is the inverse of $1 - a$, as required. $\qquad\square$

The collection of invertible elements in a Banach algebra turns out to be very important.

Definition 8.11 Let A be a unital Banach algebra. The set $G = \{a \in A : a^{-1} \text{ exists}\}$ is called the *group of invertible elements* in A.

We leave it to the reader to verify that the group of invertible elements is indeed a group. (See Exercise 8.5.) In addition to being a group, it also has interesting topological properties.

Proposition 8.12 *The group of invertible elements in a unital Banach algebra is an open set.*

Proof Let A be a unital Banach algebra and let G be the group of invertible elements in A. Let $a \in G$. We will find an open ball centered at a that is contained in G. Let

$$B = \left\{ b \in A : \|a - b\| < \frac{1}{\|a^{-1}\|} \right\}.$$

Let $b \in B$. Observe that

$$b = a - (a - b) = a (1 - a^{-1}(a - b)).$$

By assumption, a is invertible. By the choice of b, we also have

$$\|a^{-1}(a - b)\| \leq \|a^{-1}\| \, \|a - b\| < 1.$$

Thus, by Proposition 8.10, we conclude that $1 - a^{-1}(a - b)$ is invertible. Consequently, both a and $1 - a^{-1}(a - b)$ are elements of the group G. Therefore,

$$b = a\left(1 - a^{-1}(a - b)\right) \in G,$$

as the product of two elements in the group. Since the choice of $b \in B$ was arbitrary, we conclude that $B \subseteq G$. Thus, B is an open set such that $a \in B$ and $B \subseteq G$. Therefore, G is an open set in A.   □

In the proof of Proposition 8.12, we saw that if $a \in A$ is invertible, and if $b \in A$ is sufficiently close to a, then b is invertible, too. In fact, thanks to Proposition 8.10, we can even provide a formula for the inverse of b:

$$\begin{aligned}
b^{-1} &= \left(1 - a^{-1}(a - b)\right)^{-1} a^{-1} \\
&= \left(1 + a^{-1}(a - b) + a^{-1}(a - b)a^{-1}(a - b) + \cdots\right)a^{-1} \\
&= a^{-1} + a^{-1}(a - b)a^{-1} + a^{-1}(a - b)a^{-1}(a - b)a^{-1} + \cdots .
\end{aligned} \tag{8.2}$$

If A is a commutative Banach algebra, then

$$b^{-1} = a^{-1} + (a^{-1})^2(a - b) + (a^{-1})^3(a - b)^2 + \cdots .$$

Definition 8.13 Suppose A is a complex unital Banach algebra with group of invertible elements G. Let $a \in A$. The *spectrum of a* is the subset of $\mathbb{C}$ defined by

$$\mathrm{Sp}(a) = \{\lambda : \lambda 1 - a \notin G\}.$$

The *resolvent of a* is the subset of $\mathbb{C}$ defined by

$$\mathrm{Res}(a) = \{\lambda : \lambda 1 - a \in G\}.$$

Example 8.14 Consider $M_n(\mathbb{C})$, the Banach algebra of $n \times n$ complex matrices, where $n \in \mathbb{N}$. Let $T \in M_n(\mathbb{C})$ be a square matrix. A complex scalar λ is in $\mathrm{Sp}(T)$ provided $\lambda I - T$ is a non-invertible matrix; i.e., when $\det(\lambda I - T) = 0$. Consequently, we have $\lambda \in \mathrm{Sp}(T)$ if and only if λ is an eigenvalue for T. Since $\mathbb{C}$ is algebraically closed, T will always have at least one eigenvalue, and so $\mathrm{Sp}(T)$ is nonempty for all $T \in M_n(\mathbb{C})$. Therefore, in this case, we can identify the spectrum of T as the collection of eigenvalues of T; that is, $\mathrm{Sp}(T) = \{\lambda_1, \ldots, \lambda_r\}$, for a finite sequence of scalars, where $r \leq n$.

When A is a finite-dimensional space of operators, the spectrum of an element is well-understood in terms of eigenvalues. In the infinite-dimensional case, however, things are more subtle.

Example 8.15 Define an operator T on $C[0, 1]$ by

$$Tf(x) = x f(x), \quad f \in C[0, 1], \quad x \in [0, 1].$$

We saw in Example 6.3 that T has no eigenvalues. Even so, the spectrum of T is not empty. In fact, for any $\lambda \in [0, 1]$, the operator $\lambda 1 - T$ is not invertible. If $f \in C[0, 1]$,

then

$$(\lambda 1 - T)f(x) = \lambda f(x) - xf(x) = (\lambda - x)f(x), \quad x \in [0,1].$$

The inverse operator would have to be

$$Sg(x) = \frac{g(x)}{\lambda - x}, \quad g \in C[0,1], \quad x \in [0,1],$$

but S is not a bounded operator on $C[0,1]$ if $\lambda \in [0,1]$. Therefore, the spectrum of T is $\mathrm{Sp}(T) = [0,1]$. (Note that S is bounded if $\lambda \notin [0,1]$.)

Example 8.16 Let X be a Banach space and suppose $T \in \mathcal{L}(X)$. By definition, the complex scalar λ is in $\mathrm{Sp}(T)$ whenever $\lambda 1 - T$ is not invertible. Recall that an operator is not invertible either because it is not injective or not surjective (or both). There are three (not necessarily disjoint) possibilities:

(1) Suppose $\lambda 1 - T$ is not one-to-one. In this case, there exists a nonzero $x \in X$ such that $(\lambda 1 - T)(x) = 0$, and so $Tx = \lambda x$. It follows that λ is an eigenvalue of T.
(2) Suppose $\lambda 1 - T$ is one-to-one, but $(\lambda 1 - T)(X)$ is a proper closed linear subspace of X. By the Hahn–Banach Theorem, there is some $x^* \in X^*$ such that $x^* \neq 0$, but $x^*((\lambda 1 - T)(X)) = 0$. (See Exercise 5.20.) It follows that $(T^* x^*)(x) = (\lambda x^*)(x)$ for every $x \in X$, and so λ is an eigenvalue of T^*.
(3) Suppose $\lambda 1 - T$ is one-to-one, but $\lambda 1 - T$ is not bounded below; that is, there does not exist a constant $c > 0$ such that

$$\|\lambda x - Tx\| \geq c \|x\|$$

for all $x \in X$. Therefore, we can find a sequence $(x_n)_{n=1}^{\infty}$ in X such that $\|x_n\| = 1$ for all $n \in \mathbb{N}$ and such that

$$\lim_{n \to \infty} \|\lambda x_n - Tx_n\| = 0. \tag{8.3}$$

(See Exercise 1.12.) When λ satisfies (8.3) for some sequence $(x_n)_{n=1}^{\infty}$, we call it an *approximate eigenvalue* of T.

We conclude that if $\lambda \in \mathrm{Sp}(T)$, then λ is either an eigenvalue of T or T^*, or an approximate eigenvalue of T. We will reserve the term *approximate eigenvalue* to describe those $\lambda \in \mathrm{Sp}(T)$ which are not eigenvalues of T. It may happen, however, that λ is both an approximate eigenvalue of T and an eigenvalue of T^*. (See Example 8.18.)

Definition 8.17 Let $T \in \mathcal{L}(X)$ for a Banach space X. The *point spectrum* of T is the set of all eigenvalues of T. The *approximate point spectrum* of T is the set of all approximate eigenvalues of T.

Example 8.18   Let us revisit Example 8.15. Suppose $T \in \mathcal{L}(C[0,1])$ is defined by the formula:

$$Tf(x) = x f(x), \quad f \in C[0,1], \quad x \in [0,1].$$

We already mentioned that T has no eigenvalues. We now identify the eigenvalues of T^*. We start by computing $T^* : M[0,1] \to M[0,1]$. If $\mu \in M[0,1]$, then

$$T^*\mu(f) = \mu(Tf) = \int_0^1 x f(x) \mu(dx).$$

Therefore, $T^*\mu(dx) = x \mu(dx)$. Suppose that λ is an eigenvalue of T^*. Then there exists some $\mu \in M[0,1]$ such that $T^*\mu(dx) = \lambda \mu(dx)$. Then for every Borel set A,

$$\int_A x \mu(dx) = \int_A \lambda \mu(dx).$$

This implies $x = \lambda$ a.e.(μ), which means $\mu = \delta_\lambda$, the Dirac measure at λ. Therefore, $T^*\delta_\lambda = \lambda \delta_\lambda$ for each $\lambda \in [0,1]$, and so each $\lambda \in [0,1]$ is an eigenvalue of T^*.

We established in Example 8.15 that $\mathrm{Sp}(T) = [0,1]$ and we have shown that each point in $[0,1]$ is an eigenvalue of T^*. We now show that each point in $[0,1]$ is also an approximate eigenvalue of T. For each $\lambda \in [0,1]$, we need a sequence $(f_n)_{n=1}^\infty$ of continuous functions such that $\|f_n\|_\infty = 1$ and $\|\lambda f_n - Tf_n\|_\infty \to 0$ as $n \to \infty$. There are many such sequences. It suffices to find a sequence $(f_n)_{n=1}^\infty$ for which f_n is a function that peaks with value 1 at λ and then decreases to zero, where f_n decreases more rapidly to zero as n increases.

As an example, let $\lambda \in [0,1]$ be given and for each $n \in \mathbb{N}$, define $f_n \in C[0,1]$ as follows:

$$f_n(x) = \begin{cases} 0 & \text{if } 0 \leq x < \lambda - \frac{1}{n}, \\ n(x - \lambda) + 1 & \text{if } \lambda - \frac{1}{n} \leq x < \lambda, \\ n(\lambda - x) + 1 & \text{if } \lambda \leq x < \lambda + \frac{1}{n}, \\ 0 & \text{if } \lambda + \frac{1}{n} \leq x \leq 1. \end{cases}$$

For each $n \in \mathbb{N}$, the function f_n is continuous and $\|f_n\|_\infty = 1$. If $|x - \lambda| > 1/n$, then $f_n(x) = 0$. On the other hand, if $|x - \lambda| < 1/n$, then

$$|(\lambda - x) f_n(x)| \leq |\lambda - x| (n|\lambda - x| + 1) < \frac{1}{n} \cdot \left(n \cdot \frac{1}{n} + 1 \right) = \frac{2}{n}.$$

Therefore,

$$\|\lambda f_n - Tf_n\|_\infty = \sup_{x \in [0,1]} |\lambda f_n(x) - x f_n(x)| \leq \frac{2}{n}.$$

Naturally, this tends to zero, and it follows that λ is an approximate eigenvalue for the operator T.

Theorem 8.19   *Let A be a complex unital Banach algebra. If $a \in A$, then $\mathrm{Sp}(a)$ is a nonempty compact set.*

Proof We will first show that $\mathrm{Sp}(a)$ is compact. Let G denote the set of invertible elements in A. By Proposition 8.12, G is an open set in A. Define a map $\rho : \mathbb{C} \to A$ by

$$\rho(\lambda) = \lambda 1 - a, \quad \lambda \in \mathbb{C}.$$

Then ρ is a continuous map, and so $\rho^{-1}(G)$ is an open set in $\mathbb{C}$. Consequently,

$$\mathrm{Sp}(a) = \{\lambda : \lambda 1 - a \notin G\} = \mathbb{C} \setminus \rho^{-1}(G)$$

is closed in $\mathbb{C}$.

If $|\lambda| > \|a\|$, then $1 - (a/\lambda)$ is invertible, by Proposition 8.10. Thus, $\lambda 1 - a$ is invertible, as well. Consequently, if $\lambda 1 - a$ is not invertible, then $|\lambda| \leq \|a\|$. Therefore, $\mathrm{Sp}(a) \subseteq \|a\| B_{\mathbb{C}}$, and so is compact by the Heine–BorelTheorem.

Now we show that $\mathrm{Sp}(a)$ is nonempty. Assume to the contrary that $\mathrm{Sp}(a) = \emptyset$. Then $\lambda 1 - a$ is invertible for all $\lambda \in \mathbb{C}$. Let $\phi \in A^*$ and define $f : \mathbb{C} \to \mathbb{C}$ by

$$f(\lambda) = \phi((\lambda 1 - a)^{-1}), \quad \lambda \in \mathbb{C}. \tag{8.4}$$

(We call f the *resolvent function*.)

We claim that f is analytic on $\mathbb{C}$ (i.e., f is an *entire function*). Let $\lambda \in \mathbb{C}$. We will show that there is a neighborhood of λ in which f has a power series expansion. Suppose $\mu \in \mathbb{C}$ is such that

$$|\mu - \lambda| < \|(\lambda 1 - a)^{-1}\|^{-1}. \tag{8.5}$$

Using a bit of algebra:

$$\mu 1 - a = (\lambda 1 - a) + (\mu - \lambda)1 = (\lambda 1 - a)(1 + (\mu - \lambda)(\lambda 1 - a)^{-1}).$$

By assumption, $\mu 1 - a$ is invertible, and so

$$(\mu 1 - a)^{-1} = \left(1 + (\mu - \lambda)(\lambda 1 - a)^{-1}\right)^{-1}(\lambda 1 - a)^{-1}$$

Because of (8.5), we know $\|(\mu - \lambda)(\lambda 1 - a)^{-1}\| < 1$, and thus (by Proposition 8.10) there is a Neumann series for the inverse of $1 + (\mu - \lambda)(\lambda 1 - a)^{-1}$:

$$\left(1 + (\mu - \lambda)(\lambda 1 - a)^{-1}\right)^{-1} = \sum_{n=0}^{\infty} (-1)^n (\mu - \lambda)^n (\lambda 1 - a)^{-n}.$$

Therefore,

$$(\mu 1 - a)^{-1} = \sum_{n=0}^{\infty} (-1)^n (\mu - \lambda)^n (\lambda 1 - a)^{-n-1}.$$

Since $\phi \in A^*$,

$$f(\mu) = f(\lambda) + \sum_{n=1}^{\infty} (-1)^n (\mu - \lambda)^n \phi((\lambda 1 - a)^{-n-1}).$$

It follows that f is analytic on $\mathbb{C}$.

We claim that f is a bounded function. To that end, suppose that $|\lambda| > \|a\|$. Then $1 - a/\lambda$ is invertible, by Proposition 8.10. Hence, $\lambda 1 - a$ is invertible and has Neumann series

$$(\lambda 1 - a)^{-1} = \lambda^{-1} \sum_{n=0}^{\infty} \lambda^{-n} a^n.$$

Thus, for $|\lambda| > \|a\|$,

$$f(\lambda) = \phi((\lambda 1 - a)^{-1}) = \lambda^{-1} \sum_{n=0}^{\infty} \lambda^{-n} \phi(a^n).$$

Consequently, for $|\lambda| > \|a\|$,

$$|f(\lambda)| \leq \frac{1}{|\lambda|} \sum_{n=0}^{\infty} |\lambda|^{-n} \|\phi\| \|a\|^n = \frac{\|\phi\|}{|\lambda|} \sum_{n=0}^{\infty} \left(\frac{\|a\|}{|\lambda|} \right)^n.$$

Since $\|a\|/|\lambda| < 1$, by assumption, the geometric series converges, and so

$$|f(\lambda)| \leq \frac{\|\phi\|}{|\lambda|} \cdot \frac{1}{1 - \frac{\|a\|}{|\lambda|}} = \frac{\|\phi\|}{|\lambda| - \|a\|}. \tag{8.6}$$

Because f is analytic on $\mathbb{C}$, there exists some $M \geq 0$ such that $|f(\lambda)| \leq M$ for all $|\lambda| \leq 2\|a\|$. On the other hand, if $|\lambda| \geq 2\|a\|$, then $|f(\lambda)| \leq \|\phi\|/\|a\|$, because of (8.6). It follows that f is bounded by $\max\{M, \|\phi\|/\|a\|\}$.

We have established that f is a bounded entire function. Therefore, by Liouville's Theorem (Theorem B.14), f is constant. In fact, since

$$\lim_{|\lambda| \to \infty} \frac{\|\phi\|}{|\lambda| - \|a\|} = 0,$$

we conclude that $f = 0$.

We have established that $f(\lambda) = 0$ for all $\lambda \in \mathbb{C}$. The choice of $\phi \in A^*$ in (8.4) was arbitrary, and so we conclude that $\phi((\lambda 1 - a)^{-1}) = 0$ for all $\phi \in A^*$. The only way this can happen is if $(\lambda 1 - a)^{-1} = 0$ (because of the Hahn–Banach Theorem). This is a contradiction, and so we conclude that $\mathrm{Sp}(a)$ is nonempty. $\qquad \square$

For the next theorem, we recall that an algebra is a *field* if it is commutative and if every nonzero element is invertible.

Theorem 8.20 (Gelfand–Mazur Theorem) *Suppose A is a complex Banach algebra. If A is a field, then A is isometrically isomorphic to $\mathbb{C}$.*

Proof Let $a \in A$. By Theorem 8.19, $\mathrm{Sp}(a)$ is nonempty. Thus there exists some $\lambda \in \mathbb{C}$ such that $\lambda 1 - a$ is not invertible. By assumption, the only noninvertible element is 0, and so $\lambda 1 - a = 0$. Therefore, $a = \lambda 1$, and the result follows. $\qquad \square$

Theorem 8.20 remains true if we replace field with *skew-field* (also known as a *noncommutative field* or *division algebra*). If A is a *real* Banach algebra which is a skew-field, then A is $\mathbb{R}$, $\mathbb{C}$, or $\mathbb{H}$. Here, we view $\mathbb{C}$ as a real Banach algebra. The symbol $\mathbb{H}$ denotes the *quaternions*. (We use $\mathbb{H}$ in honor of the Irish mathematician Sir William Hamilton, who first described the quaternions in 1843.) We remark that $\mathbb{H}$ is not a complex Banach algebra because it does not satisfy property *(iv)* of Definition 8.1.

The Gelfand–Mazur Theorem dates back to 1938, when Mazur stated in an article that every normed division algebra over $\mathbb{R}$ is either $\mathbb{R}$, $\mathbb{C}$, or $\mathbb{H}$ [25]. His original proof did not appear until much later, in 1973 [37]. The theorem now called the Gelfand–Mazur Theorem was proved by Gelfand in 1941 [14].

Definition 8.21 Let A be a unital Banach algebra. If $a \in A$, then the number

$$r(a) = \max\{|\lambda| : \lambda \in \mathrm{Sp}(a)\}$$

is called the *spectral radius* of a.

By Theorem 8.19, the spectrum of a is nonempty and compact, and so we know there is some $\lambda \in \mathrm{Sp}(a)$ which achieves the maximum in Definition 8.21. We also know that $r(a) \le \|a\|$, because if $\lambda > \|a\|$, then $\|a\|/\lambda < 1$, and so Proposition 8.10 would imply that $\lambda 1 - a$ is invertible, which is a contradiction.

Lemma 8.22 *Let A be a unital Banach algebra. If $a \in A$, then $r(a) \le \|a^n\|^{1/n}$ for all $n \in \mathbb{N}$.*

Proof It will suffice to show that if $\lambda \in \mathrm{Sp}(a)$, then $\lambda^n \in \mathrm{Sp}(a^n)$ for all $n \in \mathbb{N}$. (Because then $|\lambda|^n \le \|a^n\|$.)

Let $n \in \mathbb{N}$ and $\lambda \in \mathrm{Sp}(a)$. Observe that

$$\lambda^n 1 - a^n = (\lambda 1 - a)(\lambda^{n-1} 1 + \lambda^{n-2} a + \cdots + \lambda a^{n-2} + a^{n-1})$$
$$= (\lambda^{n-1} 1 + \lambda^{n-2} a + \cdots + \lambda a^{n-2} + a^{n-1})(\lambda 1 - a).$$

Suppose $\lambda^n 1 - a^n$ is invertible. Then

$$1 = (\lambda 1 - a)(\lambda^{n-1} 1 + \lambda^{n-2} a + \cdots + \lambda a^{n-2} + a^{n-1})(\lambda^n 1 - a^n)^{-1}$$
$$= (\lambda^n 1 - a^n)^{-1}(\lambda^{n-1} 1 + \lambda^{n-2} a + \cdots + \lambda a^{n-2} + a^{n-1})(\lambda 1 - a).$$

From this, however, it follows that $\lambda 1 - a$ is invertible. More precisely, we have

$$(\lambda 1 - a)^{-1} = (\lambda^n 1 - a^n)^{-1}(\lambda^{n-1} 1 + \lambda^{n-2} a + \cdots + \lambda a^{n-2} + a^{n-1}).$$

(See Exercise 8.1.) This contradicts the assumption that $\lambda \in \mathrm{Sp}(a)$. Thus, $\lambda^n 1 - a^n$ is not invertible, and so $\lambda^n \in \mathrm{Sp}(a^n)$, as required. $\square$

We now come to a key result, one which provides a formula for computing the spectral radius.

Theorem 8.23 (Spectral Radius Formula) *Let A be a unital Banach algebra. If $a \in A$, then*

$$r(a) = \lim_{n \to \infty} \|a^n\|^{1/n} = \inf_{n \in \mathbb{N}} \|a^n\|^{1/n}.$$

Proof By Lemma 8.22, we know that

$$r(a) \le \inf_{n \in \mathbb{N}} \|a^n\|^{1/n} \le \liminf_{n \to \infty} \|a^n\|^{1/n}.$$

It will suffice, therefore, to show that $\limsup\limits_{n\to\infty} \|a^n\|^{1/n} \le r(a)$.

Suppose $\phi \in A^*$ and let

$$f(\lambda) = \phi((\lambda 1 - a)^{-1}), \quad \lambda \in \mathbb{C}. \tag{8.7}$$

(This is the resolvent function defined in (8.4).) We know from the proof of Theorem 8.19 that f is an analytic function in the region $|\lambda| > r(a)$. (There will be a singularity at any $\lambda \in \mathrm{Sp}(a)$.)

Now define a new function:

$$F(\xi) = \phi((1 - \xi a)^{-1}), \quad \xi \in \mathbb{C}.$$

Since $F(\xi) = \xi^{-1} f(\xi^{-1})$, we conclude that F is an analytic function on the open disk $B = \{\xi : |\xi| < 1/r(a)\}$, except possibly at the origin $\xi = 0$. We will demonstrate, however, that F has a power series on B centered at $\xi = 0$.

Suppose that $\xi \in \mathbb{C}$ is such that $|\xi| < 1/\|a\|$. Then $\|\xi a\| < 1$, and so (by Proposition 8.10) there is a Neumann series for $(1 - \xi a)^{-1}$:

$$(1 - \xi a)^{-1} = 1 + \xi a + \xi^2 a^2 + \cdots = \sum_{n=0}^{\infty} \xi^n a^n.$$

Then, for any $|\xi| < 1/\|a\|$,

$$F(\xi) = \phi((1 - \xi a)^{-1}) = \phi\left(\sum_{n=0}^{\infty} \xi^n a^n\right) = \sum_{n=0}^{\infty} \xi^n \phi(a^n).$$

The power series expansion is unique, and so we conclude that

$$F(\xi) = \sum_{n=0}^{\infty} \xi^n \phi(a^n), \quad \xi \in B. \tag{8.8}$$

Since the series in (8.1.6) converges for all $\xi \in B$, we observe that

$$\lim_{n\to\infty} |\xi^n \phi(a^n)| = 0$$

for each $\xi \in B$. Let $\rho \in \mathbb{R}$ be such that $\rho > r(a)$, but otherwise arbitrary. Then $1/\rho \in B$, and so

$$\sup_{n\in\mathbb{N}} |\phi(\rho^{-n} a^n)| = \sup_{n\in\mathbb{N}} |\rho^{-n} \phi(a^n)| < \infty.$$

This is true for all $\phi \in A^*$, and so by the Uniform Boundedness Principle, there exists a constant C_ρ (that depends on the choice of ρ) such that $C_\rho > 0$ and

$$\|\rho^{-n} a^n\| \le C_\rho, \quad n \in \mathbb{N}.$$

Then, for all $n \in \mathbb{N}$, we have that $\|a^n\| \le C_\rho \rho^n$. In particular, we have

$$\|a^n\|^{1/n} \le C_\rho^{1/n} \rho.$$

Therefore,

$$\limsup_{n\to\infty} \|a^n\|^{1/n} \le \rho.$$

The choice of $\rho > r(a)$ was arbitrary, and so

$$\limsup_{n\to\infty} \|a^n\|^{1/n} \le r(a).$$

It follows that

$$\limsup_{n\to\infty} \|a^n\|^{1/n} \le r(a) \le \inf_{n\in\mathbb{N}} \|a^n\|^{1/n} \le \liminf_{n\to\infty} \|a^n\|^{1/n}.$$

Since it is always the case that $\liminf_{n\to\infty} \|a^n\|^{1/n} \le \limsup_{n\to\infty} \|a^n\|^{1/n}$, the proof is complete. □

Observe that if a is such that $r(a) = 0$, then (by Theorem 8.23) it must be the case that $\lim_{n\to\infty} \|a^n\|^{1/n} = 0$. This motivates the next definition.

Definition 8.24 Let A be a unital Banach algebra. If $a \in A$ is such that $r(a) = 0$, then a is called *quasinilpotent*.

Let us now return our attention to compact operators on a Banach space. Let X be an infinite-dimensional Banach space and suppose $K \in \mathcal{L}(X)$ is a compact operator. We know that all eigenvalues of K are in $\mathrm{Sp}(K)$. From Theorem 6.37, we also know that the only possible limit point of the eigenvalues of K is 0. Certainly, K cannot be invertible, and so $0 \in \mathrm{Sp}(K)$. We are left with the question: *What other elements of the spectrum are not eigenvalues?* This leads us to a classical theorem, due to Fredholm.

Theorem 8.25 (Fredholm Alternative) *Let K be a compact operator on the Banach space X. If λ is a nonzero scalar, then either $\lambda 1 - K$ is invertible or λ is an eigenvalue of K.*

Proof Let λ be a nonzero scalar and assume that λ is not an eigenvalue of K. Then, by assumption, $\ker(\lambda 1 - K) = \{0\}$. Thus, by the Rank-Nullity Theorem (Theorem 6.33),

$$\dim(X/\mathrm{ran}(\lambda 1 - K)) = \dim(\ker(\lambda 1 - K)) = 0.$$

It follows that $\mathrm{ran}(\lambda 1 - K) = X$, and so $\lambda 1 - K$ is a surjection. By the initial assumption, $\lambda 1 - K$ is an injection, and therefore $\lambda 1 - K$ is invertible. □

Example 8.26 Recall the Volterra operator from Examples 3.41 and 6.27. The Volterra operator is a map $V : L_2(0, 1) \to L_2(0, 1)$ defined by

$$Vf(x) = \int_0^x f(t)\,dt, \quad f \in L_2(0, 1), \ x \in [0, 1].$$

Since V is a Hilbert–Schmidt operator, it is compact, and so $0 \in \mathrm{Sp}(V)$. In Example 6.27, it was established that V has no eigenvalues. By Theorem 8.25 (the

Fredholm Alternative), it follows that $\mathrm{Sp}(V) = \{0\}$. Therefore, $r(V) = 0$ and V is quasinilpotent.

Example 8.27 Suppose R and L are the right and left shift operators, respectively. Then

$$R\xi = (0, \xi_1, \xi_2, \dots) \quad \text{and} \quad L\xi = (\xi_2, \xi_3, \dots),$$

where $\xi = (\xi_k)_{k=1}^{\infty}$ is any sequence indexed by the natural numbers. If $1 < p < \infty$, then $R : \ell_p \to \ell_p$ is a bounded linear operator. More precisely, $\|R\xi\|_p = \|\xi\|_p$ for all $\xi \in \ell_p$, so that $\|R\| = 1$.

We wish to identify $\mathrm{Sp}(R)$. We have that $\|R^n\xi\|_p = \|\xi\|_p$ for all $\xi \in \ell_p$, and so it follows that $\|R^n\| = 1$ for all $n \in \mathbb{N}$. By the Spectral Radius Formula (Theorem 8.23), we conclude that $r(R) = 1$, and hence $\mathrm{Sp}(R) \subseteq \overline{\mathbb{D}}$. (Recall that $\mathbb{D}$ denotes the open unit disk in $\mathbb{C}$.) It is easy to see that R has no eigenvalues, and so we must look elsewhere to find elements in the spectrum of R.

Let q be the conjugate exponent for p. Then $\ell_p^* = \ell_q$. We will compute the eigenvalues of $R^* : \ell_q \to \ell_q$. The first step is to identify R^*. For ease of notation, denote the dual action of ℓ_q on ℓ_p by $\xi^*(\xi) = \langle \xi, \xi^* \rangle$ for all $\xi \in \ell_p$ and $\xi^* \in \ell_q$. A moments thought will reveal that

$$\langle R\xi, \xi^* \rangle = \langle \xi, L\xi^* \rangle, \quad \xi \in \ell_p, \ \xi^* \in \ell_q,$$

and hence $R^* = L$.

Suppose that $\lambda \in \mathbb{C}$ is an eigenvalue of $L : \ell_q \to \ell_q$. Then there is some $\xi^* \in \ell_q$ such that $L\xi^* = \lambda\xi^*$. Suppose $\xi^* = (\xi_n^*)_{n=1}^{\infty}$. Then

$$(\xi_2^*, \xi_3^*, \xi_4^*, \dots) = (\lambda\xi_1^*, \lambda\xi_2^*, \lambda\xi_3^*, \dots).$$

Therefore, for all $n \in \mathbb{N}$, we have $\xi_{n+1}^* = \lambda\xi_n^*$, and consequently $\xi_{n+1}^* = \lambda^n \xi_1^*$. It follows that

$$\xi^* = (\xi_1^*, \lambda\xi_1^*, \lambda^2\xi_1^*, \dots) = \xi_1^* (1, \lambda, \lambda^2, \dots).$$

Such a sequence is in ℓ_q if and only if $|\lambda| < 1$.

We have determined that λ is an eigenvalue for $R^* = L$ if $|\lambda| < 1$. Consequently, we conclude that $\mathbb{D} \subseteq \mathrm{Sp}(R)$. By Theorem 8.19, the set $\mathrm{Sp}(R)$ is closed, and so $\overline{\mathbb{D}} \subseteq \mathrm{Sp}(R)$. Thus, we have established mutual inclusion, and hence $\mathrm{Sp}(R) = \overline{\mathbb{D}}$.

We have observed that R has no eigenvalues, and we have shown that every point in $\mathbb{D}$ is an eigenvalue of $R^* = L$. Since the spectrum of R is the closed unit disk, it must be the case that every point in $\mathbb{T} = \partial\mathbb{D}$ is an approximate eigenvalue of R. Since no $\lambda \in \mathbb{T}$ is an eigenvalue of L, it follows that if $|\lambda| = 1$, then the operator $\lambda 1 - R$ is one-to-one and does not have closed range. (See Example 8.16.)

Example 8.28 Let $p \in (1, \infty)$ and consider the Banach space $\ell_p(\mathbb{Z})$ of doubly infinite p-summable sequences. The right shift operator $R : \ell_p(\mathbb{Z}) \to \ell_p(\mathbb{Z})$ is now given by the formula

$$R(\xi) = (\xi_{n-1})_{n \in \mathbb{Z}}, \quad \xi = (\xi_n)_{n \in \mathbb{Z}} \in \ell_p(\mathbb{Z}).$$

In this case, R is actually invertible, and the inverse is the left shift operator, so $R^{-1} = L$ (where L is defined in the obvious way).

This time, neither R nor L has any eigenvalues. To see that R has no eigenvalues, suppose $\lambda \in \mathbb{C}$ is an eigenvalue of R. (Note that $\lambda \neq 0$.) Then, for some $\xi = (\xi_n)_{n \in \mathbb{Z}}$, we have $R\xi = \lambda\xi$, and so

$$(\xi_{n-1})_{n \in \mathbb{Z}} = (\lambda\xi_n)_{n \in \mathbb{Z}}.$$

Consequently, $\xi_{n-1} = \lambda\xi_n$ for all $n \in \mathbb{Z}$. From this we conclude that

$$\xi_{-n} = \lambda^n \xi_0 \quad \text{and} \quad \xi_n = \frac{1}{\lambda^n}\xi_0, \quad n \in \mathbb{N}.$$

Therefore, $\xi = \xi_0(\lambda^{-n})_{n \in \mathbb{Z}}$. Since $\xi \in \ell_p(\mathbb{Z})$, it must be that both $|\lambda| < 1$ and $\frac{1}{|\lambda|} < 1$. Naturally, this cannot happen. A similar argument shows that L has no eigenvalues.

Despite the previous remarks, it is still the case that $\|R^n\| = 1$ for all $n \in \mathbb{N}$, and consequently $r(R) = 1$. Similarly, $r(L) = 1$, and so we have both $\operatorname{Sp}(R) \subseteq \overline{\mathbb{D}}$ and $\operatorname{Sp}(R^{-1}) \subseteq \overline{\mathbb{D}}$. It is routine to show that $\lambda \in \operatorname{Sp}(R)$ if and only if $1/\lambda \in \operatorname{Sp}(R^{-1})$. (See Exercise 8.4.) Therefore, if $\lambda \in \operatorname{Sp}(R)$, then $\lambda \in \overline{\mathbb{D}}$ and $1/\lambda \in \overline{\mathbb{D}}$. The conclusion is that $\operatorname{Sp}(R) \subseteq \mathbb{T}$. In fact, $\operatorname{Sp}(R) = \mathbb{T}$, because of the previous example. (Consider sequences ξ with $\xi_{-n} = 0$ for all $n \geq 0$.)

Example 8.29 In Example 7.29, we saw that, by Parseval's Identity, there is an isometric isomorphism between the Hilbert spaces $L_2 = L_2([0, 2\pi), \frac{d\theta}{2\pi})$ and $\ell_2(\mathbb{Z})$, given by the Fourier transform $f \mapsto (\hat{f}(n))_{n \in \mathbb{Z}}$. Define the *multiplier operator* $M : L_2 \to L_2$ by

$$Mf(\theta) = e^{i\theta} f(\theta), \quad f \in L_2, \quad \theta \in [0, 2\pi).$$

With regards to the isometric isomorphism, the multiplier operator M on L_2 corresponds to the shift operator R on $\ell_2(\mathbb{Z})$. (See Exercise 8.10.)

Remark 8.30 In Example 7.29, we used the symbol $\mathbb{T}$ to denote the interval $[0, 2\pi)$. In the context of complex analysis, however, we adopt the convention that $\mathbb{T}$ denotes the unit circle $\partial\mathbb{D} = \{z \in \mathbb{C} : |z| = 1\}$. (This usage indicates the origin of the name *torus* for the symbol $\mathbb{T}$.) Although this may seem to be an overuse of notation, the two sets are readily identifiable, since the unit circle can be written as

$$\{z \in \mathbb{C} : |z| = 1\} = \{e^{i\theta} : \theta \in [0, 2\pi)\}.$$

In fact, some authors prefer to use the symbol $\mathbb{T}$ to denote the unit interval $[0, 1)$, making use of the identification $\partial\mathbb{D} = \{e^{2\pi i\theta} : \theta \in [0, 1)\}$. For the remainder of this text, however, we will use $\mathbb{T}$ to mean the unit circle.

8.2 Commutative Algebras

In this section (as before), we consider complex Banach algebras that are unital. Shortly, we will impose on our Banach algebras the additional restriction of commutativity. This added structure leads to some remarkable consequences. We begin, however, by making some definitions in the general context of an algebra.

Definition 8.31 A nonempty linear subspace I of an algebra A is called a *left ideal* of A if $ax \in I$ whenever $a \in A$ and $x \in I$. Similarly, I is called a *right ideal* if $xa \in I$ whenever $a \in A$ and $x \in I$. If I is both a left ideal and a right ideal, it is called a *two-sided ideal*. In any case, I is called *proper* if $I \neq A$.

If an ideal contains the identity element of A, then it must contain every element of A. Thus, a proper ideal does not contain the identity element. Similarly, a proper ideal cannot contain an element that is invertible. Notice that an ideal necessarily contains 0, and $\{0\}$ is always a proper two-sided ideal in any unital algebra.

Proposition 8.32 *If I is a proper closed two-sided ideal in a unital Banach algebra A, then A/I is a unital Banach algebra, called the quotient algebra.*

Proof We know that A/I is a Banach space, by Proposition 3.47. We now define a multiplication on A/I:

$$(a + I) \cdot (b + I) = ab + I, \quad \{a, b\} \subseteq A.$$

With this multiplication, it is clear that $1 + I$ is an identity in A/I, where 1 is the identity in A. If $x \in I$ and $\|1 - x\| < 1$, then x is invertible (by Proposition 8.10), which contradicts the fact that I is a proper ideal. Therefore, $\|1 + I\| = 1$, and so A/I is unital.

It remains to show that $\|ab + I\| \leq \|a + I\| \|b + I\|$ for all a and b in A. Let a and b be fixed elements in A. The norm on A/I is an infimum, and so for each $n \in \mathbb{N}$, we may select $x_n \in a + I$ and $y_n \in b + I$ so that

$$\|x_n\| < \|a + I\| + \frac{1}{n} \quad \text{and} \quad \|y_n\| < \|b + I\| + \frac{1}{n}.$$

For each $n \in \mathbb{N}$, we have $x_n y_n \in ab + I$, and hence

$$\|ab + I\| \leq \|x_n y_n\| < \left(\|a + I\| + \frac{1}{n} \right) \left(\|b + I\| + \frac{1}{n} \right).$$

Since this bound holds for all $n \in \mathbb{N}$, we may take the limit as $n \to \infty$. It follows that $\|ab + I\| \leq \|a + I\| \|b + I\|$, as required.   □

Example 8.33 Let H be an infinite-dimensional *separable* Hilbert space. The space $\mathcal{L}(H)$ of operators on H is a Banach algebra, and the subspace $\mathcal{K}(H)$ of compact operators on H is a closed two-sided ideal. (See Theorem 6.17.) The quotient algebra $\mathcal{L}(H)/\mathcal{K}(H)$ is called the *Calkin algebra*, after the mathematician J. W. Calkin.

The Calkin algebra is significant because it is not isomorphic to an algebra of operators on a *separable* Hilbert space (even though H is itself separable) [5]. This is remarkable because the Calkin algebra is also a C^*-*algebra*, and from the *Gelfand–Naimark Theorem* it is known that every C^*-algebra is isomorphic to an algebra of operators on some Hilbert space. The Gelfand-Naimark Theorem dates back to the work of Gelfand and Naimark in 1943 [16].

For the remainder of this section, we will suppose that all Banach algebras are commutative. We remind the reader that an algebra A is *commutative* if $ab = ba$ for all a and b in A. (See Definition 8.5.) In particular, this implies that any ideal is necessarily two-sided. Consequently, when A is a commutative algebra, a two-sided ideal is called an *ideal*.

Definition 8.34 Let A be a commutative complex unital Banach algebra. A linear functional $\phi \colon A \to \mathbb{C}$ is called a *multiplicative linear functional* if it is a *ring homomorphism*; that is, if $\phi(1) = 1$ and $\phi(xy) = \phi(x)\phi(y)$ for all x and y in A.

The multiplicative property of a multiplicative linear functional ϕ guarantees that $\phi(1)^2 = \phi(1)$, and so it must be that either $\phi(1) = 1$ or $\phi(1) = 0$. We insist on the condition that $\phi(1) = 1$ in order to disqualify the trivial linear functional $\phi = 0$.

Theorem 8.35 *If ϕ is a multiplicative linear functional on a commutative complex unital Banach algebra, then ϕ is continuous and $\|\phi\| = 1$.*

Proof By assumption, ϕ is a multiplicative linear functional, and so $\phi(1) = 1$. It follows that $\|\phi\| \geq 1$. Suppose there exists some $x \in A$ such that $\|x\| \leq 1$, but $|\phi(x)| > 1$. Let $\alpha = \phi(x)$. By definition, $\phi(\alpha^{-1}x) = 1$, and so $\phi(1 - \alpha^{-1}x) = 0$. On the other hand, $|\alpha| > 1$, and so $\|\alpha^{-1}x\| < 1$. By Proposition 8.10, we have that $1 - \alpha^{-1}x$ is invertible, and so

$$\phi(1 - \alpha^{-1}x) \cdot \phi((1 - \alpha^{-1}x)^{-1}) = 1.$$

This is a contradiction, because $\phi(1 - \alpha^{-1}x) = 0$. Consequently, it must be the case that $\|\phi\| = 1$. $\qquad\square$

Definition 8.36 Let A be a commutative algebra. A proper ideal I is said to be *maximal* in A if $I = J$ whenever J is a proper ideal such that $I \subseteq J$.

Example 8.37 Suppose ϕ is a multiplicative linear functional on a commutative complex unital Banach algebra A. We claim that the kernel of ϕ is a closed maximal ideal in A. Recall that the kernel of ϕ is the set $\ker\phi = \{x : \phi(x) = 0\}$. It is a consequence of the multiplicative property of ϕ that $\ker\phi$ is an ideal. It is a proper ideal because $1 \notin \ker\phi$. (This is why we assumed $\phi(1) = 1$ in the definition of a multiplicative linear functional.)

By Proposition 3.49, the quotient $A/\ker\phi$ is isomorphic to $\mathbb{C}$. This means that the codimension of $\ker\phi$ is one and, in particular, this tells us that $\ker\phi$ is a maximal ideal. (See Exercise 8.11.)

Proposition 8.38 (Krull's Theorem) *Any proper ideal is contained in a maximal ideal. In particular, any unital algebra has a maximal ideal.*

Proof If $(J_i)_{i \in I}$ is a chain of proper ideals, then $J = \bigcup_{i \in I} J_i$ is also a proper ideal. To see this, simply note that $1 \notin J_i$ for any $i \in I$, and so $1 \notin J$. Therefore, every chain of proper ideals has an upper bound, and so the result follows from Zorn's Lemma.

To show every unital algebra has a maximal ideal, observe that $\{0\}$ is a proper ideal, and so must be contained in a maximal ideal.   □

We are working within the context of Banach algebras, but Krull's Theorem remains true in the more general setting of ring theory. Specifically, Krull's Theorem asserts the existence of maximal ideals in any unital ring. The proof is the same, and relies on Zorn's Lemma. In fact, like Zorn's Lemma, Krull's Theorem is equivalent to the Axiom of Choice [18]. Krull's Theorem is named after Wolfgang Krull, who proved the general version of the result in 1929 [22].

Proposition 8.39 *Every maximal ideal in a commutative unital Banach algebra is closed.*

Proof Suppose J is a maximal ideal in a commutative unital Banach algebra A. By the continuity of multiplication, $\overline{J}$ is also an ideal. We need only show that $\overline{J}$ is a proper ideal. It will suffice to show that $1 \notin \overline{J}$. We will show this by showing that $d(1, J) = 1$. First, observe that $d(1, J) \leq 1$, because $0 \in J$. Suppose that $d(1, J) < 1$. Then there exists an element $x \in J$ such that $d(1, x) = \|1 - x\| < 1$. This implies (by Proposition 8.10) that x is invertible, which contradicts the assumption that J is a proper ideal. Therefore, $d(1, J) = 1$, and so $1 \notin \overline{J}$. Consequently, $\overline{J}$ is a proper ideal containing the maximal ideal J, and therefore $J = \overline{J}$.   □

Theorem 8.40 *Each maximal ideal in a commutative complex unital Banach algebra is the kernel of some multiplicative linear functional.*

Proof Let J be a maximal ideal in a commutative complex unital Banach algebra A. By Proposition 8.39, the set J is a closed subset of A. It follows that A/J is a unital Banach algebra, by Proposition 8.32.

By assumption, J is a proper ideal in A, and so there exists some $x \in A$ such that $x \notin J$. Define a subset of A by

$$J[x] = \{ax + y : a \in A, \ y \in J\}.$$

Then $J[x]$ is an ideal which is strictly larger than J, and so $A = J[x]$.

Let $\pi : A \to A/J$ be the quotient map. Since $J[x] = A$, it must be that $1 = ax + y$ for some $a \in A$ and $y \in J$. Therefore, $\pi(1) = \pi(ax + y)$, and consequently

$$1 + J = ax + J = (a + J)(x + J).$$

It follows that $(x + J)^{-1}$ exists whenever $x \notin J$, and so every nonzero element of A/J is invertible. By the Gelfand–Mazur Theorem (Theorem 8.20), we conclude that A/J is isometrically isomorphic to $\mathbb{C}$. Denote this isometric isomorphism by $i : A/J \to \mathbb{C}$. Then $\phi = i \circ \pi$ is a multiplicative linear functional on A and $\ker \phi = J$.   □

Since the proof of the above theorem uses the Gelfand–Mazur Theorem, it only works for complex Banach algebras. Indeed, this theorem is not true (in general) for real Banach algebras.

Through Example 8.37 and Theorem 8.40, we have established a correspondence between the multiplicative linear functionals on a commutative complex unital Banach algebra A and the maximal ideals of A.

Definition 8.41 Let A be a commutative complex unital Banach algebra. The *spectrum* of A, which we denote by $\Sigma(A)$, is the collection of all multiplicative linear functionals on A.

One must be careful to not confuse the spectrum of an algebra $\Sigma(A)$ with the spectrum of an element $\mathrm{Sp}(x)$, where x is an element in the algebra A. We will discover shortly the reason behind this naming. Due to the correspondence between multiplicative linear functionals and maximal ideals, the set $\Sigma(A)$ is sometimes called the *maximal ideal space* of A.

Theorem 8.42 *The spectrum of a commutative complex unital Banach algebra is a nonempty w^*-compact set.*

Proof Let A be a commutative complex unital Banach algebra. By Proposition 8.38, there exists a maximal ideal in A. By Theorem 8.40, there is some multiplicative linear functional for which this maximal ideal is the kernel. Therefore, $\Sigma(A)$ is nonempty.

It remains to show that $\Sigma(A)$ is w^*-compact. From Theorem 8.35, we deduce that $\Sigma(A) \subseteq B_{A^*}$. By Theorem 5.39 (the Banach-Alaoglu Theorem), we know that B_{A^*} is a w^*-compact set. Thus, it suffices to show that $\Sigma(A)$ is w^*-closed. Observe that

$$\Sigma(A) = \{\phi \in B_{A^*} : \phi(1) = 1, \ \phi(xy) = \phi(x)\phi(y) \text{ for all } \{x, y\} \subseteq A\} \qquad (8.9)$$

$$= B_{A^*} \cap \{\phi : \phi(1) = 1\} \cap \left(\bigcap_{x \in A} \bigcap_{y \in A} \{\phi : \phi(xy) = \phi(x)\phi(y)\} \right). \qquad (8.10)$$

All of these sets are closed in the weak*-topology, and so the result follows. □

Definition 8.43 Let A be a commutative complex unital Banach algebra. If $x \in A$, then we define the *Gelfand transform of x* to be the (complex-valued) continuous function $\hat{x} \in C(\Sigma(A))$ given by

$$\hat{x}(\phi) = \phi(x), \quad \phi \in \Sigma(A).$$

The Gelfand transform of x is well-defined, because $\Sigma(A)$ is a w^*-compact subset of B_{A^*}. Observe also that the map $x \mapsto \hat{x}$, which is called *the Gelfand transform*, is a norm-decreasing algebra homomorphism.

The next proposition explains why we call $\Sigma(A)$ the *spectrum* of A.

Proposition 8.44 *Let A be a commutative complex unital Banach algebra and let $\Sigma = \Sigma(A)$. If $x \in A$, then $\mathrm{Sp}(x) = \hat{x}(\Sigma)$ and*

$$\|\hat{x}\|_{C(\Sigma)} = r(x) = \lim_{n \to \infty} \|x^n\|^{1/n}.$$

Proof Suppose $\lambda \in \hat{x}(\Sigma)$. Then there exists some $\phi \in \Sigma$ such that

$$\phi(x) = \hat{x}(\phi) = \lambda = \phi(\lambda 1).$$

Consequently, $\phi(\lambda 1 - x) = 0$, and so $\lambda 1 - x$ is not invertible. (If a is invertible, then $\phi(a)\phi(a^{-1}) = 1$, and so $\phi(a) \neq 0$.) It follows that $\lambda \in \mathrm{Sp}(x)$.

Now suppose $\lambda \in \mathrm{Sp}(x)$. Then $\lambda 1 - x$ is not invertible, and hence $(\lambda 1 - a)A$ is a proper ideal. By Proposition 8.38, there exists a maximal ideal J containing the ideal $(\lambda 1 - x)A$. By Theorem 8.40, there exists some multiplicative linear functional ϕ such that $J = \ker\phi$. We conclude that $\phi(\lambda 1 - x) = 0$, and so $\lambda = \phi(x) = \hat{x}(\phi)$, as required.

The fact that $\|\hat{x}\|_{C(\Sigma)} = r(x)$ follows from the identification $\mathrm{Sp}(x) = \hat{x}(\Sigma)$. The rest of the proposition is the Spectral Radius Formula. (See Theorem 8.23.) $\square$

Remark 8.45 To emphasize the relationship between $\Sigma(A)$ and $\mathrm{Sp}(x)$, where x is an element in the Banach algebra A, the spectrum of x is sometimes denoted $\sigma(x)$.

Example 8.46 Once again, consider the Volterra operator:

$$Vf(x) = \int_0^x f(t)\,dt, \quad f \in C[0,1], \ x \in [0,1].$$

In Example 8.26, we saw that $\mathrm{Sp}(V) = \{0\}$. (This was a result of the Fredholm Alternative, because V is a compact operator with no eigenvalues.)

Let A denote the algebra given by the closure of all polynomials in V; that is,

$$A = \overline{\left\{ \sum_{k=0}^n a_k V^k : (a_0, \dots, a_n) \in \mathbb{C}^{n+1}, \ n \in \mathbb{N} \right\}}.$$

We know that A is a commutative Banach algebra. By Proposition 8.44, we have $\widehat{V}(\Sigma(A)) = \mathrm{Sp}(V) = \{0\}$. It follows that $\phi(V) = 0$ for all multiplicative linear functionals on A. Therefore, $\Sigma(A)$ contains only one element ϕ and

$$\phi\left(\sum_{k=0}^n a_k V^k \right) = a_0,$$

for all $(a_0, \dots, a_n) \in \mathbb{C}^{n+1}$, where $n \in \mathbb{N}$.

Theorem 8.47 *Let A be a commutative complex unital Banach algebra and let $\Sigma = \Sigma(A)$. The Gelfand transform is an algebra homomorphism between A and a subalgebra of $C(\Sigma)$. Furthermore, the Gelfand transform is an isometry if and only if $\|x^2\| = \|x\|^2$ for all $x \in A$.*

Proof It is clear the Gelfand transform is an algebra homomorphism onto its image. We show that it is an isometry if and only if $\|x^2\| = \|x\|^2$ for all $x \in A$.

First, assume $\|x^2\| = \|x\|^2$ for all $x \in A$. By induction, $\|x^{2^k}\| = \|x\|^{2^k}$ for all $x \in A$. Therefore, by Proposition 8.44,

$$\|\hat{x}\|_{C(\Sigma)} = \lim_{k \to \infty} \|x^{2^k}\|^{1/2^k} = \|x\|.$$

Thus, the Gelfand transform is an isometry.

Conversely, assume $\|\hat{x}\|_{C(\Sigma)} = \|x\|$ for all $x \in A$. By Theorem 8.23 and Proposition 8.44,

$$\|x\| = \|\hat{x}\|_{C(\Sigma)} = r(x) = \inf_{n \in \mathbb{N}} \|x^n\|^{1/n} \le \|x^2\|^{1/2},$$

and hence $\|x\|^2 \le \|x^2\|$. By submultiplicativity of the norm, $\|x^2\| \le \|x\|^2$. Therefore, $\|x\|^2 = \|x^2\|$, as required.   □

We now consider some examples of Banach algebras, and we identify the spectrum in each case. In each of the following examples, we will be looking at a Banach algebra with the supremum (or essential supremum) norm, and so it is a trivial calculation to show it satisfies the relationship $\|x\|^2 = \|x^2\|$ for every x in the algebra. Consequently, in each case, the Gelfand transform is an isometry (by Theorem 8.47).

Example 8.48 Suppose K is a compact Hausdorff space. We would like to identify $\Sigma = \Sigma(C(K))$, the spectrum of $C(K)$. Here, $C(K)$ denotes the space of complex-valued continuous functions on K. Observe that, for any $s \in K$, the map

$$\delta_s(f) = f(s), \quad f \in C(K),$$

is a multiplicative linear functional. We will show that, in fact, all multiplicative linear functionals on $C(K)$ can be achieved as point evaluation.

Suppose that $\phi \in \Sigma \setminus \{\delta_s : s \in K\}$, which is a w^*-open set in Σ. There exists, then, a w^*-neighborhood W of ϕ so that $W \cap \{\delta_s : s \in K\} = \emptyset$.

The set W is a neighborhood of ϕ in the w^*-topology, and so there exists an $n \in \mathbb{N}$, an $\epsilon > 0$, and functions $\{f_1, \dots, f_n\} \subseteq C(K)$ so that

$$\{\mu \in C(K)^* : |\mu(f_1) - \phi(f_1)| < \epsilon, \dots, |\mu(f_n) - \phi(f_n)| < \epsilon\} \subseteq W.$$

If $s \in K$, then $\delta_s \notin W$, and so there exists some $j \in \{1, \dots, n\}$ such that

$$|f_j(s) - \phi(f_j)| = |\delta_s(f_j) - \phi(f_j)| \ge \epsilon.$$

For each $j \in \{1, \dots, n\}$, define a function $g_j \in C(K)$ by $g_j = f_j - \phi(f_j)1$, where 1 is the unit element in the algebra $C(K)$. (Notice that $1 = \chi_K$, the function that is constantly 1 for all $s \in K$.) By definition, $\phi(g_j) = 0$ for all $j \in \{1, \dots, n\}$. Now define $g \in C(K)$ by

$$g = \sum_{j=1}^{n} g_j \overline{g_j} = \sum_{j=1}^{n} |g_j|^2.$$

For each $s \in K$, there exists some $j \in \{1, \dots, n\}$ such that $|g_j(s)| \ge \epsilon$. Therefore, $|g(s)| \ge \epsilon^2$ for all $s \in K$, and so g is invertible (in the algebra $C(K)$).

On the other hand,

$$\phi(g) = \sum_{j=1}^{n} \phi(g_j)\phi(\overline{g_j}) = 0,$$

which implies that g is not invertible. We have derived a contradiction, and so it must be that $\Sigma = \{\delta_s : s \in K\}$. By Lemma 5.66, we conclude that Σ is homeomorphic to K. Therefore, the spectrum of $C(K)$ is in fact K. Certainly, then, in this example, we have that $C(K)$ is isometrically isomorphic to $C(\Sigma)$.

Example 8.49 Consider the sequence space ℓ_∞ of bounded complex-valued sequences. We define a multiplication in ℓ_∞ component-wise, so that $\xi \cdot \eta$ is the sequence with components

$$(\xi \cdot \eta)_n = \xi_n \eta_n, \quad n \in \mathbb{N},$$

where $\xi = (\xi_n)_{n=1}^\infty$ and $\eta = (\eta_n)_{n=1}^\infty$ are sequences in ℓ_∞.

We again wish to identify $\Sigma = \Sigma(\ell_\infty)$. For each $n \in \mathbb{N}$, the projection onto the nth coordinate $\xi \mapsto \xi_n$ determines a multiplicative linear functional. Consequently, we obtain $\mathbb{N} \subseteq \Sigma$. This inclusion cannot be equality, however, because $\mathbb{N}$ is not a w^*-compact set. In fact, the spectrum of ℓ_∞ is the set $\Sigma = \beta\mathbb{N}$, the Stone–Čech compactification of $\mathbb{N}$. The proof of this fact is beyond the scope of our current discussion, and so we will omit it. We remark, however, that it relies on the Axiom of Choice. (See Sect. B.1 for the definition of the Stone–Čech compactification.)

In this example, too, the algebra ℓ_∞ is isometrically isomorphic to $C(\Sigma)$.

Example 8.50 Consider the space $L_\infty(0, 1)$ of (equivalence classes of) essentially bounded measurable functions on the unit interval $[0, 1]$. We denote the spectrum by $\Omega = \Sigma(L_\infty(0, 1))$. Certainly, $\|f^2\|_\infty = \|f\|_\infty^2$ for all $f \in L_\infty(0, 1)$, and so the Gelfand transform is an isometry (by Theorem 8.47). It turns out that the Gelfand transform is actually an isomorphism, and so $L_\infty(0, 1)$ is identical (as a Banach algebra) to $C(\Omega)$. The space Ω is called the *Stone space* of the measure algebra, but it is very difficult to describe. Unlike Example 8.48, we cannot realize the multiplicative linear functionals in this case as point evaluation, because there is no (constructive) way to define the value of $f \in L_\infty(0, 1)$ at a given point in $[0, 1]$ (because f represents a class of functions that are equal almost everywhere).

Theorem 8.47 is a statement about complex Banach algebras. Ultimately, it relies on the Gelfand–Mazur Theorem and requires the underlying scalar field to be $\mathbb{C}$. The arguments given here cannot be applied to real Banach algebras; however, a similar theorem does hold for Banach algebras over $\mathbb{R}$. (See Theorem 4.2.5 in [2].)

8.3 The Wiener Algebra

Consider the space $\ell_1(\mathbb{Z})$ of doubly infinite absolutely summable sequences. We will equip $\ell_1(\mathbb{Z})$ with a multiplication akin to the one in Example 8.6(c), making $\ell_1(\mathbb{Z})$ into a *convolution algebra*.

Suppose $\xi = (\xi_j)_{j\in\mathbb{Z}}$ and $\eta = (\eta_j)_{j\in\mathbb{Z}}$ are sequences in $\ell_1(\mathbb{Z})$. Define the product of ξ and η to be the sequence $\xi * \eta = ((\xi * \eta)_k)_{k\in\mathbb{Z}}$ of coefficients of the formal Laurent series

$$\sum_{k=-\infty}^{\infty} (\xi * \eta)_k \, t^k = \left(\sum_{j=-\infty}^{\infty} \xi_j \, t^j \right) \left(\sum_{j=-\infty}^{\infty} \eta_j \, t^j \right). \tag{8.11}$$

Then

$$(\xi * \eta)_k = \sum_{j=-\infty}^{\infty} \xi_{k-j}\,\eta_j, \quad k \in \mathbb{Z}. \tag{8.12}$$

We must verify that this multiplication is well-defined; i.e., that $\xi * \eta$ is in fact an absolutely summable sequence. From (8.12), we have

$$\|\xi * \eta\|_1 = \sum_{k=-\infty}^{\infty} \left| \sum_{j=-\infty}^{\infty} \xi_{k-j}\,\eta_j \right| \le \sum_{k=-\infty}^{\infty} \left(\sum_{j=-\infty}^{\infty} |\xi_{k-j}|\,|\eta_j| \right).$$

For a fixed $j \in \mathbb{Z}$, we have $\sum_{k=-\infty}^{\infty} |\xi_{k-j}| = \|\xi\|_1$. Consequently, by Fubini's Theorem,

$$\|\xi * \eta\|_1 \le \sum_{j=-\infty}^{\infty} |\eta_j| \left(\sum_{k=-\infty}^{\infty} |\xi_{k-j}| \right) = \sum_{j=-\infty}^{\infty} |\eta_j|\,\|\xi\|_1 = \|\eta\|_1\,\|\xi\|_1.$$

Therefore, $\xi * \eta \in \ell_1(\mathbb{Z})$, and so the multiplication is well-defined.

We have also verified that the norm is submultiplicative, and so equipped with this multiplication, the space $\ell_1(\mathbb{Z})$ becomes a commutative Banach algebra. This Banach algebra is called the *Wiener algebra* (after Norbert Wiener), and is denoted by W. We call W a *convolution algebra* because the multiplication $*$ is called a *convolution*. Furthermore, we say that $\xi * \eta$ is the *convolution of ξ with η*.

We now identify $\Sigma(W)$, the spectrum of W. It will be convenient to identify W with the space of formal Laurent series, so that t^k corresponds to the sequence with 1 in the kth position ($k \in \mathbb{Z}$), and zero elsewhere. This makes sense because of how we defined the multiplication in W.

We wish to identify the multiplicative linear functionals ϕ on W. Suppose ϕ is a multiplicative linear functional. Observe that ϕ is completely determined by what it does on t. If $\phi(t) = \lambda$, then $\phi(t^k) = \lambda^k$ for all $k \in \mathbb{Z}$. Therefore,

$$\phi(\xi) = \sum_{k=-\infty}^{\infty} \lambda^k \xi_k, \quad \xi \in W. \tag{8.13}$$

The series in (8.13) is doubly infinite. As a result, it will converge for all $\xi \in W$ only if both $|\lambda| \le 1$ (for the series with $k \ge 0$) and $|\lambda| \ge 1$ (for the series with $k \le 0$). Consequently, $|\lambda| = 1$, and so $\lambda \in \mathbb{T}$, the unit circle in $\mathbb{C}$.

Certainly, any $\lambda \in \mathbb{T}$ determines a multiplicative linear functional according to (8.13). Thus, the linear functionals that are multiplicative linear functionals are those given by $\phi(t) = \lambda$ for $\lambda \in \mathbb{T}$.

For each $\lambda \in \mathbb{T}$, let ϕ_λ denote the multiplicative linear functional determined by $\phi_\lambda(t) = \lambda$. Then $\Sigma = \{\phi_\lambda : \lambda \in \mathbb{T}\}$. We may thus identify Σ with $\mathbb{T}$. (The argument is similar to the proof of Lemma 5.66.) Notice that $\phi_\lambda(\xi) = \phi_1(\eta)$, where the sequence $\eta = (\eta_k)_{k\in\mathbb{Z}}$ is defined so that $\eta_k = \lambda^k \xi_k$ for each $k \in \mathbb{Z}$.

Since Σ is homeomorphic to $\mathbb{T}$, we can embed W into $C(\mathbb{T})$ via the Gelfand transform. Suppose that $\xi \in W$. Then $\hat{\xi} \in C(\mathbb{T})$ via the map

$$\hat{\xi}(\lambda) = \hat{\xi}(\phi_\lambda) = \sum_{k=-\infty}^{\infty} \lambda^k \xi_k, \quad \lambda \in \mathbb{T}.$$

For every $\lambda \in \mathbb{T}$, there is a $\theta \in [0, 2\pi)$ so that $\lambda = e^{i\theta}$. Therefore, we define a function $f : [0, 2\pi) \to \mathbb{C}$ by $f(\theta) = \hat{\xi}(e^{i\theta})$, and hence

$$f(\theta) = \sum_{k=-\infty}^{\infty} \xi_k e^{ik\theta}, \quad \theta \in [0, 2\pi). \tag{8.14}$$

Consequently, each $\xi \in W$ corresponds to an $f \in C(\mathbb{T})$ with absolutely convergent Fourier series.

Motivated by the above discussion, we let $A(\mathbb{T})$ denote the collection of all functions on $\mathbb{T}$ with absolutely convergent Fourier series, and we provide it with the norm

$$\|f\|_{A(\mathbb{T})} = \sum_{k=-\infty}^{\infty} |\hat{f}(k)|, \quad f \in A(\mathbb{T}), \tag{8.15}$$

where $\hat{f}(k)$ is the kth Fourier coefficient of f. (See (4.2.1).) Then $A(\mathbb{T})$ is a Banach algebra, and W is homeomorphic to $A(\mathbb{T})$ via the Gelfand transform. Since it is isometrically isomorphic to W (as an algebra) we also call $A(\mathbb{T})$ the *Wiener Algebra*.

Remark 8.51 In the above discussion, we once again made use of the correspondence between $\mathbb{T}$ and $[0, 2\pi)$. In particular, we identified the continuous functions on $\mathbb{T}$ with the continuous functions on $[0, 2\pi)$. It is for this reason that we can say $f \in C(\mathbb{T})$ when it is actually an element of $C[0, 2\pi)$.

The next theorem gives an indication of the power behind the tools we have developed. The conclusion of the theorem is not at all obvious, but the proof itself is quite simple (and short). One should not mistake the brevity of the proof as a mark of triviality—it is brief because of the power behind our machinery.

Theorem 8.52 (Wiener Inversion Theorem) *Let $f \in C(\mathbb{T})$ have an absolutely convergent Fourier series. If $f(\theta) \neq 0$ for any $\theta \in [0, 2\pi)$, then $1/f$ has an absolutely converging Fourier series.*

Proof By assumption, $f \in A(\mathbb{T})$. There exists some $a \in W$ such that $\hat{a} = f$. By Proposition 8.44,

$$f(\mathbb{T}) = \hat{a}(\mathbb{T}) = \text{Sp}(a).$$

Since $0 \notin f(\mathbb{T})$, it follows that $0 \notin \text{Sp}(a)$, and so a is invertible in W. Therefore, f is invertible in $A(\mathbb{T})$, and the result follows. $\square$

Exercises

Exercise 8.1 Let A be a unital Banach algebra. Show that if $a \in A$ has an inverse, then it must be unique. Show that if a has both a left-inverse and a right-inverse, then a is invertible.

Exercise 8.2 Let A be a noncommutative unital Banach algebra. Suppose x and y are elements in A such that both xy and yx are invertible. Show that both x and y are invertible and show that $(xy)^{-1} = y^{-1}x^{-1}$.

Exercise 8.3 Let A be a unital Banach algebra with identity element 1. Suppose $a \in A$ is such that $\sum_{n=0}^{\infty} \|a\|^n < \infty$. Show that $1 - a$ is an invertible element and that $(1 - a)^{-1} = \sum_{n=0}^{\infty} a^n$.

Exercise 8.4 Let X be a Banach space and suppose $T \in \mathcal{L}(X)$ is invertible. Show $\lambda \in \mathrm{Sp}(T)$ if and only if $1/\lambda \in \mathrm{Sp}(T^{-1})$.

Exercise 8.5 Let A be a unital Banach algebra. Show that the set of invertible elements of A is a group.

Exercise 8.6 Let A be a commutative unital Banach algebra and let $G(A)$ be the group of invertible elements of A. Show that $x \in G(A)$ if and only if $\phi(x) \neq 0$ for all multiplicative linear functionals ϕ.

Exercise 8.7 Let A be a Banach algebra. An element $x \in A$ is called a *topological divisor of zero* if there exists a sequence $(y_n)_{n=1}^{\infty}$ in A with $\|y_n\| = 1$ for all $n \in \mathbb{N}$ such that $\lim_{n \to \infty} x y_n = 0 = \lim_{n \to \infty} y_n x$. Find an example of a Banach algebra with a topological divisor of zero that is not a divisor of zero.

Exercise 8.8 Let A be a unital Banach algebra and let $G(A)$ be the group of invertible elements. Show that every $x \in \partial G(A)$ is a topological divisor of zero, where $\partial G(A)$ is the boundary of $G(A)$.

Exercise 8.9 Let X be a commutative Banach algebra. A proper ideal I of X is called *prime* if for any elements a and b in X the product $ab \in I$ implies that either $a \in I$ or $b \in I$. Show that in a commutative unital Banach algebra any maximal ideal is prime.

Exercise 8.10 Let M be the multiplier operator on $L_2 = L_2([0, 2\pi), \frac{d\theta}{2\pi})$ given by the formula $Mf(\theta) = e^{i\theta} f(\theta)$ for all $f \in L_2$ and $\theta \in [0, 2\pi)$. Show that the Fourier transform of the multiplier operator satisfies the equation $\widehat{Mf}(n) = \hat{f}(n - 1)$ for all $f \in L_2$ and each $n \in \mathbb{Z}$. Explicitly write out the relationship between M and the shift operator R on $\ell_2(\mathbb{Z})$. (Recall that $\hat{f}(n) = \int_0^{2\pi} f(\theta) e^{-in\theta} \frac{d\theta}{2\pi}$ for all $n \in \mathbb{Z}$.)

Exercise 8.11 If N is a subspace of a vector space X, then the *codimension* of N in X is the dimension of X/N. Show that an ideal of a complex commutative unital Banach algebra X is maximal if and only if it has codimension 1. (*Hint:* See the proof of Theorem 8.40.)

Exercise 8.12 Let A be a complex unital Banach algebra. Suppose there exists some $M < \infty$ such that $\|x\|\|y\| \leq M\|xy\|$ for all x and y in A. Show that $A = \mathbb{C}$.

Exercise 8.13 Let A be a Banach algebra and suppose $a \in A$. Use Fekete's Lemma (Exercise 3.10) to show that

$$\lim_{n \to \infty} \|a^n\|^{1/n} = \inf_{n \in \mathbb{N}} \|a^n\|^{1/n}.$$

(*Hint:* Let $\theta_n = \log \|a^n\|$.)

Exercise 8.14 Let $A = \ell_1(\mathbb{Z}_+)$ be the convolution algebra from Example 8.6(c). Recall that the multiplication in A is given by the formula $(a * b)_k = \sum_{j=0}^{k} a_{k-j} b_j$ for all $k \in \mathbb{Z}_+$, where $a = (a_j)_{j=0}^{\infty}$ and $b = (b_j)_{j=0}^{\infty}$ are sequences in A. Identify $\Sigma = \Sigma(A)$, the spectrum of A, and show that $C(\Sigma)$ is the collection of all functions having an absolutely convergent Taylor series in $\overline{\mathbb{D}}$. (*Hint:* A is a subalgebra of W.)

Exercise 8.15 If f has an absolutely convergent Taylor series in $\overline{\mathbb{D}}$, and if $f(z) \neq 0$ for any $z \in \overline{\mathbb{D}}$, then show that $1/f$ has an absolutely convergent Taylor series in $\overline{\mathbb{D}}$. (*Hint:* Use the previous problem.)

Appendix A
Basics of Measure Theory

In this appendix, we give a brief overview of the basics of measure theory. For a detailed discussion of measure theory, a good source is *Real and Complex Analysis* by W. Rudin [33].

A.1 Measurability

Definition A.1 Let X be a set. A collection $\mathcal{A}$ of subsets of X is called a σ-*algebra of sets in* X if the following three statements are true:

(a) The set X is in $\mathcal{A}$.
(b) If $A \in \mathcal{A}$, then $A^c \in \mathcal{A}$, where A^c is the complement of A in X.
(c) If $A_n \in \mathcal{A}$ for all $n \in \mathbb{N}$, and if $A = \bigcup_{n=1}^{\infty} A_n$, then $A \in \mathcal{A}$.

A set X together with a σ-algebra $\mathcal{A}$ is called a *measurable space* and is denoted by $(X, \mathcal{A})$, or simply by X when there is no risk of confusion. The elements $A \in \mathcal{A}$ are known as *measurable sets*.

From (a) and (b), we can easily see that the empty set is always measurable; that is, $\emptyset \in \mathcal{A}$. If $A_n \in \mathcal{A}$ for all $n \in \mathbb{N}$, then $\bigcap_{n=1}^{\infty} A_n$ is always measurable, by (b) and (c), because

$$\bigcap_{n=1}^{\infty} A_n = \left(\bigcup_{n=1}^{\infty} A_n^c \right)^c.$$

Among other things, we can now conclude that, for measurable sets A and B, the set $A \backslash B = A \cap B^c$ is measurable.

The σ in the name σ-algebra refers to the countability assumption in (c) of Definition A.1. If we require only finite unions of measurable sets to be in $\mathcal{A}$, then $\mathcal{A}$ is called an *algebra of sets in* X.

Definition A.2 Let $(X, \mathcal{A})$ be a measurable space. A scalar-valued function f with domain X is called *measurable* if $f^{-1}(V) \in \mathcal{A}$ whenever V is an open set in the scalar field.

© Springer Science+Business Media, LLC 2014
A. Bowers, N. J. Kalton, *An Introductory Course in Functional Analysis*,
Universitext, DOI 10.1007/978-1-4939-1945-1

Example A.3 Let $(X, \mathcal{A})$ be a measurable space. If $E \subseteq X$, then the *characteristic function* (or *indicator function*) of E is the function defined by

$$\chi_E(x) = \begin{cases} 1 & \text{if } x \in E, \\ 0 & \text{if } x \notin E. \end{cases}$$

The characteristic function of E is measurable if and only if $E \in \mathcal{A}$. The characteristic function χ_E is also denoted by 1_E.

Suppose u and v are real-valued functions. Then $f = u + iv$ is a complex measurable function if and only if u and v are real measurable functions. Furthermore, if f and g are measurable functions, then so is $f + g$.

Measurable functions are very well-behaved. For example, if $(f_n)_{n=1}^{\infty}$ is a sequence of real-valued measurable functions, then the functions

$$\sup_{n \in \mathbb{N}} f_n \quad \text{and} \quad \limsup_{n \to \infty} f_n$$

are measurable whenever the function values are finite. In particular, if $(f_n)_{n=1}^{\infty}$ is a convergent sequence of scalar-valued measurable functions, then the limit of the sequence is also a scalar-valued measurable function.

Definition A.4 A *simple function* is any linear combination of characteristic functions; that is

$$\sum_{j=1}^{n} a_j \chi_{E_j},$$

where $n \in \mathbb{N}$, $a_1, \dots, a_n$ are scalars, and $E_1, \dots, E_n$ are measurable sets.

Certainly, any simple function is measurable, since it is a finite sum of measurable functions.

The next theorem is key to the further development of the subject.

Theorem A.5 *Let $(X, \mathcal{A})$ be a measurable space. If f is a positive measurable function, then there exists a sequence $(s_n)_{n=1}^{\infty}$ of simple functions such that:*

$$0 \le s_1 \le s_2 \le \cdots \le f,$$

and

$$f(x) = \lim_{n \to \infty} s_n(x), \quad x \in X.$$

A.2 Positive Measures and Integration

In this section, we will consider functions defined on a σ-algebra $\mathcal{A}$. Such functions are called *set functions*. For now, we will restrict our attention to set functions taking values in the interval $[0, \infty]$.

Definition A.6   Let $(X, \mathcal{A})$ be a measurable space. A set function $\mu : \mathcal{A} \to [0, \infty]$ is said to be *countably additive* if

$$\mu\left(\bigcup_{j=1}^{\infty} A_j\right) = \sum_{j=1}^{\infty} \mu(A_j), \tag{A.1}$$

whenever $(A_j)_{j=1}^{\infty}$ is a sequence of pairwise disjoint measurable sets. By *pairwise disjoint* we mean $A_j \cap A_k = \emptyset$ whenever $j \neq k$.

A countably additive set function $\mu : \mathcal{A} \to [0, \infty]$ such that $\mu(\emptyset) = 0$ is called a *positive measure on* $\mathcal{A}$. In such a case, we call the triple $(X, \mathcal{A}, \mu)$ a *positive measure space*. If the σ-algebra is understood, we often write (X, μ) for the positive measure space and say μ is a positive measure on X.

Notice that we allow μ to attain infinite values. The assumption that $\mu(\emptyset) = 0$ is to avoid the trivial case where the set function is always ∞. Equivalently, we could assume that there exists some $A \in \mathcal{A}$ such that $\mu(A) < \infty$.

Example A.7   The *Borel σ-algebra* on $\mathbb{R}$ is the smallest σ-algebra that contains all of the open intervals in $\mathbb{R}$. A measure defined on the Borel σ-algebra on $\mathbb{R}$ is called a *Borel measure* on $\mathbb{R}$.

Let μ be a positive measure on a σ-algebra. A measurable set that has zero measure with respect to μ is called a *μ-null set* or a *μ-measure-zero set*. A subset of a μ-null set is called *μ-negligible*. A μ-negligible set may or may not be measurable. If every μ-negligible set is measurable (and hence a μ-null set), then the measure μ is called *complete*. (That is, μ is complete if every subset of a μ-null set is μ-measurable and has μ-measure zero.)

Example A.8 (Lebesgue measure)   Let $\mathcal{B}$ be the Borel σ-algebra on $\mathbb{R}$ and let m be a positive Borel measure on $\mathcal{B}$ defined so that $m((a, b)) = b - a$ whenever a and b are real numbers such that $a < b$. (Such a measure can be shown to exist.) We now define a complete measure on $\mathbb{R}$ that agrees with m for all Borel measurable sets in $\mathcal{B}$. In order to do this, we will enlarge our σ-algebra.

The *Lebesgue σ-algebra* on $\mathbb{R}$ is the collection of all sets of the form $B \cup N$, where $B \in \mathcal{B}$ and N is an m-negligible set. Define a measure λ on this collection of sets by the rule $\lambda(B \cup N) = m(B)$ for any m-measurable set B and any m-negligible set N. It can be shown that λ is a complete countably additive measure on the Lebesgue σ-algebra on $\mathbb{R}$. The measure λ is called *Lebesgue measure* on $\mathbb{R}$.

The Lebesgue σ-algebra on $\mathbb{R}$ is the smallest σ-algebra on $\mathbb{R}$ that contains the Borel measurable sets and all subsets of m-measure-zero sets. Consequently, all Borel measurable sets are Lebesgue measurable, but there are Lebesgue measurable sets that are not Borel measurable.

Example A.9 (Counting measure)   Another classic example of a positive measure is the so-called *counting measure* on $\mathbb{N}$. We let $\mathcal{A}$ be the *power set*, which is the family of all subsets of $\mathbb{N}$, often denoted $\mathcal{P}(\mathbb{N})$ or $2^{\mathbb{N}}$. For any subset $A \subseteq \mathbb{N}$, we define $n(A)$ to be the size (or cardinality) of the set A. Then n is a positive measure on $\mathbb{N}$ and $n(\mathbb{N}) = \infty$.

Definition A.10 A positive measure that attains only finite values is called a *finite measure*. If μ is a positive measure on X such that $\mu(X) = 1$, then we call μ a *probability measure*.

Example A.11 The restriction of the measure λ from Example A.8 to the unit interval $[0, 1]$ is a probability measure.

Let us make some obvious comments about positive measures. Suppose $(X, \mathcal{A}, \mu)$ is a positive measure space. If A and B are measurable sets such that $A \subseteq B$, then $\mu(A) \leq \mu(B)$. Also, observe that if μ is a finite positive measure, then $\mu(\cdot)/\mu(X)$ is a probability measure.

Definition A.12 Let $(X, \mathcal{A}, \mu)$ be a positive measure space and let $s = \sum_{j=1}^{n} a_j \chi_{E_j}$ be a simple function. For any $A \in \mathcal{A}$, we define the *integral of s over A with respect to μ* to be

$$\int_A s\,d\mu = \sum_{j=1}^{n} a_j\, \mu(A \cap E_j).$$

(The value of this integral does not depend on the representation of s that is used.)

If $f : X \to \mathbb{R}$ is a *nonnegative* measurable function, then we define the *integral of f over A with respect to μ* to be

$$\int_A f\,d\mu = \sup \left\{ \int_A s\,d\mu \right\},$$

where the supremum is taken over all simple functions s such that $s(x) \leq f(x)$ for all $x \in X$.

In order to meaningfully extend our notion of an integral to a wider class of functions, we first define the class of functions for which our integral will exist.

Definition A.13 Let $(X, \mathcal{A}, \mu)$ be a positive measure space. A scalar-valued measurable function f is called *Lebesgue integrable* (or just *integrable*) with respect to μ if

$$\int_X |f|\,d\mu < \infty.$$

If f is integrable with respect to μ, then we write $f \in L_1(\mu)$.

Now, let $(X, \mathcal{A}, \mu)$ be a positive measure space and let $f : X \to \mathbb{R}$ be a real-valued integrable function. Define the integral of f over A by

$$\int_A f\,d\mu = \int_A f^+\,d\mu - \int_A f^-\,d\mu,$$

where

$$f^+(x) = \begin{cases} f(x) & \text{if } f(x) \geq 0, \\ 0 & \text{otherwise,} \end{cases}$$

is the *positive part of* f, and

$$f^-(x) = \begin{cases} -f(x) & \text{if } f(x) \le 0, \\ 0 & \text{otherwise,} \end{cases}$$

is the *negative part of* f. If $f : X \to \mathbb{C}$ is a complex-valued integrable function, define the integral of f over A by

$$\int_A f d\mu = \int_A \Re(f) d\mu + i \int_A \Im(f) d\mu,$$

where $\Re(f)$ and $\Im(f)$ denote the real and imaginary parts of f, respectively.

To avoid certain arithmetic difficulties which arise from the definitions above, we adopt the convention that $0 \cdot \infty = 0$. Consequently, $\int_A f\, d\mu = 0$ whenever $f(x) = 0$ for all $x \in A$, even if $\mu(A) = \infty$. Furthermore, if $\mu(A) = 0$ for a measurable set $A \in \mathcal{A}$, then $\int_A f\, d\mu = 0$ for all measurable functions f.

In some cases, it is useful to explicitly show the variable dependence in an integration, and in such a situation we write

$$\int_A f d\mu = \int_A f(x)\mu(dx).$$

A.3   Convergence Theorems and Fatou's Lemma

Definition A.14 Let $(X, \mathcal{A}, \mu)$ be a positive measure space. Two measurable functions f and g are said to be equal *almost everywhere* if $\mu\{x \in X : f(x) \ne g(x)\} = 0$. If f and g are equal almost everywhere, we write $f = g$ a.e.(μ). Sometimes we say $f(x) = g(x)$ for almost every x.

The previous definition is significant because, if $f = g$ a.e.(μ), then

$$\int_A f d\mu = \int_A g d\mu, \quad A \in \mathcal{A}.$$

This is true because the integral over a set of measure zero is always zero. Functions that are equal almost everywhere are indistinguishable from the point of view of integration. With this in mind, we extend our notion of a measurable function. Let $(X, \mathcal{A}, \mu)$ be a measure space. A function f is *defined a.e.(μ) on* X if the domain D of f is a subset of X and $X \setminus D$ is a μ-negligible set. If f is defined a.e.(μ) on X and there is a μ-null set $E \in \mathcal{A}$ such that $f^{-1}(V) \setminus E$ is measurable for every open set V, then f is called μ-*measurable*. Every measurable function (in the sense of Definition A.2) is μ-measurable and every μ-measurable function is equal almost everywhere (with respect to μ) to a measurable function.

The notion of μ-measurability allows us to extend our constructions to a wider collection of functions. We say that a μ-measurable function f is *integrable* if it is equal a.e.(μ) to a measurable function g that is integrable with respect to μ and we define $\int_A f \, d\mu$ to be $\int_A g \, d\mu$ for every measurable set A.

A sequence of measurable functions $(f_n)_{n=1}^{\infty}$ is said to converge *almost everywhere* to a function f if the set $\{x \in X : f_n(x) \not\to f(x)\}$ has μ-measure zero. In this case, we write $f_n \to f$ a.e.(μ). A function f which is the a.e.(μ)-limit of measurable functions is μ-measurable.

Theorem A.15 (Monotone Convergence Theorem) *Let $(X, \mathcal{A}, \mu)$ be a positive measure space. Suppose $(f_n)_{n=1}^{\infty}$ is a sequence of measurable functions such that*

$$0 \le f_1(x) \le f_2(x) \le f_3(x) \le \cdots$$

for almost every x. If $f_n \to f$ a.e.(μ), then f is an integrable function and

$$\int_X f d\mu = \lim_{n \to \infty} \int_X f_n \, d\mu.$$

The Monotone Convergence Theorem is a very useful tool, and one of the better known consequences of it is Fatou's Lemma, named after the mathematician Pierre Fatou.

Theorem A.16 (Fatou's Lemma) *Let $(X, \mathcal{A}, \mu)$ be a positive measure space. If $(f_n)_{n=1}^{\infty}$ is a sequence of scalar-valued measurable functions, then*

$$\int_X \left(\liminf_{n \to \infty} |f_n| \right) d\mu \le \liminf_{n \to \infty} \int_X |f_n| \, d\mu.$$

Perhaps the most important use of Fatou's Lemma is in proving the next theorem, which is one of the cornerstones of measure theory.

Theorem A.17 (Lebesgue's Dominated Convergence Theorem) *Let $(X, \mathcal{A}, \mu)$ be a positive measure space and suppose $(f_n)_{n=1}^{\infty}$ is a sequence of scalar-valued measurable functions that converge almost everywhere to f. If there exists a function $g \in L_1(\mu)$ such that $|f_n| \le g$ a.e.(μ) for all $n \in \mathbb{N}$, then $f \in L_1(\mu)$ and*

$$\lim_{n \to \infty} \int_X |f_n - f| \, d\mu = 0.$$

In particular,

$$\int_X f d\mu = \lim_{n \to \infty} \int_X f_n \, d\mu.$$

The last equality follows from the fact that $| \int_X f \, d\mu | \le \int_X |f| \, d\mu$ whenever f is an integrable function.

The next result is a direct consequence of Lebesgue's Dominated Convergence Theorem.

Corollary A.18 (Bounded Convergence Theorem) *Let $(X, \mathcal{A}, \mu)$ be a positive finite measure space and suppose $(f_n)_{n=1}^{\infty}$ is a sequence of uniformly bounded scalar-valued measurable functions. If $f_n \to f$ a.e.(μ), then $f \in L_1(\mu)$ and*

$$\int_X f d\mu = \lim_{n \to \infty} \int_X f_n \, d\mu.$$

A.4 Complex Measures and Absolute Continuity

Definition A.19 Let $(X, \mathcal{A})$ be a measurable space. A countably additive set function $\mu : \mathcal{A} \to \mathbb{C}$ is called a *complex measure*. When we say μ is countably additive, we mean

$$\mu \left(\bigcup_{j=1}^{\infty} A_j \right) = \sum_{j=1}^{\infty} \mu(A_j), \tag{A.2}$$

whenever $(A_j)_{j=1}^{\infty}$ is a sequence of pairwise disjoint measurable sets in $\mathcal{A}$, where the series in (A.2) is absolutely convergent. (Compare to Definition A.6). A complex measure which takes values in $\mathbb{R}$ is called a *real measure* or a *signed measure*.

If μ is a complex measure on $\mathcal{A}$, then the triple $(X, \mathcal{A}, \mu)$ is called a (complex) *measure space*. If the σ-algebra is understood, then we may write (X, μ).

More generally, we can define a measure to be a countably additive set function that takes values in any topological vector space, so long as the convergence of the series in (A.2) makes sense. For our purposes, such generality is not necessary, and so we will confine ourselves to complex measures and positive measures.

There is a significant difference between positive measures and complex measures. In Definition A.19, we require that a complex measure μ be finite; that is, $|\mu(A)| < \infty$ for all $A \in \mathcal{A}$. This was not a requirement for a positive measure.

Definition A.20 Let μ be a complex measure on the σ-algebra $\mathcal{A}$. We define the *total variation measure of* μ to be the set function $|\mu| : \mathcal{A} \to \mathbb{R}$ given by

$$|\mu|(E) = \sup \left\{ \sum_{j=1}^{n} |\mu(E_j)| \right\},$$

where the supremum is taken over all *finite* sequences of pairwise-disjoint measurable sets $(E_j)_{j=1}^{n}$, for all $n \in \mathbb{N}$, such that $E = \bigcup_{j=1}^{n} E_j$.

As the name suggests, the total variation measure is a measure. Additionally, it is a finite measure—a fact that we state in the next theorem.

Theorem A.21 *If $(X, \mathcal{A}, \mu)$ is a complex measure space, then $|\mu|$ is a positive measure on $\mathcal{A}$ and $|\mu|(X) < \infty$.*

There is, naturally, a relationship between a measure and its total variation measure. For example, if $|\mu|(E) = 0$ for some set $E \in \mathcal{A}$, it must be the case that $\mu(E) = 0$. This is an instance of a property called *absolute continuity*.

Definition A.22 Let μ be a complex measure on $\mathcal{A}$ and suppose λ is a positive measure on $\mathcal{A}$. We say that μ is *absolutely continuous* with respect to λ if, for all

$A \in \mathcal{A}$, we have that $\mu(A) = 0$ whenever $\lambda(A) = 0$. When μ is absolutely continuous with respect to λ, we write $\mu \ll \lambda$.

Shortly, we will state the Radon–Nikodým Theorem, one of the key results in measure theory. Before that, however, we introduce a definition.

Definition A.23 Let $(X, \mathcal{A})$ be a measurable space. A positive measure μ on $\mathcal{A}$ is said to be *σ-finite* if there exists a countable sequence $(E_j)_{j=1}^{\infty}$ of measurable sets such that $X = \bigcup_{j=1}^{\infty} E_j$ and $\mu(E_j) < \infty$ for each $j \in \mathbb{N}$.

Important examples of σ-finite measures include Lebesgue measure on $\mathbb{R}$ and counting measure on $\mathbb{N}$. (See Examples A.8 and A.9, respectively.) One can also have a counting measure on $\mathbb{R}$, but it is not σ-finite.

Theorem A.24 (Radon–Nikodým Theorem) *Suppose $(X, \mathcal{A})$ is a measurable space and let λ be a positive σ-finite measure on $\mathcal{A}$. If μ is a complex measure on $\mathcal{A}$ that is absolutely continuous with respect to λ, then there exists a unique $g \in L_1(\lambda)$ such that*

$$\mu(E) = \int_E g(x) \lambda(dx), \quad E \in \mathcal{A}. \tag{A.3}$$

The equation in (A.3) is sometimes written $\mu(dx) = g(x) \lambda(dx)$ or $d\mu = g \, d\lambda$. The function $g \in L_1(\lambda)$ is called the *Radon–Nikodým derivative* of μ with respect to λ and is sometimes denoted $d\mu/d\lambda$. When we say the Radon–Nikodým derivative is *unique*, we mean up to a set of measure zero. That is, if g and h are both Radon–Nikodým derivatives of μ with respect to λ, then $g = h$ a.e.(λ) (and consequently a.e.(μ)).

The σ-finite assumption on μ in Theorem A.24 cannot be relaxed.

We seek to define an integral with respect to a complex measure. To that end, let $(X, \mathcal{A}, \mu)$ be a complex measure space. Since $\mu \ll |\mu|$, there exists (by the Radon–Nikodým Theorem) a unique (up to sets of measure zero) function $g \in L_1(|\mu|)$ such that $d\mu = g \, d|\mu|$. We can, therefore, define the integral of a measurable function $f : X \to \mathbb{C}$ by

$$\int_E f d\mu = \int_E fg \, d|\mu|, \quad E \in \mathcal{A}. \tag{A.4}$$

More can be said about the Radon–Nikodým derivative of μ with respect to $|\mu|$. The following proposition is a corollary of Theorem A.24 and is often called the *polar representation* or *polar decomposition* of μ.

Proposition A.25 *If $(X, \mathcal{A}, \mu)$ is a complex measure space, then there exists a measurable function g such that $|g(x)| = 1$ for all $x \in X$ and such that $d\mu = g \, d|\mu|$.*

As a consequence of Proposition A.25 and the definition in (A.4), versions of the Monotone Convergence Theorem, Fatou's Lemma, and Lebesgue's Dominated Convergence Theorem hold for integrals with respect to complex measures. We apply the existing theorems to the positive finite measure $|\mu|$ and observe that

$$\left| \int_X f \, d\mu \right| \le \int_X |f| \, d|\mu|,$$

for all measurable functions f on X.

The Radon–Nikodým Theorem is named after Johann Radon, who proved the theorem for $\mathbb{R}^n$ in 1913 ($n \in \mathbb{N}$), and for Otton Nikodým, who proved the theorem for the general case in 1930 [28].

A.5 L_p-spaces

In this section, we consider a measure space $(X, \mathcal{A}, \mu)$, where μ is a positive measure. We will identify certain spaces of measurable functions on X. For $p \in [1, \infty)$, let

$$L_p(\mu) = \left\{ f \text{ a measurable function} : \int_X |f|^p \, d\mu < \infty \right\}.$$

This is the space of *p-integrable functions*, also known as L_p-*functions*, on X. For each $p \in [1, \infty)$, we let

$$\|f\|_p = \left(\int_X |f|^p \, d\mu \right)^{1/p},$$

where f is a μ-measurable function on X. Observe that $\|f\|_p < \infty$ if and only if $f \in L_p(\mu)$. For the case $p = \infty$, let

$$\|f\|_\infty = \inf \left\{ K : \mu(|f| > K) = 0 \right\},$$

for all measurable functions f. The quantity $\|f\|_\infty$ is called the *essential supremum norm* of f, and is the smallest number having the property that $|f| \le \|f\|_\infty$ a.e.(μ).

The set

$$L_\infty(\mu) = \{ f \text{ a measurable function} : \|f\|_\infty < \infty \}$$

is the space of *essentially bounded measurable functions* on X.

Theorem A.26 *Let (X, μ) be a positive measure space. If $1 \le p \le \infty$, then $\| \cdot \|_p$ is a complete norm on $L_p(\mu)$. In particular, $L_p(\mu)$ is a Banach space.*

In fact, the L_p-spaces are collections of *equivalence classes* of measurable functions. Two functions f and g in $L_p(\mu)$ are considered equivalent if $f = g$ a.e.(μ). In spite of this, we will generally speak of the elements in L_p-spaces as functions, rather than equivalence classes of functions.

The proof of Theorem A.26 relies heavily on the following fundamental inequality, which provides the triangle inequality for $L_p(\mu)$.

Theorem A.27 (Minkowski's Inequality) *Let* (X, μ) *be a positive measure space. If f and g are in $L_p(\mu)$, where $1 \le p \le \infty$, then $f + g \in L_p(\mu)$ and*

$$\|f + g\|_p \le \|f\|_p + \|g\|_p.$$

Before we give another theorem, we must introduce a definition.

Definition A.28 If $1 < p < \infty$ and $\frac{1}{p} + \frac{1}{q} = 1$, then q is called the *conjugate exponent* of p. The conjugate exponent of $p = 1$ is defined to be $q = \infty$ (and *vice versa*).

If $p \in [1, \infty]$ and q is the conjugate exponent of p, then we can also say that q is conjugate to p, or even that p and q are conjugate to each other. If $p \in (1, \infty)$ and q is conjugate to p, it is sometimes convenient to write $q = \frac{p}{p-1}$.

The following theorem, which is ubiquitous in measure theory, can be seen as a generalization of the Cauchy–Schwarz inequality.

Theorem A.29 (Hölder's Inequality) *Let (X, μ) be a positive measure space. Suppose $1 \le p < \infty$ and let q be conjugate to p. If $f \in L_p(\mu)$ and $g \in L_q(\mu)$, then $fg \in L_1(\mu)$ and*

$$\|fg\|_1 \le \|f\|_p \|g\|_q.$$

Hölder's Inequality ensures that, given any $g \in L_q(\mu)$, the map

$$f \mapsto \int_X f(x) \, g(x) \, \mu(dx), \quad f \in L_p(\mu),$$

defines a bounded linear functional on $L_p(\mu)$ whenever $1 \le p < \infty$. It turns out that any bounded linear functional on $L_p(\mu)$ can be achieved in this way, which is the content of the next theorem.

Theorem A.30 *Let (X, μ) be a positive σ-finite measure space. If $1 \le p < \infty$, and if q is conjugate to p, then $L_p(\mu)^* = L_q(\mu)$.*

Frequently, in order to prove something about L_p-spaces, it is sufficient to prove it for simple functions. This is a consequence of the next theorem, which follows from Theorem A.5 and Theorem A.15 (the Monotone Convergence Theorem).

Theorem A.31 (Density of Simple Functions) *Let (X, μ) be a positive measure space. The set of simple functions is dense in $L_p(\mu)$ whenever $1 \le p \le \infty$.*

A.6 Borel Measurability and Measures

Definition A.32 Suppose X is a topological space. The smallest σ-algebra on X containing the open sets in X is called the *Borel σ-algebra*, or the *Borel field*, on X. A function which is measurable with respect to the Borel σ-algebra is called a *Borel* *measurable function*, or a *Borel function*. A measure on the Borel σ-algebra is called a *Borel measure*.

We recall that a topological space is said to be *locally compact* if every point has a compact neighborhood. Naturally, all compact spaces are locally compact, but the converse need not be true. For example, the real line $\mathbb{R}$ with its standard topology is locally compact, but not compact.

If X is a locally compact Hausdorff space, then we denote by $C_0(X)$ the collection of all continuous functions that *vanish at infinity*. We say f *vanishes at infinity* if for every $\varepsilon > 0$, there exists a compact set K such that $|f(x)| < \varepsilon$ for all $x \notin K$. The set $C_0(X)$ is a Banach space under the supremum norm.

Example A.33 The Banach space $C_0(\mathbb{N})$ is the classical sequence space c_0 of sequences tending to zero.

Definition A.34 Let X be a locally compact Hausdorff space and suppose μ is a positive Borel measure on X. If, for every measurable set E,

$$\mu(E) = \inf\{\mu(V) : E \subseteq V, \; V \text{ an open set}\},$$

then μ is called *outer regular*. If, for every measurable set E,

$$\mu(E) = \sup\{\mu(K) : K \subseteq E, \; K \text{ a compact set}\},$$

then μ is called *inner regular*. If μ is both inner regular and outer regular, then we call μ a *regular measure*. A complex measure is called regular if $|\mu|$ is regular.

Definition A.34 allows us to describe the dual space of $C_0(X)$.

Theorem A.35 (Riesz Representation Theorem) *Let X be a locally compact Hausdorff space. If Λ is a bounded linear functional on $C_0(X)$, then there exists a unique regular complex Borel measure μ such that*

$$\Lambda(f) = \int_X f \, d\mu, \quad f \in C_0(X),$$

and $\|\Lambda\| = |\mu|(X)$.

Theorem A.35 is named after F. Riesz, who originally proved the theorem for the special case $X = [0, 1]$ [31]. The next theorem is named after Nikolai Lusin (or Luzin), who also worked in the context of the real line [23].

Theorem A.36 (Lusin's Theorem) *Let X be a locally compact Hausdorff space and suppose μ is a finite positive regular Borel measure on X. The space $C_0(X)$ is dense in $L_p(\mu)$ for all $p \in [1, \infty)$.*

The hypotheses of Lusin's Theorem can be relaxed somewhat. Indeed, the theorem holds for any positive Borel measures that are finite on compact sets. Note that Lusin's Theorem is not true when $p = \infty$. Convergence in the L_∞-norm is the same as uniform convergence, and the uniform limit of a sequence of continuous functions is continuous; however, functions in $L_\infty(\mu)$ need not be continuous.

Proposition A.37 *If X is a locally compact metrizable space, then any finite Borel measure on X is necessarily regular.*

Again, the Borel measure in question need only be finite on compact sets for the conclusion to hold. In light of Proposition A.37, Theorem A.35 is often stated for a metrizable space X, in which case the term "regular" is omitted from the conclusion. (This was done, for example, in Theorem 2.20 of the current text.)

A.7   Product Measures

Let $(X, \mathcal{A})$ and $(Y, \mathcal{B})$ be two measurable spaces. A *measurable rectangle* in $X \times Y$ is any set of the form $A \times B$, where $A \in \mathcal{A}$ and $B \in \mathcal{B}$. We denote by $\sigma(\mathcal{A} \times \mathcal{B})$ the smallest σ-algebra containing all measurable rectangles in $X \times Y$.

Proposition A.38 *Let $(X, \mathcal{A})$ and $(Y, \mathcal{B})$ be two measurable spaces. If a scalar-valued function f on $X \times Y$ is $\sigma(\mathcal{A} \times \mathcal{B})$-measurable, then the map $x \mapsto f(x, y)$ is $\mathcal{A}$-measurable for all $y \in Y$, and $y \mapsto f(x, y)$ is $\mathcal{B}$-measurable for all $x \in X$.*

Now let $(X, \mathcal{A}, \mu)$ and $(Y, \mathcal{B}, \nu)$ be two σ-finite measure spaces. We define the *product measure* on $\sigma(\mathcal{A} \times \mathcal{B})$ to be the set function $\mu \times \nu$ given by the formula

$$(\mu \times \nu)(Q) = \int_X \left(\int_Y \chi_Q(x, y) \, \nu(dy) \right) \mu(dx)$$

$$= \int_Y \left(\int_X \chi_Q(x, y) \, \mu(dx) \right) \nu(dy),$$

for all $Q \in \sigma(\mathcal{A} \times \mathcal{B})$. As the name implies, $\mu \times \nu$ is a measure on $\sigma(\mathcal{A} \times \mathcal{B})$. Furthermore, the measure $\mu \times \nu$ is such that $(\mu \times \nu)(E \times F) = \mu(E)\nu(F)$ for all $E \in \mathcal{A}$ and $F \in \mathcal{B}$.

The fundamental result in this section is known as Fubini's Theorem. It is named after Guido Fubini, who proved a version of the theorem in 1907 [13].

Theorem A.39 (Fubini's Theorem) *Let $(X, \mathcal{A}, \mu)$ and $(Y, \mathcal{B}, \nu)$ be two σ-finite measure spaces. If $f \in L_1(\mu \times \nu)$, then*

$$\int_{X \times Y} f \, d(\mu \times \nu) = \int_X \left(\int_Y f(x, y) \, \nu(dy) \right) \mu(dx)$$

$$= \int_Y \left(\int_X f(x, y) \, \mu(dx) \right) \nu(dy).$$

The same conclusion holds if f is a $\sigma(\mathcal{A} \times \mathcal{B})$-measurable function such that

$$\int_X \left(\int_Y |f(x, y)| \, \nu(dy) \right) \mu(dx) < \infty.$$

Fubini's Theorem is sometimes called the Fubini-Tonelli theorem, after Leonida Tonelli, who proved a version of Theorem A.39 in 1909 [35].

Appendix B
Results From Other Areas of Mathematics

Throughout the course of this text, we have invoked important results (usually by name), sometimes without explicitly writing out the statement of the result being used. Many of these come from measure theory, and so appear in Appendix 8.3. We include the rest here, for easy reference. For proofs of these theorems, as well as more discussion on these topics, see (for example) *Topology: a first course* by James Munkres [27] and *Functions of One Complex Variable* by John Conway [7].

B.1 The One-Point and Stone-Čech Compactifications

We will give only a brief discussion of the necessary topological concepts. For more, we refer the interested reader to *Topology: a first course*, by James Munkres [27].

We start by defining the *one-point compactification* of a locally compact Hausdorff space that is not compact. (Recall that a space is *locally compact* if every point has a compact neighborhood containing it.)

Definition B.1 Let X be a locally compact Hausdorff space that is not compact. Adjoin to X an element ∞, called the *point at infinity*, to form a set $Y = X \cup \{\infty\}$. Define a topology on Y by declaring a set U to be open in Y if either U is open in X or $U = Y \setminus K$, where K is compact in X. The space Y is called the *one-point compactification of X*.

The classic example of a one-point compactification is $\mathbb{N} \cup \{\infty\}$, the one-point compactification of the natural numbers, where $\mathbb{N}$ is given the discrete topology. Notice that $\mathbb{N}$ is locally compact but not compact. The space $\mathbb{N} \cup \{\infty\}$ is homeomorphic to the subspace $\{1/n : n \in \mathbb{N}\} \cup \{0\}$ of $\mathbb{R}$. (See Exercise 2.15.)

© Springer Science+Business Media, LLC 2014
A. Bowers, N. J. Kalton, *An Introductory Course in Functional Analysis,*
Universitext, DOI 10.1007/978-1-4939-1945-1

Theorem B.2 *Let X be a locally compact Hausdorff space that is not compact and let Y be its one-point compactification. Then Y is a compact Hausdorff space, $Y \setminus X$ contains exactly one point, X is a subspace of Y, and Y is the closure of X (in Y).*

In general, the one-point compactification of a locally compact metric space is not itself metrizable; however, there are some circumstances under which a metric does exist.

Theorem B.3 *Let X be a locally compact Hausdorff space that is not compact. The one-point compactification of X is metrizable if and only if X is second countable.*

In particular, any locally compact separable metric space that is not compact (such as $\mathbb{N}$) will have a metrizable one-point compactification. We will not provide a proof of Theorem B.3. The inquisitive reader is directed to Theorem 3.44 in [3].

The next type of compactification we wish to define is the Stone-Čech compactification. Before introducing this concept, however, we require some background. We begin by restating Tychonoff's Theorem (Theorem 5.38), which we will use presently.

Theorem B.4 (Tychonoff's Theorem) *Let J be an index set. If $\{K_\alpha\}_{\alpha \in J}$ is a collection of compact topological spaces, then $\prod_{\alpha \in J} K_\alpha$ is compact in the product topology.*

Tychonoff's Theorem is a statement about compactness, and while we do not provide a proof of this theorem, we can profit from some knowledge of the methods used within. The standard proof of Tychonoff's Theorem uses an alternate characterization of compactness, one for which we need a definition.

Definition B.5 (Finite Intersection Property) Let X be a topological space. A collection $\mathcal{C}$ of subsets of X is said to satisfy the *Finite Intersection Property* if for every finite subcollection $\{C_1, \ldots, C_n\}$ of $\mathcal{C}$, the intersection $C_1 \cap \cdots \cap C_n$ is nonempty.

We now state an alternative formulation of compactness.

Theorem B.6 *A topological space X is compact if and only if every collection $\mathcal{C}$ of closed sets in X satisfying the Finite Intersection Property is such that the intersection $\bigcap_{C \in \mathcal{C}} C$ of all elements in $\mathcal{C}$ is nonempty.*

The above theorem leads directly to the following useful corollary.

Corollary B.7 (Nested Interval Property) *Let X be a topological space. If $(C_n)_{n=1}^{\infty}$ is a sequence of nonempty compact sets such that $C_{n+1} \subseteq C_n$ for all $n \in \mathbb{N}$, then the intersection $\bigcap_{n=1}^{\infty} C_n$ of all sets in the sequence is nonempty.*

A space X is called *completely regular* if all singletons (single-point sets) in X are closed and if for each $x \in X$ and each closed subset A not containing x there exists a bounded continuous function $f : X \to [0, 1]$ such that $f(x) = 1$ and $f(A) = \{0\}$.

In particular, if X is completely regular, then the collection of bounded real-valued continuous functions on X *separates the points of X*. That is, whenever $x \neq y$, there is some continuous function $f : X \to [0, 1]$ such that $f(x) = 1$ but $f(y) = 0$.

We remark here that a space X is called *regular* if any closed subset A and point $x \notin A$ can be separated by open neighborhoods. That is to say, there exist disjoint open neighborhoods containing A and x, respectively. A completely regular space is regular, but there exist regular spaces that are not completely regular.

Proposition B.8 *A topological space X is completely regular if and only if it is homeomorphic to a subspace of a compact Hausdorff space.*

A *compactification* of a space X is a compact Hausdorff space Y that contains X as a dense subset. (An example is the one-point compactification given earlier.) In order for a noncompact space X to have a compactification, it is both necessary and sufficient that X be completely regular. There are generally many compactifications for a completely regular space X. The one we consider now is the Stone-Čech compactification.

Let X be a completely regular space. Let $\{f_\alpha\}_{\alpha \in J}$ be the collection of all bounded continuous real-valued functions on X, indexed by some (possibly uncountable) set J. For each $\alpha \in J$, let I_α be any closed interval containing the range of f_α, say

$$I_\alpha = \left[\inf_{x \in X} f_\alpha(x), \ \sup_{x \in X} f_\alpha(x) \right],$$

or

$$I_\alpha = \left[- \| f_\alpha \|_\infty, \ \| f_\alpha \|_\infty \right].$$

Each of the intervals I_α is compact in $\mathbb{R}$, and so, by Tychonoff's Theorem, the product $\prod_{\alpha \in J} I_\alpha$ is a compact space.

Define a map $\phi : X \to \prod_{\alpha \in J} I_\alpha$ by

$$\phi(x) = (f_\alpha(x))_{\alpha \in J}, \quad x \in X.$$

It can be shown that ϕ is an embedding. (This follows from the complete regularity of X, because the collection $\{f_\alpha\}_{\alpha \in J}$ separates the points of X. See Theorem 4.4.2 in [27] for the details.) We conclude that $\overline{\phi(X)}$ is a compact Hausdorff space.

Let $A = \overline{\phi(X)} \backslash \phi(X)$ and define a space Y by $Y = X \cup A$. We have a bijective mapping $\Phi : Y \to \overline{\phi(X)}$ given by the rule

$$\Phi(y) = \begin{cases} \phi(y) & \text{if } y \in X, \\ y & \text{if } y \in A. \end{cases}$$

Define a topology on Y by letting U be open in Y whenever $\Phi(U)$ is open in $\overline{\phi(X)}$. It follows that Φ is a homeomorphism, and X is a subspace of the compact Hausdorff space Y.

The topological space Y constructed above is known as the *Stone-Čech compactification* of X, and is generally denoted $\beta(X)$. It may seem that $\beta(X)$ is not well-defined, since one could construct it with a different choice of sets; however, the Stone-Čech compactification is unique up to homeomorphism (and a homeomorphism can be chosen so that it is the identity on X).

While a full discussion of the Stone-Čech compactification is beyond the scope of this appendix, we will state a very important property that it possesses.

Theorem B.9 *Let X be a completely regular space with Stone-Čech compactification $\beta(X)$. Any bounded continuous real-valued function on X can be uniquely extended to a continuous real-valued function on $\beta(X)$.*

The Stone-Čech compactification is named after Marshall Harvey Stone and Eduard Čech. Tychonoff's Theorem is named after Andrey Nikolayevich Tychonoff. It is not clear that these theorems are named quite as they should be, a fact which was remarked upon by Walter Rudin in his *Functional Analysis* [34, Note to Appendix A]:

> ... Thus it appears that Čech proved the Tychonoff theorem, whereas Tychonoff found the Čech compactification—a good illustration of the historical reliability of mathematical nomenclature.

B.2   Complex Analysis

The subject of complex analysis is overflowing with fantastic and improbable theorems. Unfortunately, we only encounter a small portion of the subject in our current undertaking. Let us begin with some definitions. A complex-valued function is called *differentiable* at z_0 in the complex plane $\mathbb{C}$ if

$$f'(z_0) = \lim_{z \to z_0} \frac{f(z) - f(z_0)}{z - z_0}$$

exists (and thus is a complex number). In this context, the notation $z \to z_0$ means that $|z - z_0| \to 0$. If f is differentiable at every point in a set, then we say f is differentiable in the set (or on the set). A complex-valued function is called *holomorphic* (or *analytic*) if it is differentiable in a neighborhood of every point in its domain. The next theorem is one of the key results of complex analysis.

Theorem B.10 *A holomorphic function is infinitely differentiable.*

The next theorem is a frequently cited result and is used to show, for example, that the disk algebra $A(\mathbb{D})$ is a subalgebra of $C(\mathbb{T})$, the space of continuous functions on the unit circle. (See Example 8.6(b).)

Theorem B.11 (Maximum Modulus Theorem) *Let D be a bounded open set in the complex plane. If f is a holomorphic function in D that is extendable to a continuous function on $\overline{D}$, then*

$$\max_{z \in \overline{D}} |f(z)| = \max_{z \in \partial D} |f(z)|.$$

Holomorphic functions display a variety of interesting and useful properties. One such property is stated in the next theorem.

Theorem B.12 (Cauchy's Integral Formula) *Let $f : D \to \mathbb{C}$ be a holomorphic function on the simply connected domain D. If γ is a closed curve in D, then $\int_\gamma f(z)\,dz = 0$.*

The integral appearing in Theorem B.12 is a standard line integral over a path γ. When we say D is *simply connected*, we mean that any two points in D can be connected by a path, and that any path connecting those two points can be continuously transformed into any other. A less precise way of saying that is to say D has no "holes" in it.

Cauchy's Integral Formula has a converse, which is named after the mathematician and engineer Giacinto Morera.

Theorem B.13 (Morera's Theorem) *Let D be a connected open set in the complex plane. If $f : D \to \mathbb{C}$ is a continuous function such that $\int_\gamma f(z)\,dz = 0$ for every closed piecewise continuously differentiable curve γ in D, then f is holomorphic on D.*

Morera's Theorem is used to show (among other things) that the uniform limit of holomorphic functions is holomorphic: Suppose $(f_n)_{n=1}^\infty$ is a uniformly convergent sequence of holomorphic functions, and suppose $f = \lim_{n \to \infty} f_n$. By Cauchy's Integral Formula, if γ is any continuously differentiable closed curve, then we have $\int_\gamma f_n(z)\,dz = 0$ for all $n \in \mathbb{N}$. Since $\int_\gamma f(z)\,dz = \lim_{n \to \infty} \int_\gamma f_n(z)\,dz = 0$ (by uniform convergence), it follows from Morera's Theorem that f is holomorphic.

A function that is holomorphic on all of $\mathbb{C}$ is called an *entire* function. There are some truly remarkable theorems related to entire functions.

Theorem B.14 (Liouville's Theorem) *A bounded entire function is constant.*

According to Liouville's Theorem, if an entire function is not constant, then it must be unbounded. In fact, something even stronger can be said.

Theorem B.15 (Picard's Lesser Theorem) *If an entire function is not constant, then its image is the entire complex plane, with the possible exception of one point.*

The classic example of such a non-constant function is $f(z) = e^z$. The range of this function is $\mathbb{C} \backslash \{0\}$. Theorem B.15 was proved by Charles Émile Picard in 1879. Another significant theorem due to Picard concerns *essential singularities*.

Theorem B.16 (Picard's Greater Theorem) *Let f be an analytic function with an essential singularity at z_0. On any neighborhood of z_0, the function f will attain every value of $\mathbb{C}$, with the possible exception of one point, infinitely often.*

A function is said to have a *singularity* at the point z_0 if f is analytic in a neighborhood containing z_0, except $f(z_0)$ does not exist. If $\lim_{z \to z_0} f(z)$ exists, then we call it a *removable singularity*. A classic example is the function $f(z) = \frac{\sin z}{z}$, which has a removable singularity at $z = 0$.

A singularity z_0 is a *pole* of f if there is some $n \in \mathbb{N}$ such that

$$\lim_{z \to z_0} (z - z_0)^n f(z) \tag{B.5}$$

exists. We call z_0 a *pole of order n* if n is the smallest integer such that (B.5) exists. A simple example of a function with a polar singularity of order $n \in \mathbb{N}$ at $z = 0$ is $f(z) = \frac{1}{z^n}$.

If the limit in (B.5) does not exist for any $n \in \mathbb{N}$, we call z_0 an *essential singularity* of f. Examples of functions with an essential singularity at $z = 0$ are $f(z) = e^{1/z}$ and $f(z) = \sin(1/z)$.

An alternate characterization of singularities can be obtained from the Laurent series of the function f. A function f has a *Laurent series about z_0* if there exists a doubly infinite sequence of scalars $(a_n)_{n \in \mathbb{Z}}$ such that

$$f(z) = \sum_{n=-\infty}^{\infty} a_n(z - z_0)^n.$$

There is a formula to compute a_n for $n \in \mathbb{Z}$:

$$a_n = \frac{1}{2\pi i} \int_\gamma \frac{f(z)}{(z - z_0)^{n+1}} \, dz,$$

where γ is a circle centered at z_0 and contained in an annular region in which f is holomorphic. Any function with a singularity at z_0 will have a Laurent series there. Such a function has a pole of order $n \in \mathbb{N}$ at z_0 if $a_{-n} \neq 0$, but $a_{-k} = 0$ for all $k > n$. If $a_{-k} \neq 0$ for infinitely many $k \in \mathbb{N}$, then f has an essential singularity at z_0.

Augustin-Louis Cauchy did fundamental research in complex analysis in the first half of the nineteenth century. He and Joseph Liouville, after whom Theorem B.14 is named, were contemporaries. Charles Émile Picard came later, not being born until 1856, only one year before Cauchy passed away (on 23 May 1857).

References

1. L. Alaoglu, Weak topologies of normed linear spaces. Ann. Math. (2), **41**, 252–267 (1940)
2. F. Albiac, N.J. Kalton, *Topics in Banach Space Theory*, vol. 233 of Graduate Texts in Mathematics (Springer, New York, 2006)
3. C.D. Aliprantis, K.C. Border, *Infinite Dimensional Analysis: A Hitchhiker's Guide*, 3rd edn. (Springer, Berlin, 2006)
4. J.L. Bell, D.H. Fremlin, A geometric form of the axiom of choice. Fundam. Math. **77**, 167–170 (1972)
5. J.W. Calkin, Two-sided ideals and congruences in the ring of bounded operators in Hilbert space. Ann. Math. (2) **42**, 839–873 (1941)
6. P.G. Casazza, A tribute to Nigel J. Kalton (1946–2010). Notices Am. Math. Soc. **59**, 942–951 (2012). (Peter G. Casazza, coordinating editor)
7. J.B. Conway, *Functions of One Complex Variable*, vol. 11 of Graduate Texts in Mathematics, 2nd edn. (Springer-Verlag, New York, 1978)
8. J.B. Conway, *A Course in Operator Theory*, vol. 21 of Graduate Studies in Mathematics (American Mathematical Society, Providence, 2000)
9. P. Enflo, A counterexample to the approximation problem in Banach spaces. Acta Math. **130**, 309–317 (1973)
10. M. Fekete, Über die Verteilung der Wurzeln bei gewissen algebraischen Gleichungen mit ganzzahligen Koeffizienten. Math. Zeitschrift **17**, 228–249 (1923)
11. E. Fischer, Sur la convergence en moyenne. C. R. Acad. Sci. (Paris) **144**, 1022–1024 (1907)
12. I. Fredholm, Sur une classe d'équations fonctionnelles. Acta. Math. **27**, 365–390 (1903)
13. G. Fubini, Sugli integrali multipli. Rom. Accad. Lincei. Rend. (5) **16**(1), 608–614 (1907)
14. I. Gelfand, Normierte Ringe. Rec. Math. N.S. (Mat. Sbornik) **9**(51), 3–24 (1941)
15. I. Gelfand, A. Kolmogoroff, On rings of continuous functions on topological spaces. Dokl. Akad. Nauk. SSSR **22**, 11–15 (1939)
16. I. Gelfand, M. Neumark, On the imbedding of normed rings into the ring of operators in Hilbert space. Rec. Math. N.S. (Mat. Sbornik) **12**(54), 197–213 (1943)
17. A. Grothendieck, Produits tensoriels topologiques et espaces nucléaires. Mem. Am. Math. Soc. **1955**, 140 (1955)
18. W. Hodges, Krull Implies Zorn. J. London Math. Soc. (2) **19**, 285–287 (1979)
19. R.C. James, Characterizations of Reflexivity. Studia Math. **23**, 205–216 (1963/1964)
20. Y. Katznelson, *An Introduction to Harmonic Analysis*, Cambridge Mathematical Library, 3rd edn. (Cambridge University Press, Cambridge, 2004)
21. M. Krein, D. Milman, On extreme points of regular convex sets. Studia Math. **9**, 133–138 (1940)
22. W. Krull, Idealtheorie in Ringen ohne Endlichkeitsbedingung. Math. Ann. **101**, 729–744 (1929)
23. N. Lusin, Sur les propriétés des fonctions mesurables. C. R. Acad. Sci. (Paris) **154**, 1688–1690 (1912)

© Springer Science+Business Media, LLC 2014
A. Bowers, N. J. Kalton, *An Introductory Course in Functional Analysis*,
Universitext, DOI 10.1007/978-1-4939-1945-1

24. R.D. Mauldin (ed.), *The Scottish Book*, Birkhäuser Boston, Mass., 1981. Mathematics from the Scottish Café, Including selected papers presented at the Scottish Book Conference held at North Texas State University, Denton, Tex. (May 1979)
25. S. Mazur, Sur les anneaux linéaires. C. R. Acad. Sci. (Paris) **207**, 1025–1027 (1938)
26. G.H. Moore, *Zermelo's Axiom of Choice. Its Origins, Development, and Influence*, vol. 8 of Studies in the History of Mathematics and Physical Sciences (Springer-Verlag, New York, 1982)
27. J.R. Munkres, *Topology: A First Course* (Prentice-Hall Inc., Englewood Cliffs, 1975)
28. O. Nikodym, Sur une généralisation des intégrales de M. J. Radon. Fundam. Math. **15**, 131–179 (1930)
29. J.C. Oxtoby, *Measure and Category: A Survey of the Analogies Between Topological and Measure Spaces*, vol. 2 of Graduate Texts in Mathematics, 2nd edn. (Springer-Verlag, New York, 1980)
30. F. Riesz, Sur les systèmes orthogonaux de fonctions. C. R. Acad. Sci. (Paris) **144**, 615–619 (1907)
31. F. Riesz, Sur les opérations fonctionnelles linéaires. C. R. Acad. Sci. (Paris) **149**, 974–977 (1909)
32. J.W. Roberts, A compact convex set with no extreme points. Studia Math. **60**, 255–266 (1977)
33. W. Rudin, *Real and Complex Analysis*, 3rd edn. (McGraw-Hill Book Co., New York, 1987)
34. W. Rudin, *Functional Analysis*. International Series in Pure and Applied Mathematics, 2nd edn. (McGraw-Hill Inc., New York, 1991)
35. L. Tonelli, Sull' integrazione per parti. Rom. Accad. Lincei. Rend. (5). **18**(2), 246–253 (1909)
36. M. Väth, The dual space of L_∞ is L_1. Indag. Math. N.S. **9**, 619–625 (1998)
37. W. Żelazko, *Banach Algebras* (Elsevier Publishing Co., Amsterdam, 1973). (Translated from the Polish by Marcin E. Kuczma.)

Index

A

$A(\mathbb{D})$, *disk algebra* 182
$A(\mathbb{T})$, *Wiener Algebra* 204
abelian group, 45
absolutely continuous
    function, 179
    measure, 213
absolutely convex set, 93
absorbent, 86
adjoint
    Banach space adjoint, 50, 108
    Hilbert space adjoint, 158
    of a matrix, 50
Alaoglu, Leonidas, 106
almost convergent sequence, 59
almost everywhere
    convergence, 211
    equality, 211
almost open, 72
analytic function, *holomorphic function* 222
annihilator, 57
anti-isomorphism, 157
approximate eigenvalue, 187
approximate point spectrum, 187
Approximation Problem, 140
associative algebra, 181
Axiom of Choice, 31, 198

B

$\beta(X)$, *Stone-Čech compactification* 221
Baire Category Theorem, 61, 64
Baire, René-Louis, 61
balanced set, 86
Banach algebra, 181
    commutative, 196
Banach limit, 42
Banach space, 2

Banach, Stefan, 7, 106
Banach–Mazur Characterization of Separable
    Spaces, 74
Banach–Stone Theorem, 121
Banach-Alaoglu Theorem, 105, 199
Banach-Steinhaus Theorem, 66
Banach-Tarski Paradox, 31
base for a topology, 84
Bessel's Inequality, 160
bidual, 49
bijection, 5
bilinear map, 43, 151
Borel
    σ-algebra, 209
    field, 216
    measurable function, 216
    measure, 216
    measure on $\mathbb{R}$, 209
    σ-algebra, 216
bounded
    function, 8
    sequence, 7
    set, 65, 127
Bounded Convergence Theorem, 136, 212
Bounded Inverse Theorem, 74
bounded linear map, 2
bounded sequence, 14

C

c_0, 15
    has no extreme points, 113
c, 36
$C(K)$, 20, 182, 201
    completeness of, 26
    separability for $K = [0, 1]$, 38
$C_0(X)$, 216
Calkin algebra, 196

© Springer Science+Business Media, LLC 2014
A. Bowers, N. J. Kalton, *An Introductory Course in Functional Analysis,*
Universitext, DOI 10.1007/978-1-4939-1945-1